Fossil Hydrocarbons

Fossil Hydrocarbons

CHEMISTRY AND TECHNOLOGY

Norbert Berkowitz
School of Mining and Petroleum Engineering
University of Alberta
Edmonton, Alberta
Canada

ACADEMIC PRESS
San Diego London Boston New York Sydney Tokyo Toronto

This book is printed on acid-free paper. ♾

Academic Press
a division of Harcourt Brace & Company
525 B Street, Suite 1900, San Diego, California 92101-4495, USA
http://www.apnet.com

Academic Press Limited
24-28 Oval Road, London NW1 7DX, UK
http://www.hbuk.co.uk/ap/

Library of Congress Cataloging-in-Publication Data

Fossil hydrocarbons : chemistry and technology / [edited] by Norbert Berkowitz.
p. cm.
Includes bibliographical references and index.
ISBN 0-12-091090-X (alk. paper)
1. Hydrocarbons. 2. Fossil fuels. I. Berkowitz, N. (Norbert). date.
TP343.F65 1997
553.2--dc21 97-23439
CIP

PRINTED IN THE UNITED STATES OF AMERICA
97 98 99 00 01 02 EB 9 8 7 6 5 4 3 2 1

For Sheila

CONTENTS

PREFACE

The decade of 1973–1983, in which most of the Western world moved from economic turmoil and panic created by an oil crisis to blissfully "putting it all behind us", illustrates how easily we persuade ourselves to forget what should have taught us a profound lesson—and how cavalierly we face the need to secure long-term supplies of liquid fuels.

The oil crisis abated not because of how we responded to it, but because the major Middle East oil producers raised output in expectation of recouping revenues that had fallen victim to the Iraq–Iran war. This generated an oil glut that halved crude oil prices and allowed us to return to the *status quo ante*. Laissez-faire economic policies once again allowed profligate use and/or export of diminishing indigenous reserves of gas and conventional oil. Alternative sources—the heavy fossil hydrocarbons that could help us to attain reasonable energy self-sufficiency—were once again consigned to the dim recesses of our collective minds. Research and development, which outlined and sometimes defined the new technologies through which self-sufficiency could be achieved, was abruptly discontinued. And development of future crude oil supplies once again became centered on distant sources over which we have little, if any, control. All this occurred, despite the demonstration by major commercial ventures (in particular, South Africa's coal-based SASOL complexes and production of "synthetic" light crudes from Alberta's oil sands) of what is technically possible and could be competitively accomplished.

It is difficult to understand a mindset that allows such energy "policies"—and, not coincidentally, reflects a deplorable disregard for macroeconomics on which all national well-being ultimately depends—as anything other than an attitude of *apres mois la deluge*.

Even academia is not immune to that malaise. Instruction in petroleum engineering at universities and technical colleges is rarely augmented by the study of heavy fossil hydrocarbons, the existence of which is, as a rule, acknowledged only when deemed to be relevant to instruction in rock mechanics, mining engineering, and mineral preparation. And between petroleum devotees who generally don't much care about these "other" resources and a dwindling band of professionals who do, we have thus promoted two solitudes—each sustained by a technical jargon that suggests differences where few exist, and each side seemingly incapable of speaking to the other.

In these circumstances, it seemed to me pertinent to draw attention to the fact that the entrenched dichotomy between petroleum hydrocarbons and coal, which in no small measure shaped popular attitudes about energy, is technically inadmissable and to observe that the heavy fossil hydrocarbons—all much more abundant than natural gas and conventional crudes and distributed more equitably across the globe—offer attractive sources of synthetic gas and light oils even though the required conversion technologies are sometimes, as in the matter of coal liquefaction, still far from fully developed.

The format of this book, which discusses the fossil hydrocarbons under common topic headings rather than by type, reflects my objectives. The first section (Chapters 1–7) thus opens with a review of indicators that support the underlying concept and then considers source materials, biosources, metamorphic histories, host rock geology and geochemistry, classification, and molecular structure. Chapters 8–10 focus on preparation, processing, and conversion technologies. Finally, Chapter 11 examines some of the environmental issues that arise from production, processing, and use of fossil hydrocarbons. Each topic is augmented by end-of-chapter notes that I deemed to be helpful, but did not want to insert into the main text (where they might have proved disruptive), and for each topic I have sought to provide a reasonably detailed bibliography for the interested reader to consult. To assist such reference, I have, wherever possible, made use of English-language literature—even though this might, at first glance, distort the scene and not give proper recognition to the outstanding contributions made in many other countries and reported in other languages. Because I wanted to retain some historical flavor that traced the development of the relevant science and technologies as well as give credit where due, I have also, as far as possible, stayed with the original literature rather than cite more recent sources that added little to what had long been known.

The use of the term *fossil hydrocarbons* in the title and throughout the text does, of course, take liberties with chemical nomenclature. But as *petroleum hydrocarbons* include bitumens and kerogens whose oxygen contents are no lower than those of some coals, I make no apology for such indiscretion. Nor do I apologize for overtly differentiating between preparation and processing, because the former is generally concerned with physically modifying the raw hydrocarbon and the latter changes it chemically.

In preparing the text, I have drawn on open literature, on my lecture notes, and on what I have learned over the years from discussions with friends and colleagues. I must in this connection acknowledge my particular indebtedness to Dr. E. J. Wiggins, who served as Director of the Alberta Research Council and (later) as Board Member of AOSTRA (Alberta's Oil Sands Technology & Research Authority), as well as to colleagues at the University of Alberta—Dr. A. E. Mather, Professor of Chemical Engineering; the late Dr. L. G. Hepler, Professor of Chemistry; Drs. R. G. Bentson and S. M. Farouq Ali, Professors of Petroleum Engineering; and Dr. J. M. Whiting, Professor of Mining Engineering.

I must also thank the publisher, Academic Press, who encouraged me to undertake the writing of this book, and the Alberta Research Council's librarian, Ms. Nancy R. Aikman, who steered me to much helpful literature and thereby made my task so much easier.

I would, however, be terribly remiss if I did not here also specifically express my deep gratitude to my wife for her unfailing love, support, and endurance during the many months I devoted to writing. To her I dedicate this volume.

CHAPTER 1

Introduction: The Family of Fossil Hydrocarbons

Flawed classifications of living and inanimate matter are not uncommon and are usually of little concern except to the specialist. Sometimes, however, the flaws are "validated" by common usage and, in time, become counterproductive fallacies. So in the case of "petroleum": this term has come to be progressively expanded to "petroleum hydrocarbons," which include natural products as diverse as natural gas, light crudes, heavy oils, all bituminous substances, and oil-shale kerogens—but which, be definition, exclude all types of coal and thereby create an untenable distinction. The venerable role that coal has played as a primary fuel since the late 12th century [1][1]; as sources of metallurgical coke since the early 1700s; and in the mid-1700s as the trigger of an industrial revolution that changed the very course of human society—all this may provide a *historical* perspective for sometimes setting it apart from the "petroleum hydrocarbons." But as the record also makes clear, there is little technically legitimated warrant for such dichotomy [2].

Taken for what it is popularly assumed to mean, "petroleum hydrocarbons" is a semantically questionable term even though it may be sanctioned by some dictionaries: for, although kerogens are indeed oil precursors, and bituminous substances—notably bitumen in oil sands[2]—may represent microbially and/or oxidatively altered oil residues [3], neither they nor other bituminous materials (such as tars and asphaltics) are oils, as "petroleum" implies and is commonly understood to mean [4]. However, more to the point here is the untenable implicit technical meaning of the term. Designating natural gas—a variable mix of C_1–C_6 alkanes—as a petroleum hydrocarbon is certainly warranted by its composition, its common association with crude oil, and its descent from residual oily matter in late stages of kerogen catagenesis. But adoption of bituminous substances and kerogens into the petroleum hydrocarbon family can only be justified if they are deemed to be, or to be *chemically directly related* to, oil precursors. And if so, one might ask why

[1] Numbers in square brackets refer to end-of-chapter notes.

[2] Oil sands are also commonly referred to as bituminous sands or tar sands.

hydrogen-rich boghead and cannel coals, which meet this criterion by closely resembling sapropelic kerogens in their origin, developmental history, and chemistry, are excluded from an otherwise all-embracing clan; and if, on reflection, they are admitted, why orphan the equally closely related and much more abundant, albeit less H-rich, humic coals?

Differentiation between fossil hydrocarbons and choices among them are, of course, often necessitated by economic and/or supply constraints. But it is significant that where such need exists, choices almost always require resort to one or another of the heavy hydrocarbons, to bitumens, oil shales, or coals. In practice, differentiation is, in short, between these and what are *properly* termed "petroleum hydrocarbons"—i.e., natural gas and light crude oils; and where circumstances actually force resort to the "heavies," choices are always based on consideration of availability and costs [5].

Nor can it be otherwise, because fossil hydrocarbons stem from the same source materials—the entities that make up the basic fabric of living organisms—and consequently form a continuum of chemically related substances that extend from methane to anthracite. What differentiated the assemblages of source materials that over time developed into different hydrocarbon forms were primarily the *relative proportions* of the source materials; and these were determined by when and in what environments they accumulated [6]. In one form or another, organic carbon was continuously deposited from late Precambrian times, when primitive biota first appeared in ocean waters; and from early Devonian times, when terrestrial vegetation made its appearance, the locales in which organic carbon accumulated ranged therefore from alpine meadows, woodlands, and oxic swamps to disoxic lacustrine regions, paralic environments, and deep anoxic seas.

Qualitatively, a hydrocarbon continuum is indicated by general connections between different hydrocarbon forms (Table 1.1). But more convincing chemical linkages between them emerge when they are broadly arranged in order of increasing gravity, as in

natural gas—light oils—heavy oils—bituminous substances—
kerogens—sapropels—humic coals

In such serial order, the continuum is mirrored in an uninterrupted, progressive fall of the H/C ratio from 4 in the case of methane, the principal component of natural gas, to ~0.7, an average value for mature bituminous coals; and because that indicates increasing carbon aromaticity from progressive internal cyclization and dehydrogenation, it defines the nature of the transitions from gaseous to liquid, semisolid, and solid hydrocarbons.

But the continuity of the fossil hydrocarbon series is also convincingly shown in other features.

There is, for example, a remarkable similarity between constructs that pur-

TABLE 1.1 Connections among "Petroleum Hydrocarbons" and Coals[a]

Gaseous			marsh gas (CH_4)
			natural gas
Bituminous	fluid	petroleum	natural gas liquids
		hydrocarbons	crude oils
	viscous		asphalts
			bitumens
			tars
	solid	kerogens	
		sapropels	cannel coals
			boghead coals
		humic coals	lignites
			subbituminous coals
			bituminous coals
			anthracites
Waxy		mineral waxes	

[a] Adapted from R. R. F. Kinghorn, *An Introduction to the Physics and Chemistry of Petroleum,* Wiley & Sons, New York, 1983.

port to depict average molecular structures of bituminous substances, kerogens, and coals and to show macromolecular, pseudo-crystallographic ordering in them [7].

There are pronounced behavioral similarities—for instance, a close parallel between thermal degradation of kerogen (during catagenesis) and coal (during carbonization), with both yielding H-rich liquids (*oils* or *tars*) in amounts determined by their H/C ratios or hydrogen contents [8], and both leaving correspondingly H-depleted solid residues.

And although the behavior of coal is profoundly influenced by its solidity and rank-dependent porosity, its responses to chemical processing—to thermal cracking and hydrogenation—are much the same as those of other heavy fossil hydrocarbons.

As might indeed be expected, these similarities make for *interchangeability* between bituminous substances, kerogen-rich oil shales and coal, and allow virtually identical processing techniques to transform any one of them into more useful, lighter members of the series. Regardless of whether the feed is a heavy oil, oil residuum, bitumen, oil shale, or coal, such transformation always entails some particular form of carbon rejection or H-addition—in one case increasing the H/C ratio by pyrolytically abstracting carbon as "coke," CO, and/or CO_2, and in the other raising it by inserting externally sources H into the feed [9].

These procedures, summarized in Chapters 9 and 10, make it technically feasible to convert natural gas into an almost pure form of carbon [10] and, more important, to convert heavy oils, bituminous substances, oil shale kerogens, or coals into light transportation fuels. They also allow transforming heavy hydrocarbons into a "substitute" or "synthetic" natural gas (SNG), or into a syngas from which an extraordinarily wide range of hydrocarbon liquids and industrial chemicals can be produced by Fischer–Tropsch techniques (see Chapter 10). The aromaticity of the feed will, as a rule, only determine the *severity* of processing—that is, the *extent* of carbon rejection or H-addition, and in practice, this rarely means much more than selecting suitable processing regimes [11].

This given, questions of whether or when any of these options might be exercised can generally only be answered in light of prevailing economic circumstances.

NOTES

[1] Authentic documentary evidence places the first use of coal as a heating fuel in late 12th-century England, but there are indications that it was occasionally also used as such by the Roman legions in Britain during the 1st century.

[2] Episodal uses of coal other than as primary fuel are a matter of record. In the mid-1800s, it began to be gasified and thereby converted into a domestic fuel gas. In the 1920s, it commenced service as a source of syngas needed for production of gasolines and diesel and aviation fuels. And by the early 1930s, it had established itself as a hydrogenation feedstock for manufacturing transportation fuels, heating oils, and high-purity carbon electrodes. These activities were mostly abandoned in the early 1950s, when abundant supplies of cheap oil and natural gas almost entirely displaced coal as anything other than a primary fuel and source of metallurgical coke, and sometimes displaced it even there. Since then, coal conversion has only attracted attention in perceived crises: in the 1960s and 1970s, coal gasification commanded wide but transient interest because projections, later proven wide of the mark, anticipated serious shortages of acceptably priced natural gas; and intensive work on transforming coal into liquid hydrocarbons lasted no longer than the crippling economic dislocations that followed the 1973 oil crisis, but were soon "remedied" by such events as the Iraq–Iran war—a conflict that, by the convoluted economic policies of an international oil cartel (OPEC), caused an oversupply of oil and a consequent oil-price collapse that is likely to continue until well into the 21st century. That coal conversion can neverthelsss remain attractive is demonstrated by some 20 commercial plants in Europe and Asia that currently produce ammonia (for fertilizer use) from coal-based syngas.

Parenthetically it is also worth observing that oft-repeated "technical" justifications for the dichotomy between "petroleum hydrocarbons" and coal are more contentious than real. Arguments that coal cannot meet the multifaceted needs of modern societies, or meet them as easily or conveniently as natural gas and/or petroleum, seem to ignore advances in coal processing since the mid-1940s. And the contention that coal is so much dirtier than oil and gas, and therefore environmentally "unfriendly," discounts what is required to prepare oil and gas for environmentally acceptable use, and ignores impressive advances in coal preparation (and combustion) over the past 40 or so years.

[3] Although the relevant literature seems to regard *all* bitumens to be microbially oxidized and water-washed (see Chapter 3) and designates most heavy oils in like terms, the evidence

for this view is not entirely satisfactory. There are, as suggested in Chapter 3, other possible mechanisms that could explain the their origins.

[4] Oils are generated by thermal degradation of kerogen much as tars are thermally generated from coal, and bituminous substances are widely (but not necessarily correctly) assumed to have been microbially altered much as weathered coals were abiotically altered. In neither case can a reaction product be properly classified with its precursor.

[5] That availability should often force the decision is due to a natural inequity—the fact that the so-called advanced societies tend to be well endowed with heavy fossil hydrocarbons (mainly coal), but lack the abundant natural gas and crude oils that, for the most part, occur in less developed jurisdictions.

[6] Particularly important in this context is that prior to Devonian times, lignin—an important constituent of higher terrestrial plants that then made their first massive appearance—contributed very little to the precursor masses of fossil hydrocarbons.

[7] As illustrated in Chapter 7, these constructs differ primarily in their carbon aromaticities, which increase steadily from the lighter to the heavier members of the series at the expense of aliphatic and, later, naphthenic moieties.

[8] Because of the overriding importance of H for generation of hydrocarbon liquids (see Chapter 3), their yields and compositions depend on the concentration of lipids (or lipid-like matter) in the precursor; and this implies that *oily matter increases and tends to become the lighter the farther its origin from an inland location*. In other words, light oils originate in deep or moderately deep marine conditions, kerogens and sapropelic coals in paralic and/or lacustrine environments, and humic coals on land.

[9] The many seemingly different (or differently named) processing methods turn out, on closer inspection, to be versions of basic techniques that differ in little more than operating minutiae; the rich vocabulary that characterizes petroleum preparation and processing (see Chapters 8 and 9) merely reflects the wide range of products that technical development and stimulated market demands allowed to be made from crude oil. By the same token, the much more limited process terminology relating to oil shale and coal mirrors the limited utilization of these resources. Oil shale, from which substantial quantities of (shale) oil were produced in the 19th century, is now little more than an occasional subject of a "hard look", and coal, long used as primary fuel and source of metallurgical coke, continues to be restrictively viewed as such.

[10] This is, in fact, done in production of carbon blacks, which are used as pigments in printing inks, fillers for rubber tires, etc. Such carbons are characterized by small (10–1000 nm), nearly spherical particles and bulk densities as low as 0.06 g/cm^3.

[11] For carbon rejection, the primary components of an "appropriate" regime are temperature, pressure, time, and, in some versions, catalysts. For hydrogenation, they are mainly temperature, pressure, and a suitable catalyst.

CHAPTER 2

Origins

1. THE CHEMICAL PRECURSORS

All biota, even the most primitive algae and bacteria that contributed their substance to the source materials of fossil hydrocarbons in early Paleozoic times, construct their fabric by selectively drawing on a pool of four chemically well-defined classes of matter: lipids, amino acids, carbohydrates, and lignins.

Of these, particularly important for eventual formation of oils are *lipids,* a group of closely related aliphatic hydrocarbons that include water-insoluble neutral fats, fatty acids, waxes, terpenes, and steroids. In living organisms, lipids serve primarily as sources of energy; however, during putrefaction they are hydrolyzed to long-chain carboxy acids and subsequently decarboxylated to form alkanes.

The fats of the group, mixed triglycerides, illustrate this reaction, first being saponified by aqueous NaOH to yield glycerol and the Na salts of the corresponding fatty acids, as in

$$\begin{array}{lll} H_3C{-}(CH_2)n{-}C(O){-}O{-}CH_2 & HO{-}CH_2 & \\ \qquad\qquad\qquad\qquad\quad | & \quad | & \\ H_3C{-}(CH_2)n{-}C(O){-}O{-}CH \rightarrow & HO{-}CH & + H_3C{-}(CH_2)n{-}C(O){-}O{-}, \\ \qquad\qquad\qquad\qquad\quad | & \quad | & \\ H_3C{-}(CH_2)n{-}C(O){-}O{-}CH_2 & HO{-}CH_2 & \end{array}$$

and the fatty acids—which can exist in saturated forms exemplified by palmitic ($C_{16}H_{32}O_2$) and stearic ($C_{18}H_{36}O_2$) acids or unsaturated versions such as oleic ($C_{18}H_{34}O_2$), linoleic ($C_{18}H_{32}O_2$), and linolenic ($C_{18}H_{30}O_2$) acids—then losing —COOH and yielding straight-chain alkanes [1].

However, waxes, terpenes, and steroids belonging to the group are structurally more complex and undergo correspondingly more complex changes when they decompose.

The waxes are esters of alcohols other than glycerol, contain as a rule only one —OH group, and present themselves either as sterols such as cholesterol

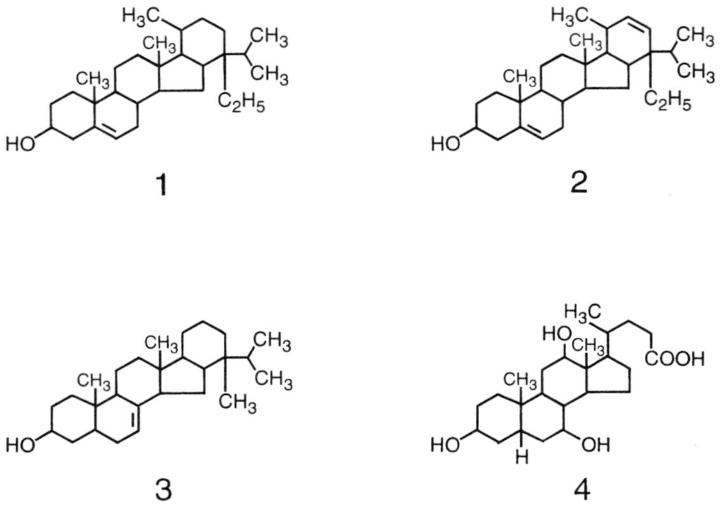

FIGURE 2.1.1 Some important natural sterols. (1) β-Sitosterol, (2) stigmasterol, (3) fungisterol, (4) cholic acid.

(Fig. 2.1.1) or as straight-chain C_{16}–C_{36} aliphatic alcohols such as cetyl alcohol, $CH_3(CH_2)_{14}CH_2OH$.

The terpenes are polymeric forms of isoprene (or 2-methyl-1,3-butadiene; see Chapter 5),

$$H_2C{=}\underset{\displaystyle CH_3}{\underset{|}{C}}{-}CH{=}CH_2,$$

a basic building block of chlorophyll [2] and of the natural gums of higher plants. In nature, they are encountered as:

1. monoterpenes (C_{10}; Fig. 2.1.2), i.e., isoprene dimers that abound in algae and in the essential oils [3] of many higher plants;
2. sesquiterpenes (C_{15}; Fig. 2.1.3) and diterpenes (C_{20}; Fig. 2.1.4), respectively comprising three and four isoprene units and, combined with phenylpropane derivatives such coniferyl alcohol, representing major components of conifer resins;
3. triterpenes (C_{30}); Fig. 2.1.5), made up of six isoprene units, often developing from squalene ($C_{30}H_{50}$; see Fig. 2.1.5) and believed to be direct precursors of petroleum hydrocarbons;
4. tetraterpenes (C_{40}), a group of carotenoid pigments represented by, and commonly present as, carotene (Fig. 2.1.6).

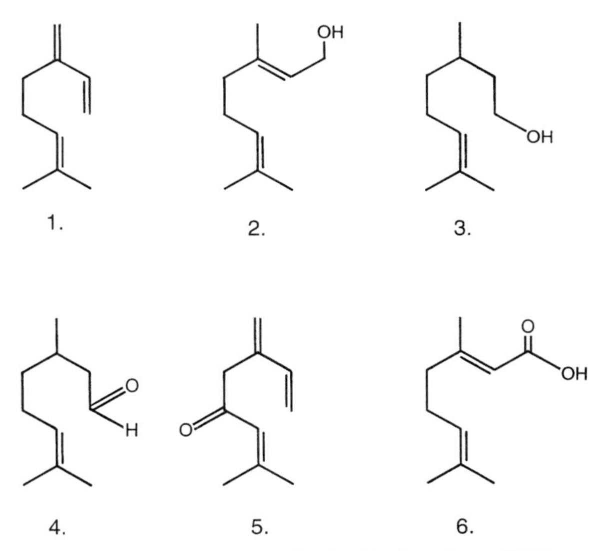

FIGURE 2.1.2 Some monoterpenes in essential oils of higher plants. (1) Myrcene, (2) geraniol, (3) citronellol, (4) citronellal, (5) myrcenone, (6) geranic acid.

A second class of compounds that contributed to source materials were *amino acids*, which are encountered in nature in acidic, basic, and neutral forms. Figure 2.1.7 exemplifies the simplest of these. Amino acids can mutually interact to form peptide linkages between them via their carboxyl and amide groups, and in that manner create extended polypeptides, as in

$$\mathrm{CH_3{-}\underset{\displaystyle NH_2}{\underset{|}{CH}}{-}COOH + HOOC{-}\underset{\displaystyle NH_2}{\underset{|}{CH}}{-}CH_3} \rightarrow$$

$$\mathrm{CH_3{-}\underset{\displaystyle COOH}{\underset{|}{CH}}{-}NH{-}CO{-}\underset{\displaystyle NH_2}{\underset{|}{CH}}{-}CH_3 + H_2O}, \text{ etc.}$$

These are termed *proteins* when their molecular weights exceed 10^4. A reverse reaction, an enzymatic hydrolysis of the peptide linkage that results in partial dissolution of a polypeptide, is shown in Fig. 2.1.8.

The vast variety of proteins is illustrated by the fact that 26 natural amino acids, each capable of reacting with itself or with any of the other 25, have been identified, and that there are therefore no less than 10^{84} possible sequences in which these 26 can be linked in a 60-acid unit.

A third contributor to source materials was *carbohydrates*, a class of compounds composed of carbon, hydrogen, and oxygen, characterized by O/C =

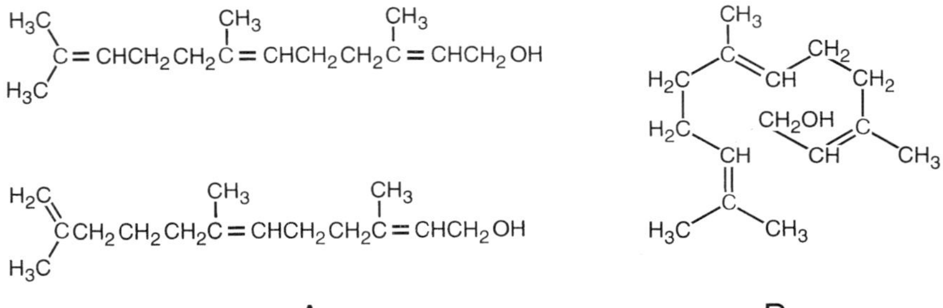

$(H_3C)_2C{=}CHCH_2CH_2C(CH_3){=}CHCH_2CH_2C(CH_3){=}CHCH_2OH$

$H_2C{=}C(H_3C)CH_2CH_2CH_2C(CH_3){=}CHCH_2CH_2C(CH_3){=}CHCH_2OH$

A B

FIGURE 2.1.3 Farnesol: an important sesquiterpene in bacteria. (A) Naturally occurring isomers of farnesol; (B) farnesol configured as precursor of dicyclic sesquiterpenes.

2, and including sugars, starches, and celluloses, the last a dominant structural material of plants.

Sugars are aldehydes or ketones of polyhydric alcohols and form two groups, viz., monosaccharides ($C_6H_{12}O_6$) exemplified by glucose and fructose, and disaccharides ($C_{12}H_{22}O_{11}$) exemplified by sucrose and β-maltose [5].

By interaction of the aldehyde —CHO or ketonic ═CO group with the alcoholic —OH function, hemiacetals or hemiketals are generated, and when the hemiacetal form of a monosaccharide further interacts with the —OH of another monosaccharide, *polysaccharides* composed of eight or more monosaccharide units can form. (Units comprised of eight or fewer monosaccharides are sometimes also referred to as oligosaccharides.)

The most important of the polysaccharides are celluloses based on glucose; in living plants, these contain up to $10–15 \times 10^3$ glucose units and possess molecular weights up to 2.4×10^6. Other polysaccharides, all closely related to celluloses and only differentiated from them by their peripheral substituents, include:

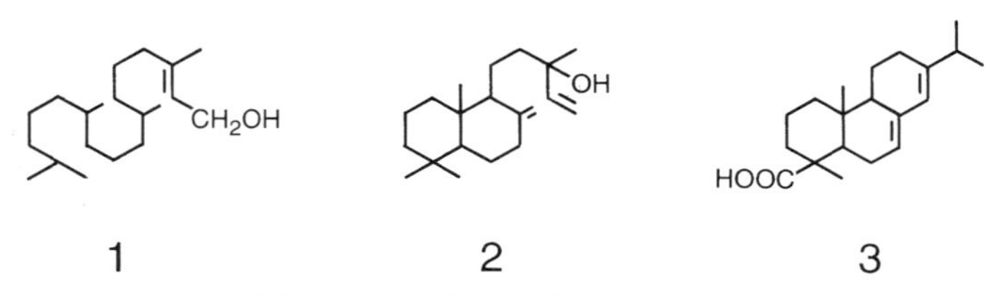

FIGURE 2.1.4 Structures of diterpenoids. (1) Acyclic: phytol; (2) dicyclic: manool; (3) tricyclic: abietic acid.

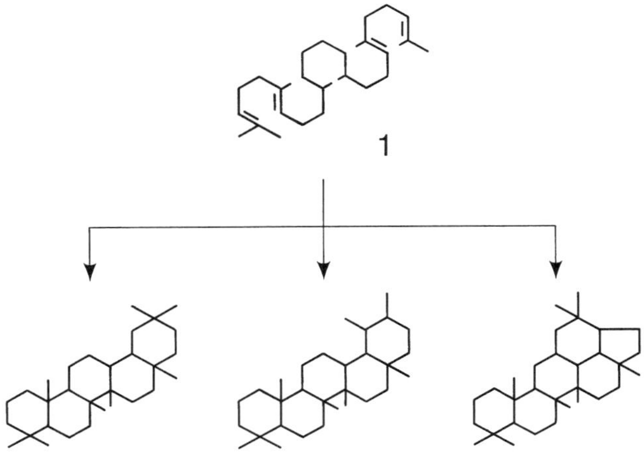

FIGURE 2.1.5 Squalene (1) and some pentacyclic triterpenoid types: (2) oleanane type, (3) ursane type, (4) lupane type.

1. alginic acid, a constituent of brown algae (*Phaeophyta*);
2. pectin, a component of bacteria and higher plants;
3. chitin, a component of some algae and of the hard outer shell of insects and crustaceans;
4. starches, characterized by the configuration of acetal linkages [6] between the monosaccharide units.

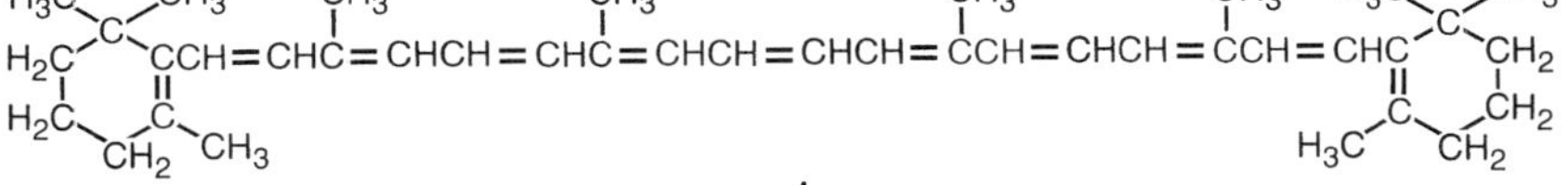

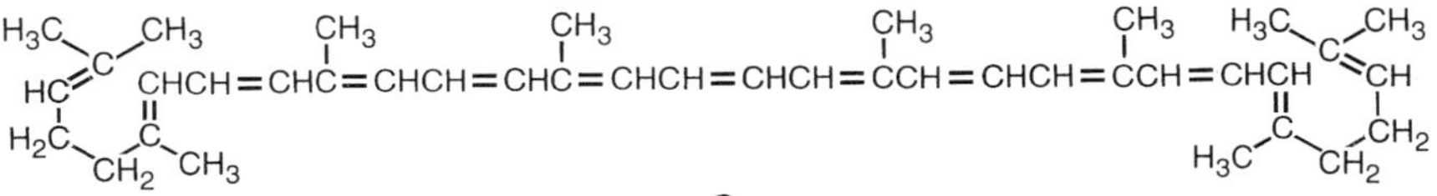

FIGURE 2.1.6 Two important carotenoids: (1) β-carotene, (2) lycopane.

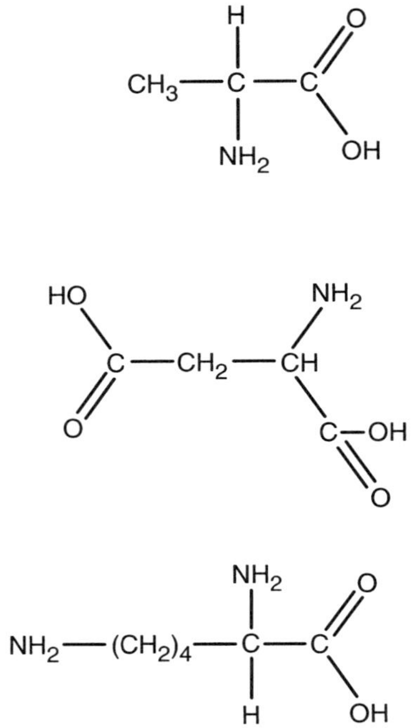

FIGURE 2.1.7 Amino acid types: 1. α-alanine (neutral), 2. aspartic acid (acidic), 3. lysine (basic)

Some of these entities are illustrated in Fig. 2.1.9, which also shows the relationship between sugars and a structure element of cellulose, and in Fig. 2.1.10.

Glycosides, which are mainly encountered as plant pigments, represent a disaccharide subset in which one unit bonded by the glucoside linkage is an alcohol. The sugar component of a glycoside is usually referred to as a glycon, and the alcohol component as an aglycon.

With massive appearance of terrestrial plants in late Devonian and Lower Carboniferous times, this pool of source materials was substantially augmented by *lignins*. These substances are characterized by an abundance of aromatic units and phenolic —OH, and are believed to be three-dimensionally cross-linked "biopolymers" of coniferyl, sinapyl, and *p*-coumaryl alcohols (see Fig. 2.1.11).

Tannins, a secondary component of higher plants, resemble glycosides and differ from them merely in that the linkage to the sugar component is an ester

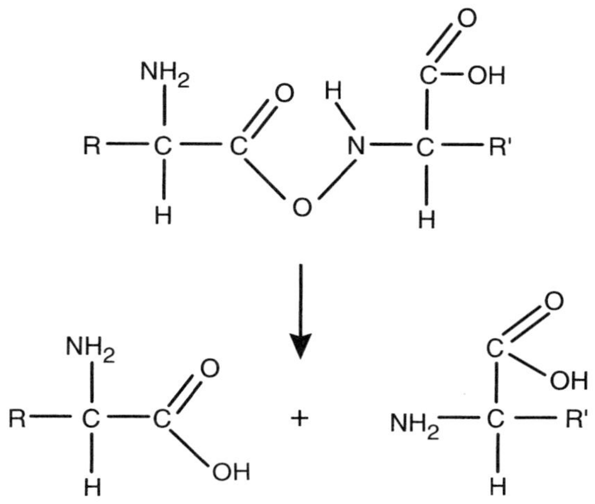

FIGURE 2.1.8 Hydrolytic dissolution of a peptide linkage.

group. The acid component of the ester is commonly gallic (a) or *m*-digallic acid (b):

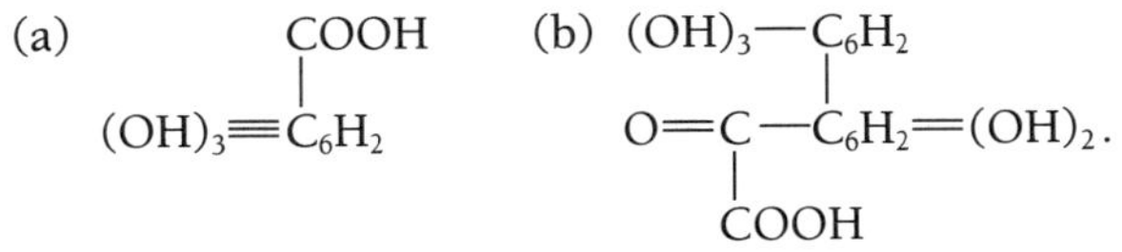

(a)

$$(OH)_3{\equiv}C_6H_2{-}COOH$$

(b)

$$(OH)_3{-}C_6H_2{-}C(=O)(COOH){-}C_6H_2{=}(OH)_2.$$

2. THE BIOSOURCES

The life forms that donated their substance to the biosources from which fossil hydrocarbons eventually formed, and consequently the compositions of the biosources, depended upon when and in what environment they flourished.

Because photosynthesis, which can be formally represented by

$$6\ CO_2 + 6\ H_2O \rightarrow C_6H_{12}O_6 + 6\ O_2,$$

will only proceed in the presence of chlorophyll, the manner in which the chemical source materials formed and then enabled life to begin is still conjectural. It is only possible to identify the earliest biota, i.e., primitive autotrophic phyto- and zooplankton [8], which appeared in the open seas some 10^9 years ago. These met their carbon requirements from CO_2 and/or CO_3^-, obtained their energy from atmospheric N_2, and thereby initiated formation of an

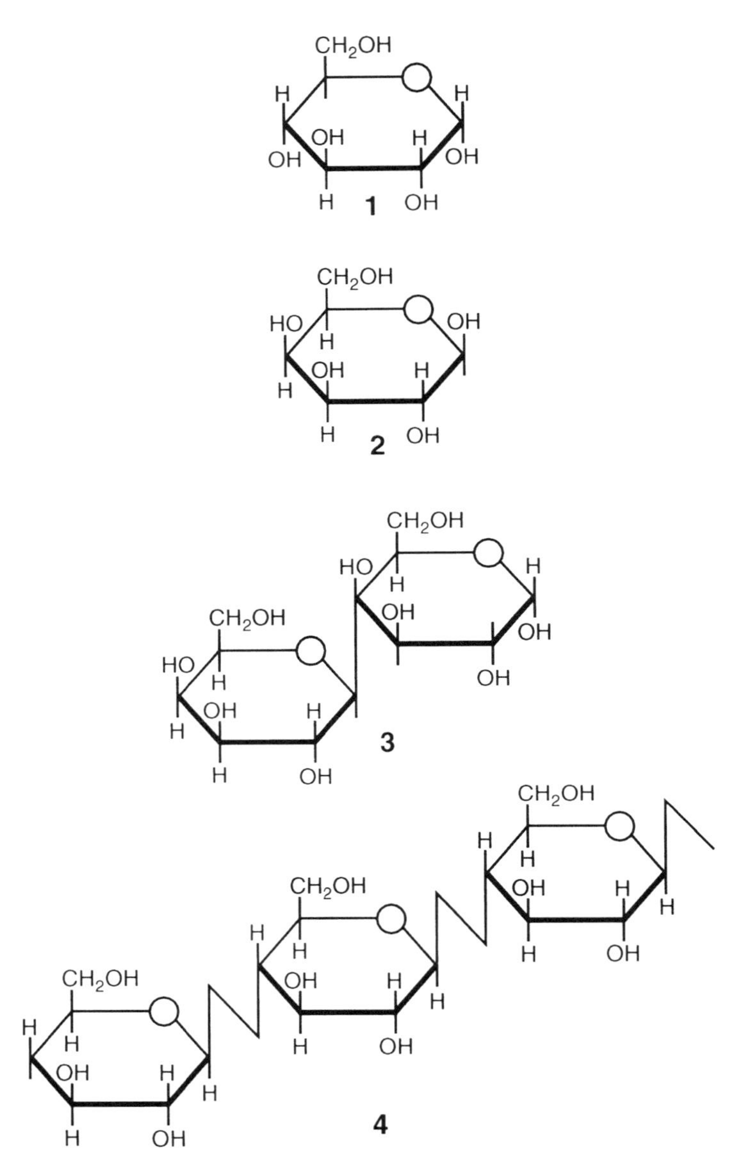

FIGURE 2.1.9 Some simple carbohydrates: (1) glucose; (2) D-galactose; (3) lactose; and (4) a structural element of cellulose.

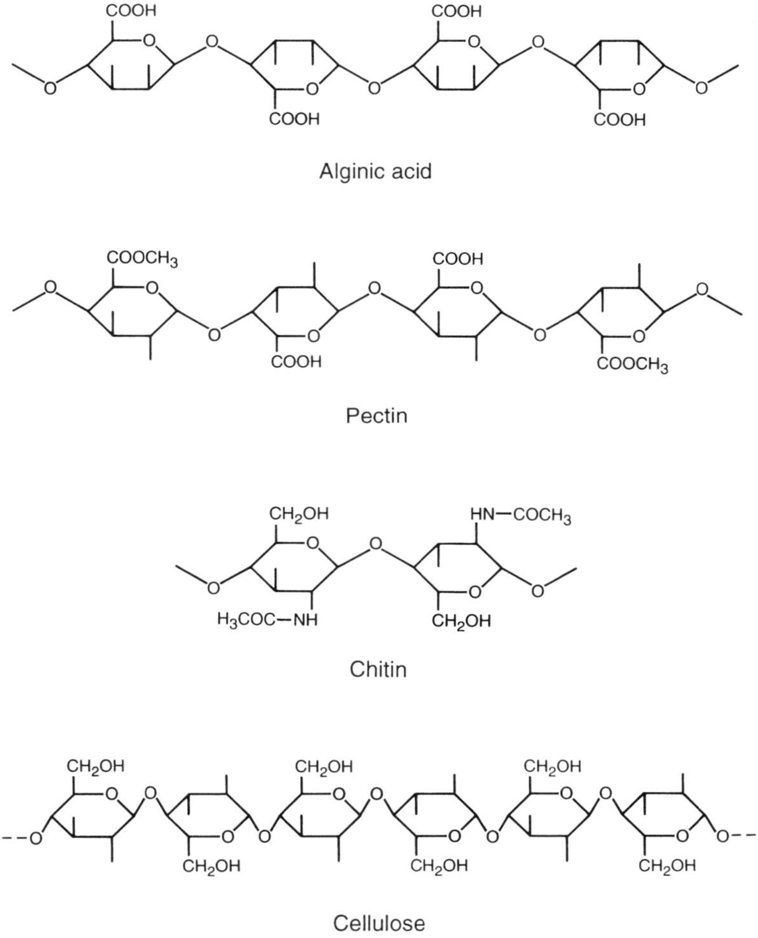

COOH
O
O
COOH
O
COOH
O
O
COOH
O

Alginic acid

$COOCH_3$
O
O
O
COOH
O
COOH
O
O
O
$COOCH_3$
O

Pectin

CH_2OH
O
O
$H_3COC—NH$
HN—$COCH_3$
O
CH_2OH
O

Chitin

CH_2OH
O
O
CH_2OH
O
O
CH_2OH
O
O
--O
O
CH_2OH
O
O
CH_2OH
O
O
CH_2OH
O--

Cellulose

FIGURE 2.1.10 Structure elements of cellulose and some closely related compounds; "open" bonds carry —OH groups

oxygen-containing atmosphere [9]. But since early Paleozoic times, evolution and topographic changes have progressively diversified the production of organic matter [10] and allowed it to proceed in formative environments that ranged from anoxic marine to oxic alpine (Table 2.2.1).

In anoxic subaquatic sedimentary environments, which usually lie several hundred kilometers off a coastline (Kruijs and Barron, 1990), biomasses formed primarily from unicellular diatoms [11] and dinoflagellates [12], but were occasionally augmented by algal matter; these accumulations betray alter-

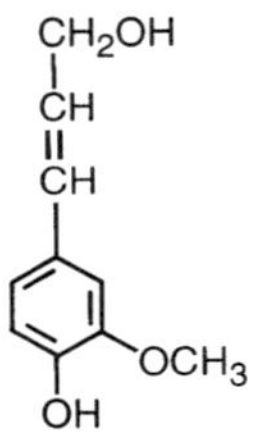

FIGURE 2.1.11 Building blocks of lignin: (1) coniferyl alcohol; (2) sinapyl alcohol; (3) *p*-coumaryl alcohol; (4) a structure element of lignin.

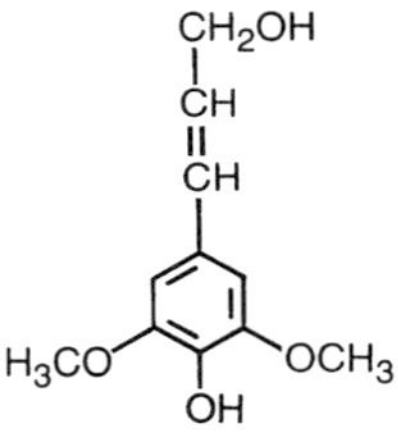

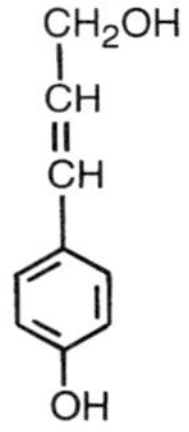

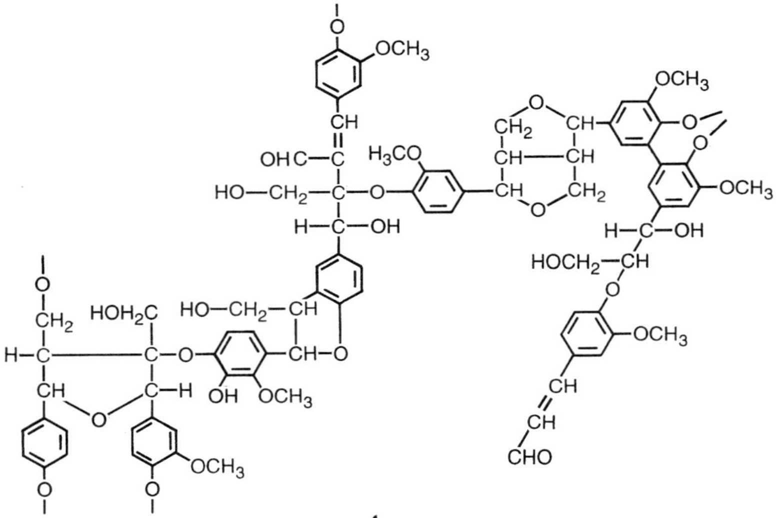

ation by marine organisms in preference for generating C_{15}, C_{17}, and C_{19} *n*-alkanes.

In paralic environments—that is, on suboxic continental shelves and in offshore, deltaic, and lacustrine waters in which biomasses were also shielded against oxidative attack and could putrefy—algal matter, supplemented by less lipid-rich seeds, pollen, and fungal spores from terrestrial vegetation, created sapropelic matter that stood compositionally between marine and terrestrial biomasses.

TABLE 2.2.1 Low-O_2 Regimes and Corresponding Biofacies[a]

Oxygen, ml/liter	Environment	Biofacies
2.0–8.0	oxic	aerobic
0.2–2.0	dysoxic	dysaerobic
[1.0–2.0 moderate]		
[0.5–1.0 severe]		
[0.2–0.5 extreme]		
0.0–0.2	suboxic	quasi-anaerobic
0.0[b]	anoxic	anaerobic

[a] Tyson and Pearson (1991).
[b] H_2S present.

In wetlands like the Florida everglades, and in other domains in which the water table lay at or near the sediment/atmosphere interface and the water was sufficiently acidic (pH 3–4) to inhibit microbial activity (Smith, 1957, 1962), biomasses were predominantly produced from putrefying reeds, primitive mosses and liverworts (*Bryophyta*), and/or ferns and tree ferns (*Pteridophyta*).

And in humid continental domains intermittently open to air, biomasses formed mainly from *Gymnospermae* and *Pteridospermeae* (the forerunners of contemporary conifers and cycads) and later, from flowering and fruit plants (*Angiospermae*). The dominant components of these biomasses were therefore derived from celluloses and lignins rather than from lipids, and alteration by abiotic O_2 and terrestrial biota was reflected in a marked preference for generating C_{27}, C_{29}, and C_{31} *n*-alkanes as well as small amounts of even-numbered C_8–C_{26} straight-chain fatty acids, mostly represented by palmitic (C_{16}) and stearic (C_{18}) acids [13].

Except where topographic features interfered, these domains merged into one another and consequently promoted a spectrum of organic matter in which components derived from celluloses and lignins gradually (and at the expense of lipids) became more prominent toward dry land.

3. DIAGENESIS

Subject to cyclical variations in the abundances of biomass precursors, formation of organic matter has continued uninterruptedly, although not uniformly across the globe, into the present [14], and this allows the major chemical changes that altered biomass compositions in different environments to be inferred from what is known about the decay of contemporary debris.

Whenever open to microbial and/or abiotic oxidative degradation, organic matter is chemically reprocessed and thereby provides energy sources for future generations of fauna and flora. But how the melange of materials that constitute a biomass is altered is determined by two site-specific factors—the biota that produced the organic carbon and the environment in which they flourished and decayed.

The extremes are, as already noted, (i) alpine, dry meadows and forests, and (ii) anoxic marine environments.

In the former situation, organic matter—derived from vegetation with 50–70% celluloses + lignins—is fully open to the atmosphere, and if oxidation is not prematurely arrested, it will promote dry rot that over time degrades the debris to CO_2, H_2O, and a fibrous charcoal-like material known as fusain [15]. Formation of a humus is substantially precluded.

However, in an anoxic marine environment in which organic matter is attacked by anaerobic microorganisms and *putrefies,* carbohydrates, proteins, and lipids are enzymatically degraded and then collectively produce a polymeric material from which lipid-rich kerogens (see Section 5) can develop. Although the minutiae of this process are not fully understood, it is known to entail (Bouska, 1981)

1. hydrolysis of cellulose, proteins, and fats to, respectively, sugars, amino acids, and fatty acids;
2. formation of mercaptans (or thiols);
3. evolution of CH_4, NH_3, H_2O, H_2S, and CO_2;
4. secondary condensation reactions that eventually produce H-rich but substantially insoluble "bituminous" matter.

Hunt (1979) has suggested that the reactions that generate this matter mainly involve the following:

1. hydrogen disproportionation, exemplified by

$$\alpha\text{-pynene} \rightarrow p\text{-cymene} + p\text{-menthane}$$

or

$$\text{abietic acid} \rightarrow \text{retene} + \text{fichtelite};$$

2. decarboxylation and reduction, exemplified by

$$\underset{\text{palmitoleic acid}}{2\,C_{15}H_{29}COOH} \rightarrow \underset{\text{pentadecane}}{C_{15}H_{32}} + \underset{\text{pentadecene}}{C_{15}H_{30}};$$

3. deamination, decarboxylation, and reduction, as in

$$\underset{\text{methionine}}{H_3C.S.(CH_2)_3(NH_2)COOH} \rightarrow \underset{\text{propane}}{C_3H_8} + \underset{\text{methyl mercaptan}}{CH_3SH} + NH_3 + CO_2;$$

4. β-carbon dealkylation and reduction, as in

$$\underset{\text{ethyl benzene}}{C_6H_5CH_2CH_3} \rightarrow \underset{\text{toluene}}{C_6H_5CH_3} + CH_4;$$

5. deformylation, as in

$$\underset{\text{benzaldehyde}}{C_6H_5CHO} \rightarrow \underset{\text{benzene}}{C_6H_6} + CO.$$

But between the extremes of oxic and anoxic environments lie paralic and continental domains in which the decay of organic matter is caused by aerobic as well as anaerobic microorganisms, and how decay proceeds is then governed by temperatures, humidity, and accessibility of the substrate to atmospheric oxygen.

When deposited in dysoxic stagnant lacustrine, deltaic, or shallow marine waters, organic matter suffers little oxidative degradation and mostly putrefies much like similar material in an anoxic or severely disoxic environment. An example is the putrefaction of spores, pollen, and leaf cuticles carried into a paralic environment by wind and/or floodwaters.

More far-reaching changes do, however, become manifest in swamps and marshlands where the organic matter is at least transiently open to the air. Under such conditions, lipids undergo little more than O_2-promoted polymerization, and pigments such as chlorophyll survive by rearranging intramolecularly into stable porphyrins (see Section 4). But celluloses are very rapidly degraded to sugars; lignins are oxidized to alkali-soluble humic acids, which slowly break down to hymatomelanic acids, fulvic acids, and, eventually, water-soluble benzenoid derivatives; glycosides are hydrolyzed to sugars and aglycons (such as sapogenins and derivatives of hydroquinone); and proteins are denatured by random scissions of their polypeptide chains to yield ill-defined slimes and free amino acids.

Over time, these processes convert the organic debris into a more or less extensively aromatized humus in which primary degradation products frequently interact further [16].

4. BIOMARKERS

The descent of fossil hydrocarbons from faunal and floral organisms, which has to this point only been *asserted*, is validated by *biomarkers* that can be traced to antecedent biota, survived diagenesis and subsequent catagenesis (see Section 5), and can now be unequivocally identified in crude oil and coal.

The most prominent biomarkers attesting to the biogenic origin of crude oil are *porphyrins* that are derived from chlorophyll [17]. The core of these

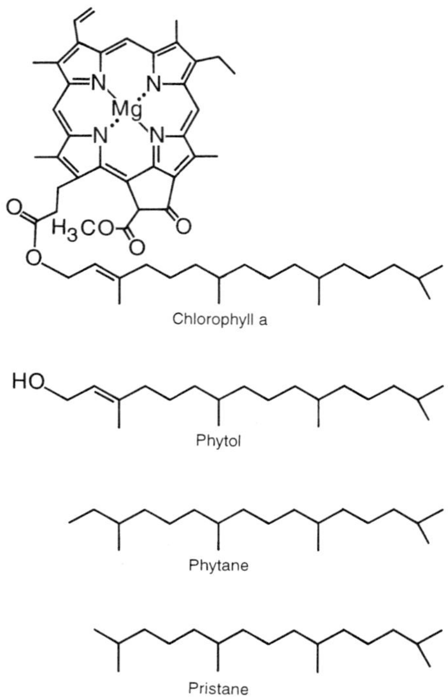

Chlorophyll a

Phytol

Phytane

Pristane

FIGURE 2.4.1 Chlorophyl-A, and three derivative compounds from its side chain.

compounds is a tetrapyrrole unit that appears in hundreds of homologs when organic matter is progressively altered during diagenesis and catagenesis. Such alteration is illustrated in Fig. 2.4.1 by chlorophyll-A and three biomarker compounds that were components of the alkanoid chain of the chlorophyll; in Fig. 2.4.2 by a porphin and Ni-chelated porphin; and in Fig. 2.4.3 by representative chlorins.

Other important biomarkers in crude oils are terpenoids, exemplified by steranes and hopanes, *n*-paraffins with odd numbers of C-atoms, and iso-branched chains.

Less direct, but also compelling evidence for biogenic origins of oil is offered by its carbon isotope ratio, which is usually written as

$$\delta^{13}\mathrm{C} = 1{,}000\{[(^{13}\mathrm{C}/^{12}\mathrm{C})_a/(^{13}\mathrm{C}/^{12}\mathrm{C})_b] - 1\},$$

where subscripts a and b indicate reference to the sample and standard. The most widely used standard is a belemnite from the Peedee Formation of South Carolina, USA [18]; as shown in Table 2.4.1, which lists some representative

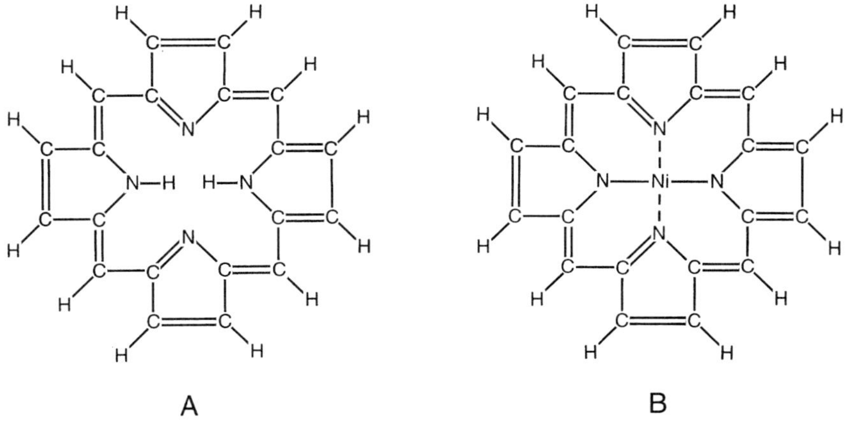

FIGURE 2.4.2 (A) a porphin and (B) a Ni chelate of a porphin.

isotope compositions, a negative value of $\delta^{13}C$ demonstrates biological or biogenic origin.

The biogenic origin of coal is even more directly proven by botanical features that can be identified in thin sections or polished surfaces of coal when viewed under a microscope. Such fossilized, but otherwise well-preserved entities or *phyterals* (Cady, 1942) include leaf fragments, woody structures, pollen grains, fungal spores, pollen, and the like, and have provided important information about paleoclimates.

5. CATAGENESIS

Diagenetic change is terminated by premature arrest of microbial and oxidative degradation of the organic material. That occurs when (i) marine accumulations are gradually buried under other sediments and eventually subjected to geothermal temperatures of 55–60°C, or (ii) decaying vegetal debris in continental wetlands and swamps is inundated by an advancing sea and covered by the silt it carries.

In both cases, catagenesis [18], the final phase in the evolution of a biomass into fossil hydrocarbons, progresses as a response to increasing overburden pressures and geothermal temperatures. But the physical status and composition of organic matter at termination of diagenesis demands differentiation between what subsequently develops under aquatic and continental conditions (see Potonié, 1908; [19]).

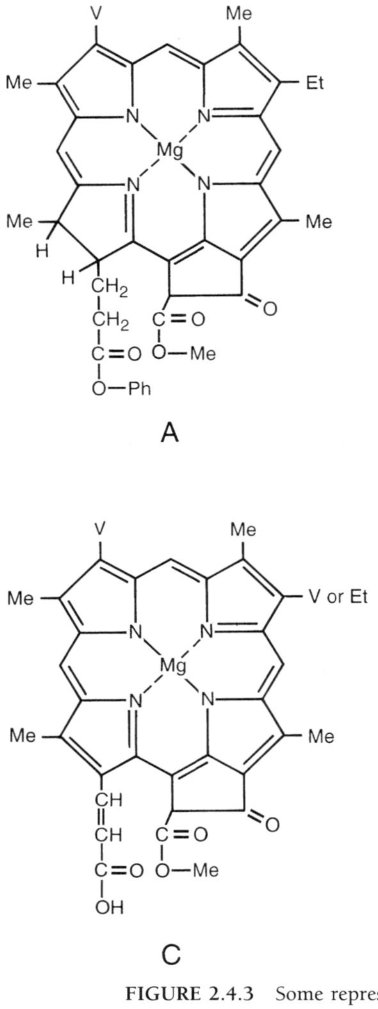

FIGURE 2.4.3 Some representative chlorin structures.

The Nature of Kerogens

In aquatic domains, chemically reworked organic matter from marine flora and fauna settles concurrently with, and into, inorganic sediments and is therefore finely disseminated throughout the sediments. In that form it is termed *kerogen,* a designation first used in reference to the substantially insolu-

TABLE 2.4.1 Typical Isotopic Compositions of Vegetation, Crude Oils, and Natural Gas

	$\delta^{13}C$, o/oo
Peedee belemite (PDB)	0
Typical limestone	+5
Marine vegetation	−10 to −18
Freshwater vegetation	−22
Plankton lipids	−30
Terrestrial vegetation	−22 to −26
Crude oils	−20 to −32
Biogenic gas	−55 to −85
Thermogenic gas	−25 to −60

ble organic carbon [21] of oil shales (Crum-Brown, 1927), but later extended to seemingly similar organic material in black shales (Breger, 1961), and now rather loosely applied to all polycondensed solid biomass-derived matter. It has been described as a high-molecular-weight nitrogenous humic substance or "geopolymer" (Hunt, 1979) that results from diagenetic reworking of biopolymers [22] and is defined as a *disseminated minor organic component of inorganic sediments* (Tissot and Welte, 1978; Schobert, 1990). The insertion of *minor* in this statement reflects recognition that even the richest of marine sediments, oil shales, rarely contain more than 6–18% kerogen.

Kerogen is ubiquitous in sedimentary rocks and carbonaceous shales, as well as in oil shales, and represents a quantity of organic carbon that exceeds by two orders of magnitude the total of all known coal, oil, and gas resources. However, kerogen compositions are as diverse as the biota and formative environments that produced them, and it is therefore proper to speak of kerogen in the plural.

A nomenclature that reflects this diversity and explicitly defines the nature of different kerogens is used by palynologists, who differentiate among five types:

1. *Algal* kerogen—morphologically recognizable marine or nonmarine algal matter, with occasional minor contributions from other sapropelic matter
2. *Amorphous* kerogen—of marine, deltaic, and/or lacustrine origin, and representing sapropelic organic matter mainly derived from plankton and other simple biota
3. *Herbaceous* kerogen—mainly of continental origin and characterized by morphologically recognizable spores, pollen, cuticles, leaf epidermis, and other discrete cell material

TABLE 2.5.1 Classification of Organic Matter in Sedimentary Rock[a]

Kerogen	Sapropelic			Humic	
	algal	amorphous	herbaceous	woody	coaly
	types I,II		type II	type III	
H/C	1.7–0.3		1.4–0.3	1.0–0.3	0.5–0.3
O/C	0.1–0.02		0.2–0.02	0.4–0.02	0.1–0.02
Org. source	marine lacustrine			terrestrial	
Maceral group[b]	liptinite (exinite)			vitrinite	inertinite
Maceral[b]	alginite	amorphous	sporinite	collinite	fusinite
			cutinite	telinite	micrinite
			resinite, etc.		
Fossil fuels	oil, sapropels		oil, gas	gas, tar	humic coal

[a] Hunt (1979).
[b] Equivalents in coals.

4. *Woody* kerogen—originating in peat swamps and composed of small fibers and/or fibrous matter with clearly expressed woody structures
5. *Coaly* kerogen—consisting of variously reworked organic material and naturally charred, oxidized, and/or fungally degraded vegetal matter

Such distinctions are important, because yields and composition of oil and/or gas that kerogens furnish on heating depend on their H/C ratios, and therefore on their origin. Algal, amorphous, and herbaceous kerogens, collectively referred to as "sapropelic" kerogens, thus tend to deliver relatively light oils [23], whereas woody and coaly kerogens—Potonié's humic matter [20]—furnish tars and hydrocarbon gases.

These aspects, and relationships between kerogen and coal, are expressed in a kerogen classification (Table 2.5.1) that shows how H/C and O/C ratios change as kerogens mature, and also indicates the equivalent coal macerals and maceral groups (Chapter 5). A similar classification, which employs a H/C vs O/C diagram to define three kerogen types (Fig. 2.5.1) and then relates these to their pyrolytic products (Table 2.5.2), has also been proposed by Tissot and Welte (1978).

However, given the definition of "kerogen" and the manner in which coal precursors were deposited, explicit designation of type III kerogen as a coal precursor is inappropriate, and adoption of three kerogen types as a framework for classification runs counter to statements (Kinghorn, 1983) that there is no discernible number of different kerogens. What determines kerogen compositions is, in fact, ultimately (i) the source materials and microbial changes wrought in them during deposition; (ii) the biochemical alteration of the

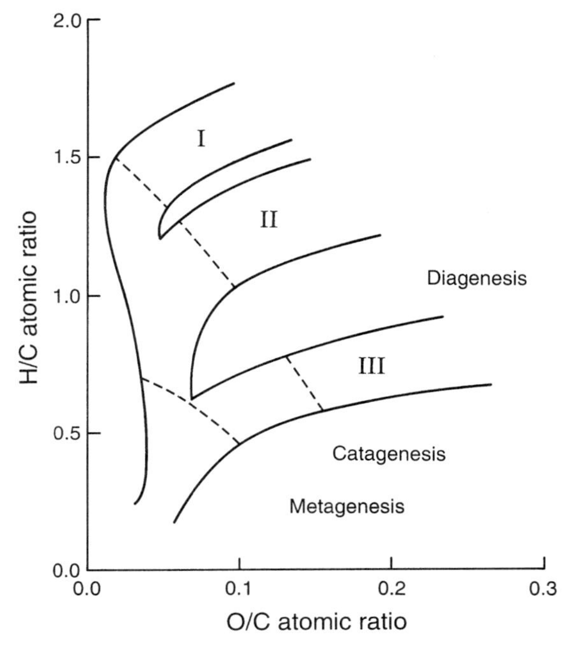

FIGURE 2.5.1 Proposed developmental pathways of the principal kerogen types. (after Tissot and Welte, 1978; reproduced with permission)

biomass during diagenesis; and (iii) subsequent thermal alteration (Hunt, 1979), and this is underscored by frequent localized horizons of *oil-producing coals* in tidal-dominated deltaic plains. The Tertiary coals of Indonesia's Mahakram delta, which consist almost wholly of vitrinite, thus show hydrogen indices up to 460 (as against ~150 for "normal" vitrinites) and yield oils in which waxy *n*-alkanes (and occasionally cyclic derivatives of resins) predominate [24]. Thompson *et al.* (1985) have suggested that such coals are typical products of consistently wet tropical climates and represent reworked ombrogenous peat that formed at proximal margins of deltaic plains. Similar inferences have been drawn from an investigation of other potential "oil source rock" coals (Bertrand, 1984) and from a study of some oil-generating Tertiary Texas (USA) coals (Mukhopadhyay and Gormly, 1989). Cooper (1990) has also drawn attention to the Jurassic coals of the Eromanga (Australia) Basin, where the precursor material accumulated in lagoons on the fringes of an inland freshwater lake: here, the detritus, which contained high proportions of spores and leaf cuticle, was sometimes augmented by algal matter and produced

TABLE 2.5.2 Schematic Development of Fossil Hydrocarbons from Precursor Organic Matter

Organic matter			
aerobic diagenesis		anaerobic diagenesis	
↓		↓	↓
kerogen III	↓	kerogen II	kerogen I
	gas		
↓		↓	↓
coals		sapropels	oils
			↓
			thermal gas

early-mature coals, which now show hydrogen indices ranging as high as 500.

The Kerogen → Hydrocarbon Transformation

Further reworking of diagenetically altered biomasses proceeds during abiotic geothermal processing between ~50 and 200°C and corresponds to the metamorphic development of coal [25]. Data for kerogens from the Lower Toarcian shales at 700 and 2500 m provide an indication of the structural changes that accompany the transformation of kerogen during catagenesis (Table 2.5.3): they reflect structural modification by thermal cracking reactions that are occasionally accelerated by tectonic upheavals or upward movement of molten magmas that augment geothermal heat. But how organic matter is affected by such changes depends on its origin or, more specifically, on its H/C or C_{ar}/C ratio.

Sapropelic Kerogens

Kerogens formed from algal, amorphous, and/or herbaceous source materials (and, as noted earlier, collectively termed sapropelic kerogens) are characterized by a preponderance of aliphatic carbon chains randomly connected by aromatic and/or naphthenic moieties; in such structures, thermolytic cleavage of C—C bonds causes elimination of aliphatic CH as gaseous and/or liquid hydrocarbons and leaves a proportionately carbon-enriched residue [26].

TABLE 2.5.3 Structural Changes in Kerogen as a Function of Depth[a,b]

	700 m			2500 m		
	C	H	O	C	H	O
Kerogen	940	1220	120	806	876	40
Hydrocarbons	11	17	—	106	180	—
Resins	14	21	1.2	33	41	1.6
Asphaltenes	13	17	1.4	22	23	1.6
MAB extract	22	31	3.7	28	35	3.4
CO_2				5		10
H_2O					102	51
H_2S					36	
	1000	1306	126.3	1000	1293	111.6

[a] Expressed as atomic balances per 1000 C atoms; N and S omitted.
[b] Tissot and Welte (1978).

Concurrent reactions, which reduce the O/C ratio of the precursor without significantly altering its H/C ratio, include (Barker, 1985):

1. thermal dehydration of hydrolyzed oils, fats, and saponified waxes to alkanes, as in

$$RCH_2CH_2OH \rightarrow RCH{=}CH_2OH + H_2O,$$

which is catalyzed by clays or by acids in solution (e.g., H_3O^+)
2. thermal decarboxylation of carboxy acids, as in

$$R(CH_2)_nCH_2COOH \rightarrow R(CH_2)_nCH_3 + CO_2;$$

3. decarboxylation of esters, which initially furnishes an alkene and acid, as in

$$RCH_2CH_2OOCR' \rightarrow RCH{=}CH_2 + HOOCR',$$

but at higher temperatures yields an alkane and CO_2 through loss of COOH; and
4. where lignin or lignin derivatives have contributed to the kerogen, condensation of phenols to polynuclear aromatics + CO_2.

Laboratory studies have shown that oils can also form by low-temperature cracking reactions catalyzed by associated mineral matter. Engler (1911) thus observed hydrocarbon liquids forming when oleic acid or its derivatives were heated over clays at 80–100°C. Frost (1945) generated saturated as well as unsaturated C_8–C_{30} hydrocarbons by heating cyclohexane, methylcyclopentane, or octyl alcohol in presence of clay, and in a similar reaction obtained benzene from cyclohexanone. Bogomolov and Panina (1961) heated oleic acid

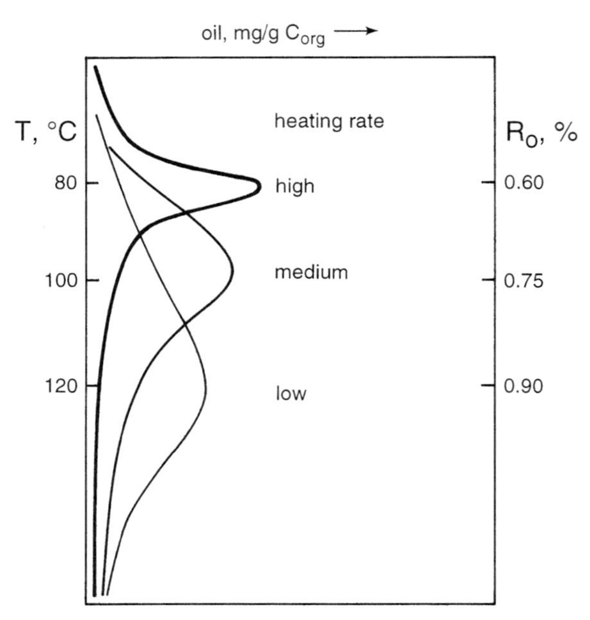

FIGURE 2.5.2 Kerogen degradation during catagenesis: the temperature zones in which oil formation is intense vary inversely with geothermal heating rates.

with clays at 150°C and thereby produced a hydrocarbon mix of which 50% boiled in the gasoline/kerosine range, while the remainder consisted of distillate oils and residua. Day and Erdman (1963) generated toluene, *m*-xylene, 2,6-dimethylnaphthalene, ionene, and other aromatics by heating β-carotene in benzene at 120°C. Heating a mix of oleic and stearic acids produced heptane and octane isomers, as well as some C_7 and C_8 naphthenes (Petrov *et al.*, 1968).

Like rates of most other chemical reactions, the rates at which kerogen will decompose will roughly double with each 10°C rise. But over very long periods of time and under high overburden pressures, such decomposition can proceed at much lower temperatures than needed in laboratory tests; there exists, therefore, an oil "window," usually at 2000–4000 m depth, in which oil formation begins at depths equivalent to ~60°C; accelerates as temperatures rise to ~120–130°C; and slowly peters out toward ~170°C. This is followed by a brief interval in which "thermal" gas (C_1–C_4) is generated by severe cracking of the residual (carbon-enriched) organic matter, and that, too, ceases near 200°C as the sediment enters a metamorphic (or metagenetic) phase (Fig. 2.5.2).

Because an accelerating oil producion (indicated by Fig. 2.5.2) implies increasingly intensive molecular cracking, the entire process is reflected in decreasing oil densities and average molecular sizes. Thus, whereas the onset of oil formation at ~60°C generates heavy, viscous oils composed of relatively large molecules such as tridecane ($C_{13}H_{28}$, b.p. 234°C, η = 1.55 mPas) or octadecane ($C_{18}H_{38}$; b.p. 317°C, η = 2.86 mPas), higher temperatures (at greater depths) generate increasing proportions of C_5–C_{10} hydrocarbons, and at 150–160°C the only detectable products are C_1–C_4 hydrocarbon gases.

Laboratory studies directed toward elucidating the mechanisms of such thermal cracking have led to the conclusion that it entails homolytic as well as heterolytic scission of carbon bonds. The former generates free radicals by which long-chain hydrocarbons are transformed into shorter *n*-and/or iso-alkanes, as in

$$H_3C{-}(CH_2)_4{-}CH_3 \rightarrow {-}C{-}C{-}C{-}C{-}^* + {-}C{-}C{-}^* \rightarrow n{-}C_4H_6 + H_2C{=}CH_2,$$

whereas the latter creates carbonium ions and negative carbanions that, in the presence of a Lewis acid, participate in reactions of the type

$$CH_3{-}H{:}CH{-}CH_2{-}CH_3 + A^+ \rightarrow CH_3{-}H^+{-}C{-}CH_2{-}CH_3 + A{:}H$$

$$\downarrow$$

$$H{:}\underset{CH_3}{\overset{CH_3}{\overset{|}{\underset{|}{C}}}}{-}CH_3 + A^+ \quad \leftarrow \quad \underset{CH_3}{\overset{CH_3}{\overset{|}{\underset{|}{C^+}}}}{-}CH_3 + A{:}H$$

In sediments in which moisture has little effect beyond slowing reaction rates, the required catalytic activity would arise from mineral matter with such Lewis-acid sites as Al^{3+} or Fe^{3+}.

Hydrocarbon liquids formed at relatively low (<100–120°C) temperatures can, of course, undergo further thermal alteration in the host rocks into which they migrated (see Chapter 3). That could, for instance, happen in a reservoir in which a nearby igneous intrusion develops higher temperatures than prevailed in the mature source rock, and where such secondary cracking occurs, the additional light hydrocarbons (including <C_7) would also cause significant natural deasphalting of the oil by slow precipitation of asphaltenes.

Humic Matter

As the proportions of aromatic carbon in kerogens increase, their ability to generate predominantly aliphatic hydrocarbons necessarily diminishes, and highly aromatized precursors can therefore furnish little more than C_1–C_3 gases. Humic matter, which behaves like an H-poor kerogen during catagenesis, is a case in point.

TABLE 2.5.4 Relationships between Coal and Oil Maturity (after Fuller, 1919, 1920)

% Carbon[a]	Petroleum quality
<71	accumulations of heavy petroleum in U.S. coastal plains and unconsolidated Tertiary formations
71–75	principal (medium) petroleum pools in Ohio, Indiana, and midcontinent states
75–79	principal light petroleum and gas pools in the Appalachian region
79–83	few commercial pools,but petroleum is of excellent (light) quality; gas occurrences frequent but minor
83–89	no commercially significant oil or gas pools
86–89	petroleum and/or gas accumulations rare

[a] Recalculated from "fixed carbon" data (NB).

Extensive aromatization of humic matter during diagenesis is facilitated by the high proportion of lignin in the plant debris and by its episodal exposure to atmospheric oxygen. The presence of lignin allows condensation and cyclization via phenolic —OH, and oxygen not only progressively destroys aliphatic and naphthenic entities, but also supports concurrent polymerization. Together, these processes create a humic mass in which aromatic carbon represents >50% of the total carbon, aliphatic structures are rarely longer than $—CH_2CH_2CH_3$, and very few peripheral functions other than $—CH_3$, $—CH_2CH_3$, —OH, and —COOH persist. Because aromatic structures are thermally much more stable than nonaromatic ones, and thermolysis of naphthenic components in aromatic/naphthenic entities (see Chapter 7) cannot produce chains longer than C_4, humic matter is thus substantially limited to furnishing C_1–C_4 hydrocarbons, CO, CO_2, and H_2O unless temperatures exceed the ~350–400°C needed for active thermal decomposition (Chapter 6).

In contrast to sapropelic kerogens, humic matter thus represents the H-poor end of a series of diagenetically altered biomasses, and develops into increasingly aromatic coal when maturing at gradually rising geothermal temperatures. By Hilt's Rule (1873), later repeatedly confirmed by more detailed measurements and expressed by other "rank" parameters (see Chapters 4 and 5), the "volatile matter" contents of coal thus fall as overburden thicknesses increase and the aromatic carbon (C_{ar}) gradually rises to >80% in anthracites.

This inverse relationship between the maturities of sapropelic kerogens and humic matter allows the use of coal for delineating oil and gas "windows" and thereby defining the depth limits beyond which a search for significant oil and/or gas accumulations would be futile. Table 2.5.4 illustrates this for major oil occurrences in the United States, and Table 2.5.5 does so with summary data for some 145 Canadian oil pools. The liquids window (~65–135°C)

TABLE 2.5.5 Relationships between Coal and Oil Maturity[a]

R_0[b] (%)	Carbon[c] (%)	Paleotemp. (°C)	Petroleum/gas occurrences
0.5–0.9	76–83	68–116	almost 90% of all crude oil; 40% reached current status at 82–96°C
>1.20	>88	>143	petroleum absent[d]
		106–177	thermal gas

[a] Hacquehard (1975).
[b] Vitrinite reflectance (see Chapter 5).
[c] Estimated from R_0 (NB).
[d] Thermal gas and petroleum and thermal gas coexist up to 135°C.

identified in the latter [27] is virtually identical with those defined by others (Landes, 1966; Tissot and Welte, 1978).

6. THE "HEAVY" HYDROCARBONS

Despite broad consensus about the origins of fossil hydrocarbons, the formative histories of heavy *sapropelic* hydrocarbons remain surrounded by several still-unresolved questions.

With respect to sapropelic *coals* (see Chapter 4), these questions concern the sites at which their precursor biomasses accumulated. Because such coals contain 8–12% H and resemble sapropelic kerogens [28], they are generally supposed to derive from pollen, fungal spores, and leaf fragments that were transported by wind and/or water from continental growth sites and deposited into substantially dysoxic paralic environments. This view is in part supported by microscopic studies that show an abundance of such fossilized plant remains embedded in the coal matrix. However, given the apparent densities of pollen grains and spores, it is no simple matter to envisage how sufficient volumes of such lightweight debris, even when loaded into relatively *still* waters, could accumulate in any one locality to enable massive development of sapropelic coal.

Also, with respect to asphaltic "petroleum hydrocarbons" such as bitumens, uncertainties arise from the fact that in few, if any, instances do these substances now reside in typical oil source rocks. Because current concepts of oil migration (Chapter 3) effectively rule out significant migration into present-day habitats on the grounds of excessive molecular size, it has been conjectured that they represent hydrocarbon precursor material whose catagenetic development was prematurely arrested by major uplift of the source strata, or that they are residua from which lighter components escaped in some manner. But in the absence of any reasonably direct evidence, neither of these notions is particularly plausible, and greater interest attaches therefore to the view that heavy oils are biodegraded, and that asphaltics represent biodegraded oils that have been stripped of residual low-molecular-weight hydrocarbons by water washing (Chapter 3). This view may explain the developmental history of Alberta's massive oil-sand deposits; however, whether it can be generalized and, in particular, account for Venezuela's Orinoco Tar Belt hydrocarbons and Utah's oil sand accumulations is uncertain. If one proceeds from constructs that model kerogen → oil transformations, it is by no means impossible that some heavy hydrocarbons migrated during the *early* stages of catagenesis (when such hydrocarbons would be generated [29]). The fact that degradation of kerogen, and consequent generation of heavy oils, begins at temperatures that vary *inversely with heating rates* (see Fig. 2.5.2)—and hence sets in under *overburden pressures that vary in like manner*—offers some direct support for this argument.

7. ABIOTIC FORMATION OF OIL

Although the biogenic origin of fossil hydrocarbons is well established, the possibility, first advanced by Mendeleef in the early 1900s, that some small oil accumulations may have been formed abiotically cannot be peremptorily dismissed. One plausible mechanism, premised on the existence of metal carbides at great depths, is an interaction with hydrothermal solutions, as in

$$FeC_2 + 2H_2O \rightarrow HC{\equiv}CH + Fe(OH)_2$$

which would at high temperatures be followed by

$$HC{\equiv}CH \rightarrow C_6H_6$$

and formation of a complex mix of other hydrocarbons.

Another abiotic path to hydrocarbons presents itself in reverse shifting,

$$CO_2 + H_2 \rightarrow CO + H_2O,$$

and subsequent Fischer–Tropsch synthesis (see Chapter 10), e.g.,

$$CO + 3H_2 \rightarrow CH_4 + H_2O.$$

But also credible, and in some respects more interesting, are two routes that, although not involving abiotic generation in the strict sense of that term, proceed by pathways other than those usually accepted for formation of hydrocarbon liquids and gases.

Hunt (1972, 1979) has noted that crustal granitic and metamorphic rocks contain ~48×10^{20} g carbon. If half of this once formed kerogen in sediments, catagenesis (metamorphism) could have converted 5×10^{20} g carbon into CH_4 and higher hydrocarbons; that total would exceed the carbon mass in all known oil accumulations by two orders of magnitude. Parenthetically, one notes in this connection that metamorphism could also generate *heavy* hydrocarbons, because temperatures above 950–1000°C promote $CH_4 \rightarrow CH{\equiv}CH$, and acetylene could then readily polymerize to oils and asphaltics.

In a similar manner, hydrocarbons could conceivably have formed in Precambrian and Phanerozoic rocks and then been distilled from them by igneous intrusions (Baker and Claypool, 1970). However, because heat from intrusions such as dykes and sills is always rapidly dissipated into surrounding strata, hydrocarbon yields from that route would be small.

NOTES

[1] Branched-chain alkanes are more often generated by thermal cracking of terpenes.

[2] Chlorophyll *per se* combines a phytyl side chain with an N-bearing porphin nucleus. From the former, a wide variety of isoprenoid compounds in sediments and petroleums can develop, and the latter is a precursor of porphyrins. See Section 4 for definitions of these compound designations.

[3] "Essential oils" are volatile oils that bestow a characteristic odor upon the plant. They are often extracted for use in perfumes, flavorings, and pharmaceuticals.

[4] This term is collectively used for sugars and their polymers and reflects the empirical class formula $C_n(H_2O)_n$, which, misleadingly, suggests a hydrated form of carbon.

[5] The structures of saccharides (Figs. 2.1.9 and 2.1.10) can be better visualized by writing them in their Fischer projections: a monosaccharide such as glucose would thus be written in the cross notation

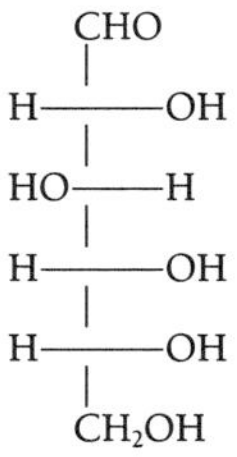

$$\begin{array}{ccc} & \mathrm{CHO} & \\ & | & \\ \mathrm{H} & - & \mathrm{OH} \\ & | & \\ \mathrm{HO} & - & \mathrm{H} \\ & | & \\ \mathrm{H} & - & \mathrm{OH} \\ & | & \\ \mathrm{H} & - & \mathrm{OH} \\ & | & \\ & \mathrm{CH_2OH} & \end{array}$$

in which each intersection is assumed to be occupied by a C atom. Di, tri, and polysaccharides can then be visualized by noting that each monosaccharide molecule contains either an aldehyde or a ketone function that allows it, analogous to the reaction between acetaldehyde and methanol,

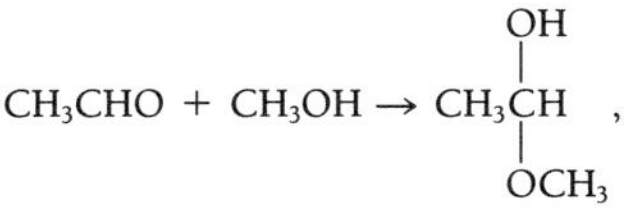

$$CH_3CHO + CH_3OH \rightarrow CH_3\underset{\substack{|\\ OCH_3}}{\overset{\substack{OH\\ |}}{C}}H \quad ,$$

to react with itself and create a cyclic hemiacetal or hemiketal (the format in which most monosaccharides exist). "Growth" into disaccharide acetals or ketals and, beyond, into oligosaccharides, etc., proceeds by further interactions with alcohols or ketones, ultimately creating naturally occurring polysaccharides with 100–3000 monosaccharide units.

[6] Acetal linkages between monosaccharide units are termed *glycoside* linkages, a glucoside linkage being an acetal group connecting two glucose units.

[7] There is some evidence that lignin assumes significantly different chemical forms in conifers and deciduous trees.

[8] These biota usually contributed their substance in a 10–15 : 1 ratio (Hunt, 1979).

[9] In this connection it does, however, merit noting that accumulation of organic carbon is often attributed to the fact that formal reversal of photosynthesis,

$$C_6H_{12}O_6 \rightarrow 6CO_2 + 6H_2O,$$

which a fully closed carbon cycle would demand, does not go to 100% completion, and that there is always some carbon leakage from the cycle.

[10] Changing land-mass configurations and climates since mid-Paleozoic times point unequivocally to a vast variety of biomass compositions. In Devonian times, the Northern Hemisphere's continents—*Laurentia* in the west and *Angara* in the east—were separated from each other and from the Southern Hemisphere's *Gondwana* landmass by the Tethys Sea, and whereas slow retreat of that ocean since L. Carboniferous (Mississippian) times toward the present limits of the Mediterranean fostered massive production of biomass from terrestrial vegetation in the Northern Hemisphere, prevailing desert conditions in Gondwanaland delayed that until Permian and Triassic times (when desert conditions prevailed in much of the *Northern* Hemisphere). Thereafter, shrinkage of a shallow Carboniferous Sea, which vertically bisected the North American continent, set the stage for abundant biomass formation in western Canada and the western United States during the Cretaceous, and fragmentation of Gondwanaland, as well topographic changes in the Americas, promoted production of biomasses from the late Mesozoic until the early Tertiary. Subject to regional climatic and topographic disparities, generation of C_{org} continued in the Quaternary—and into the present.

[11] Diatoms are unicellular microscopic algae with symmetrical siliceous exoskeletons.

[12] Protozoans characterized by possessing two flagella. Some members of this phylum cause the so-called red tide and generate highly toxic water-soluble alkaloids lethal to many aquatic species. Others, less harmful, are bioluminescent.

[13] Among lipids of higher plants, *n*-alkanes with odd numbers of C atoms are favored by a factor of 10 over those with even numbers. Identified *n*-alkanes range to dohexacontane ($C_{62}H_{126}$).

[14] A rough balance seems now to exist between primitive and more specialized producers: 50–60% of global C_{org} production is now ascribed to marine phytoplankton and bacteria (Hunt, 1979).

[15] Further reference to fusain, one of four coal lithotypes, is made in Chapters 4 and 5.

[16] Humic acids and humins are also frequently major components of lacustrine, deltaic, and coastal marine sediments—but are then assumed to be derived from wind- or water-transported (terrestrial) spores and pollen. This assumption is lent credence by the fact that marsh and swamp humus is significantly richer in aromatic moieties derived from lignin.

[17] The variety of porphyrin structures is reflected in the applicable nomenclature:

A *porphin* is an unsubstituted porphyrin.
A *porphyrin* is a porphin that has one or more substitutent side chains and that is usually chelated with a metal.
A *heme* is a porphyrin in which the chelated metal is iron.
A *chlorin* is a hydroporphyrin, a characteristic structural element of chlorophyll.

[18] A belemnite is a cephalopod resembling an octopus or squid, which first appeared in the Carboniferous era, proliferated throughout the Mesozoic, and became extinct in Eocene times.

[19] In relation to coal, catagenesis is designated as *metamorphism* or metamorphic development. However, unlike catagenesis, metamorphism is not limited to temperatures below 200°C, and therefore includes processes that petroleum geologists collectively term *metagenesis.*

[20] Early studies (Potonié, 1908) had already led to the conclusion that almost all biomass products could be divided into humic and sapropelic matter, the former resulting from decay and peatification of continental plant debris, and the latter produced by microbial attack on lipid-rich planktonic matter, algal debris, and spores in disoxic aquatic domains. Sapropelic matter (H/C = 1.3–1.7) is exemplified by coorongites, torbanites, tasmanites, and Siberian boghead coals of Jurassic age that are almost wholly made up of exinite (≡ liptinite). Petroleum geologists concerned with oil source rocks (which generally contain $<$2–3% organic matter; see Chapter 3) often use this approach by designating potentially good and poor oil sources as sapropelic or humic matter.

[21] Although routinely described as insoluble, kerogen is so only in common solvents such as acetone, benzene, or chloroform. In potent solvents, such as methylene chloride, pyridine, or ethylenediamine, it is or can be slightly soluble.

[22] This term is often applied to chemical precursors discussed in Section 1.

[23] Yields of *light* oils generally decrease quite rapidly from algal to herbaceous kerogens.

[24] Hydrogen indices and related parameters are discussed in Chapter 3.

[25] Metamorphic development or metamorphism are terms used to describe heat- and pressure-driven maturation of coal. Metamorphic development is substantially synonymous with catagenesis, but can include changes *above* 200°C and therefore encompasses what petroleum geologists define as metagenesis.

[26] The extent to which a kerogen has been transformed into hydrocarbons is measured by extactable bitumen, i.e., by the sum of hydrocarbons, resins, and asphaltenes (HC + R + A) or by HC only. If C_{org} denotes the total organic carbon, degradation of kerogen to liquid hydrocarbons is given by the bitumen ratio

$$\text{bitumen}/C_{org} = (\text{HC} + \text{R} + \text{A})/C_{org},$$

or, more simply, by the hydrocarbon ratio HC/C_{org}.

[27] The use of the vitrinite reflectivity R_o for assessing the maturity of kerogens in oil source rocks is reviewed in Chapter 3. R_o *per se* is discussed in Chapter 5.

[28] Far less abundant than their humic counterparts, sapropelic coals (or "sapropels") have attracted relatively little attention, and there is a distinct paucity of information about their geochemistry. However, their source materials, compositions, and pyrolytic behavior suggest that their precursors are more akin to oil shale kerogens than to the humic matter from which "humic" coals developed.

[29] This matter is further discussed in Section 2 of Chapter 3.

REFERENCES

Baker, D. R., and G. E. Claypool. *Amer. Assoc. Pet. Geol. Bull.* 54, 456 (1970).

Barker, C. Origin, composition and properties of petroleum. In *Enhanced Oil Recovery I: Fundamentals and Analysis,* E. C. Donaldson, G. V. Chiligarian, and T. F. Yen, eds.) 1985. New York: Elsevier.

Bertrand, P. *Org. Geochem.* 6, 481 (1984).

Bogomolov, A. I., and K. I., Panina. *Geokhim. Sbornik* 7, 174 (1961).
Bouska, V. *Geochemistry of Coal,* 1981. Amsterdam: Elsevier.
Breger, I. Kerogen. In *McGraw-Hill Encyclopedia of Science and Technology,* 1961, New York: McGraw-Hill.
Cady, C. H. *J. Geol.* **50**, 437 (1942).
Cooper, B. S., *Practical Petroleum Geochemistry,* 1990. London: Robertson Science Publishers.
Crum-Brown, A. *Geological Survey Memoirs (Scotland),* 1927. London: H. M. Stationery Office.
Day, W. C., and J. G. Erdman. *Science* **141**, 808 (1963).
Engler, C. *Petrol. Zhurnal* 7, 399 (1911/12).
Frost, A. V. *Progr. Chem.* **14**(6), 1 (1945).
Fuller, M. L., *Econ. Geol.* **14**, 536 (1919); **15**, 225 (1920).
Hacquebard, P. A. Geol. Surv. Paper No. 75-1B, 1975. Ottawa: Dept. Energy, Mines and Resources.
Hilt, C. *Zeitschr. Ver. Dtsch. Ing.* **17**, 194 (1873).
Hunt, J. M. *Amer. Assoc. Pet. Geol. Bull.* **56**, 2273, 1972.
Hunt, J. M. *Petroleum Geochemistry and Geology,* 1979. San Francisco: W. H. Freeman & Co.
Kinghorn, R. R. F. *An Introduction to the Physics and Chemistry of Petroleum,* 1983. New York: Wiley & Sons.
Kruijs, E., and E. J. Barron. *Deposition of Organic Facies* (A. Y. Huc, ed.) *Studies in Geology,* Vol. 30, p. 195, 1990. Amer. Assoc. Petr. Geol.
Landes, K. K. *Oil Gas J.,* May 2, 1966.
Mukhopadhyay, P. K., and J. R. Gormly. *Org. Geochem.* **14**, 351 (1989).
Petrov, A. A., T. V. Tikhomolova, and S. D. Pustilnikova. *Adv. Org. Chem.,* 1968. New York: Pergamon Press.
Potonié, H. *Abh. Kgl. Preuss. Geol. Landesanstalt, New Ser.* **1**, 55 (1908).
Schobert, H. H. *The Chemistry of Hydrocarbon Fuels,* 1990. London: Butterworths.
Smith, A. H. V. *Yorks. Geol. Mag.* **93**, 345 (1957); *Proc. Yorks. Geol. Soc.* **33**, 423 (1962).
Thompson, S., B. S. Cooper, R. J. Morley, and P. C. Barnard. Petroleum Geochemistry. In *Expl. Norwegian Shelf, Norwegian Petroleum Society,* 1985. London: Graham & Trotman.
Tissot, B. P., and D. H. Welte. *Petroleum Formation and Occurrence,* 1978. Berlin: Springer.
Tyson, R. V., and T. H. Pearson, eds. *Modern and Ancient Continental Shelf Anoxia; Geol. Soc. Spec. Publ.* **58** (1991).

CHAPTER 3

Host-Rock Geochemistry

The formations from which fossil hydrocarbons are now extracted are usually very different from the sediments into which their respective precursor biomasses were deposited and reprocessed, and it is therefore important to distinguish between source rocks and reservoirs. This is particularly so with respect to crude oils and gas, for which migration from source to reservoir was the norm; but to some extent, it is also true of *allochthonous* coals which developed from plant debris that had been transported—usually by floodwaters—from the original growth site. In all such instances, the reservoir can offer little information about the precursor environment beyond what can be inferred from surviving botanical features [1].

1. PETROLEUM HOST ROCKS

Source Rocks

The source rocks of conventional light and medium crude oils are fine-grained marine sediments that now present themselves as limestones, dolomites or shales, but which, when first deposited, formed permeable aggregates whose high porosities were reflected in high water contents. Thickening overburdens progressively compacted these sediments, slowly collapsed the pore space (see Table 3.1.1), and, by attendant expulsion of water, initiated subsequent hydrocarbon migration.

Source-Rock Quality

"Free," and therefore easily recoverable, organic material in a source rock can be quantitatively extracted with an appropriate solvent or a solvent mixture such as benzene/methanol/acetone (Hunt and Meinert, 1954) or 2,3-dimethyl butane/diethyl ether (Philippi, 1956), and extraction has indeed sometimes been used for preliminary assessment and classification of source rocks as

TABLE 3.1.1 Typical Variation of Source-Rock Porosity with Depth[a]

Depth		
ft	m	Porosity, %
2,000	610	27
6,560	2,000	17
9,840	3,000	9
12,120	3,700	6
16,400	5,000	4

[a] Welte (1972).

"very good/excellent" (if containing >1500 ppm organic matter), "fair/good" (if holding 150–1500 ppm), and "poor" (if yielding <150 ppm). A similar assessment can, of course, also be made with other procedures—for example, by (i) 6–36 h Soxhlet extraction with CH_2Cl_2, followed by evaporation of the solvent and the light components of the extract on a water bath in order to isolate a residual semisolid, brown oily "C_{15}+ extract"; or (ii) extracting the sample with CO_2 at 275°C under autogenic pressure and, by condensing the extract in a cold trap, generating material equivalent to the S_1 stream of the Rock-Eval pyrolysis (see later discussion).

The latter procedure does, in fact, offer a complete C_1–C_{15} hydrocarbon suite that cannot easily be obtained otherwise. However, because free organic matter cannot reflect the oil potential unless the rock has reached full maturity and (improbably) *retained all products from thermal degradation of the kerogen,* source-rock quality is more apropriately defined by (i) its total organic carbon content (TOC), which is measured by treating a comminuted rock sample with nitric acid to remove dolomite and related carbonates, combusting it in O_2, and calculating [TOC] from the volume of CO_2 thereby generated; or (ii) since [TOC] can be compromised by the presence of sulfates such as gypsum or anhydrite, by recourse to a destructive distillation technique known as Rock-Eval pyrolysis (Espitalié *et al.*, 1977).

Rock-Eval pyrolysis entails heating the powdered rock specimen in an inert atmosphere from 250 to 550°C over a 20-min period and recording two gas pulses. The first, which issues between 250 and 300°C, delivers a hydrocarbon stream (S_1) composed of free and sorbed organic material as well as (separately measured) CO_2 (S_3). The second pulse, emitted at >350°C, furnishes thermolytically generated hydrocarbons (S_2) and shows the temperature (T_{max}) at which the rate of hydrocarbon evolution peaks. S_2/TOC is then defined as the hydrogen index (HI), S_3/TOC as the oxygen index (OI), $S_1/(S_1 + S_2)$ as the production index, and $(S_1 + S_2)$ as the hydrocarbon potential [2]. HI > 250

TABLE 3.1.2 Heat Flow in Rocks

Anhydrite	λ = 11–14
Dolomite	6–14
Marl	3–6
Sandstone	4–13
Shale	3–7

is taken as indicative of an acceptable (or better) oil potential, and <250 is deemed to be a gas potential.

If the source rock contains a variety of kerogens, the relative amounts of the different types are estimated microscopically by observing natural and fluorescence colors, but this procedure is limited by its inability to quantify mixed amorphous, "coaly," and sapropelic kerogens.

SOURCE-ROCK MATURITY

The developmental status, or *maturity,* of a source rock is a function of its thermal history, and consequently reflects the historical geothermal gradient ($\Delta T/\Delta d$). Globally, this gradient averages 25°C/km (Lee and Uyeda, 1965), but can fluctuate widely from one locale to another and is, as a rule, appreciably lower in crystalline than in noncrystalline rocks. In sedimentary basins, $\Delta T/\Delta d$ lies generally between 15 and 50°C/km (Tissot and Welte, 1978), but can even here be as low as 5 or exceed 85°C. In the Landau oilfield of Germany's Rheingraben, Doebl *et al.* (1974) have thus registered 76.9°C/km, and for Indonesia's Walio oilfield, Redmond and Koesoemadinata (1976) have reported 90°C/km.

Heat flow within a formation is given by

$$Q = \lambda(\delta T/\delta d),$$

where λ is a proportionality constant and $\delta T/\delta d$ is expressed in °C cm^{-1}. But this also varies widely. Although the usually cited "representative" Q-value for oil-producing basins is 1.5×10^{-6} cal cm^{-2} s^{-1}, different rock porosities, mineral compositions, and pore-filling materials can cause λ to vary greatly from one sediment to another (see Table 3.1.2) [3].

In principle, estimates of source rock maturity are therefore more conveniently derived from Rock-Eval analyses, or from measurements of a reflectance parameter (R_o), which varies almost linearly with depth from 0.5 to 0.7% during diagenesis and early catagenesis, increases rapidly to 1.5–1.8% during oil formation, and reaches or exceeds 2.0 at onset of metagenesis near 200°C. (A meaningful value of R_o—a parameter adapted from coal petrography—requires recording the reflectance of 20–50 randomly chosen particles of associated carbonaceous rock or kerogen [4].) In either case, kerogen maturity can then be assessed by reference to models that purport to show kerogen degradation in the 60–200°C "oil window" interval. An example is a correlation of source-rock maturity indicators with vitrinite reflectance data (Tissot and Welte, 1978):

$R_o < 0.5–0.7\% \equiv$ diagenetic phase, immature source rock
$0.7 < R_o < 1.3\% \equiv$ catagenetic phase, principal zone of oil formation
$1.3 < R_o < 2.0\% \equiv$ catagenetic phase, zone of wet gas and condensate
$R_o > 2.0\% \equiv$ metagenesis, dry gas zone, only hydrocarbon is CH_4.

However, such assessments are open to question. Evaluations based Rock-Eval data necessarily assume that laboratory measurements of kerogen decomposition at ≥300°C accurately reflect progressive *in-situ* decomposition at 60–150°C—and this assumption is challenged by the need to differentiate among *labile, refractory,* and *inert* kerogen matter (Quigley *et al.,* 1987; Mann, 1990). Assessments proceeding from R_o can be seriously flawed by suppression of reflectivity by bituminous matter in amorphous H-rich kerogens (Price and Barker, 1985). And major uncertainties attach to the assumptions upon which alternative kerogen-degradation models are founded. Thus, whereas some models (Albrecht, 1970; Le Tran, 1972) make the generation of liquid and gaseous hydrocarbons linear functions of depth, others (e.g., Connan, 1974) allow little degradation before the threshold of intensive oil generation is reached, and therefore express the time/temperature relationship in terms of the Arrhenius equation:

$$\ln t = E/RT - A,$$

where T is the absolute temperature (K), t the time, and A a frequency factor [6].

Some of these difficulties, but by no means all, are skirted by kinetic models that attempt to reconstruct subsidence histories from geological and geophysical data (Tissot and Espitalié, 1975; Jüntgen and Klein, 1975; Ungerer and Pelet, 1987; Mann, 1990).

The simplest and most direct of these (Pusey, 1973) takes it as axiomatic that kerogen decomposition and generation of hydrocarbon liquids occurs only within a depth window defined by T = 60–150°C, and consequently focuses in that interval on variations of spore color with corrected strata temperatures. Inferences from this correlation can then, when required, be cross checked against the dependence of the wet-gas content or the vitrinite reflectivity on depth.

An alternative point of departure (Waples, 1984) is the rate of kerogen decomposition, which is given by

$$dq/dt = Ae^{-E/RT}$$

and assumed approximately to double with each 10°C temperature rise. The temperature history of the sediment is then considered to consist of successive time periods $t_1, t_2, \ldots t_n$, in each of which the temperature is constant, but 10°C higher than in the immediately preceding period. The thermal exposure τ in each period and the time–temperature integral (tTi) can consequently be written as

$$\tau = t2(T - 100)/100$$

$$tTi = \sum_{0 \to t} t2(T - 100)/100,$$

and if T is measured from 0°C, correlation with R_o yields an expression of the form

$$R_o = a(P + b)c,$$

where P denotes the sum of terms written in log form, and $a = 0.0525$, $b = 250$, $c = 0.21$.

If the temperature history of the source rock is known, modeling can also proceed from the assumption that kerogen decomposition involves several simultaneous first-order reactions for which

$$k = Ae^{-E/RT}.$$

The dependence of A and E on the kerogen under study is then accommodated by using an arbitrary set of 18 reactions with a common value of A, and increasing E in equal steps from 40 to 80 kcal/mol (Tissot and Espitalié, 1975); or the frequency factor A can be varied with E for each reaction by means of an expression such as

$$\log A = aE + b,$$

where a and b are constants (Ungerer *et al.*, 1988). Alternatively, a Gaussian distribution can be imposed on A or E or both (Mackenzie and Quigley, 1988; Larter, 1988).

However, because these approaches are also beset by intractable shortcomings, interest has increasingly shifted to the possibility of assessing source-rock maturity by observing color changes of spores in the kerogen or associated inorganic matrix. These changes are governed by the maximum temperature to which the rock was exposed and are comparatively insensitive to time at T_{max}. Colors range form greenish-yellow through golden yellow and shades of orange to brown, dark brown, and black (Cooper, 1990), and are codified by numbers between 1 and 5 in a thermal alteration index TAI, or between 1 and 10 in a spore color index SCI. (The latter index tends to be preferred because it has been calibrated for typical sections to yield straight lines when plotted against source rock depth.) The availability of sporomorph standards that cover a set of 20 hues precludes unduly subjective color determinations, and calibration of the SCI scale for a typical section, which covers oil generation between SCI = 3.5 and 8.5, allows the oil window to be broadly divided into "early mature" (3.5–5), "middle mature" (5–7), and "late mature" (7–8.5).

Other nonpyrolytic methods for evaluating source-rock maturity are less favored because results from them depend on the nature of the kerogen. A case in point is sterane and triterpane isomerization: this varies with temperature (Monin *et al.*, 1988; Mackenzie and Quigley, 1988) and can be conveniently expressed as an isomer ratio, but suffers from the fact that different, although equally mature, kerogens often show markedly dissimilar ratios of generated to originally present hydrocarbons. The use of such ratios therefore demands calibration for each particular kerogen.

Reservoir Rocks

In contrast to source rocks, typical reservoir rocks are variously cemented porous formations with porosities between 10 and 30% (Kinghorn, 1983). However, their pore structures tend to be complex because the total void space is made up of widely different pore shapes and sizes and commonly includes "blind" cul-de-sac pores [7]. Interest therefore centers mostly on an *effective* rock porosity p_e, which defines the capacity of the reservoir to store fluids, and on the permeability of the rock.

Unlike the absolute porosity p_a, which reflects the total void volume of the rock, effective porosity encompasses only interconnected pores and thereby offers simple means for assessing reservoir quality (Table 3.1.3). If, as is usual,

TABLE 3.1.3 Effective Porosity as Indicator of Reservoir Quality[a]

Porosity (%)	Reservoir quality
15–20	good
10–15	fair
5–10	poor
<5	negligible

[a] Levorsen (1967).

p_e is determined by mercury penetration, the method of measurement also allows assessment of pore size distributions from

$$r = -(2\sigma/p)\cos\theta = 75{,}000/p.$$

where r is the pore radius (nm), p the pressure (atm.), σ the surface tension of Hg (4.8 m cm^{-1}), and θ the contact angle of Hg (140°). However, p_e is affected by (i) the initial porosity p_i of the young formation, (ii) lithological factors (clay contents, grain size, packing density, and cementation), and (iii) structural alterations caused by weathering, percolating acid waters, fissuring, and remineralization [8]. Although the initial porosity of a young sediment with cemented grains will therefore tend to be greater than that of its freshly precipitated precursor—so that p_i(sandstone) will usually exceed p_i(carbonate)—the effective porosity of a reservoir rock can be larger or smaller than its p_i. The types of pore systems encountered in reservoir rocks are identified in Table 3.1.4.

The permeability (κ) of a reservoir rock measures its ability to allow fluid flow through interconnected pores and is defined by Darcy's law:

$$r = -(\kappa A/\eta)(dP/dL),$$

where r is the rate at which a fluid with viscosity η can flow through a test piece of cross-sectional area A (cm^2) and length L (cm). dP/dL is the pressure gradient across L. κ is expressed in darcys (D) or millidarcys (mD); in reservoir rocks it varies from as little as 0.1 to >1000 mD. However, because permeability to a gas is always much greater than to a liquid, κ is usually determined with air, in which case

$$\kappa_a = \kappa_\infty(1 + b/p),$$

where p is the pressure at which κ_a is measured, b is a constant related to the average pore diameter, and κ_∞ is obtained by extrapolating κ_a to infinite pressure. κ_a is termed the "equivalent permeability" of the reservoir rock and can

TABLE 3.1.4 Classification of Pore Systems in Reservoir Rocks

1. Primary or syndepositional
   - (a) Intergranular or interparticular—characteristic feature of sandstones
   - (b) Intragranular or intraparticular
2. Secondary or postdepositional
   - (a) Solution porosity—commonly seen in limestones
     - (i) Moldic (fabric selective)
     - (ii) Vuggy (cavernous)
   - (b) Intercrystalline—characteristic of dolomites
   - (c) Fracture porosity—can develop in any rock

TABLE 3.1.5 Permeability as Indicator of Reservoir Quality[a]

Permeability (mD)	Reservoir quality
100–1000	very good
10–100	good
1–10	fair
<1	poor

[a] Kinghorn (1983).

also be used to assess reservoir quality (Table 3.1.5). Like p_e, κ and κ_a are, of course, the resultants of an initial permeability affected by subsequent changes of rock properties.

Reservoir Fluids

Of the fluids that almost always coexist with crude oil in a reservoir and would be coproduced with oil, the most abundant is water, of which three types are recognized:

1. *Meteoric* water entered the reservoir as rain, filled porous shallow rocks or percolated through them along fractures, bedding planes, and/or permeable layers, and contained CO_2 that reacted to form carbonates and bicarbonates; it also held dissolved oxygen that interacted with sulfides to generate sulfates.
2. *Connate* water, originally identified with seawater in which marine sediments were deposited, now denotes all interstitial waters with high concentrations of Cl^- ions (mainly as Na^+Cl^-).
3. *Mixed* waters are characterized by relatively high concentrations of dissolved Cl^-, CO_3^{2-}, HCO_3^{2-}, and SO_4^{2-} and are often so salty as to merit being termed oil-field *brine*.

All three types may exist as free water in interconnected pores, or as interstitial water that can occupy as much as 50% of the total void space.

The brines are believed to have their origins in the seawaters in which the source sediments were deposited and are assumed to have migrated to the reservoir sites as the dominant components of water/oil emulsions (see the following section). The fact that their compositions vary widely and commonly differ greatly from that of typical seawater (Table 3.1.6) is attributed to compositional changes in dissolved mineral matter during migration.

Oil in a reservoir rock, which usually represents only a small fraction of

TABLE 3.1.6 Concentrations of Common Ions in Seawaters and Oil-Field Brines

	Seawater (mg/liter representative)	Oil field brines (mg/liter)	
		Average	Maximum
Na^+	11,000	43,450	120,000
K^+	350	160	1,200
Ca^{2+}	500	3,000	30,000
Mg^{2+}	1,300	1,600	30,000
Fe^{3+}	0.01	10	1,000
Cl^-	19,000	90,000	270,000
HCO_3^-		233	3,000
Organic acids		188	2,300

the total fluid volume (Levorsen, 1967) is quantitatively assessed as "oil in place" by multiplying the pore volume, which can be calculated from core analyses or electric well logs, with the oil concentration of the cores [9], and expressing the result in cubic meters (or acre-ft).

Gas in a reservoir rock is mainly composed of CH_4 and smaller amounts of C_2+ hydrocarbons that may occasionally extend to C_7, but also contains substantial proportions of inorganic contaminants, in particular H_2S, N_2, CO_2, and He. It is termed *unassociated* if it occurs alone, or *associated* if it coexists with oil; in the latter case, it will often form a gas cap that, if underlain by oil, is also considered associated. Underlain by water, it is said to be nonassociated.

Gas can also occur *dissolved* in oil; in that state, the amounts depend on the composition of the oil as well as on the temperature and pressure of the reservoir. Concentrations in crude oils have thus been found to range from a few hundred ml/m^3 to as much as 175 m^3/m^3. Conversely, because oil can also dissolve in gas, condensate gases are on occasion encountered with exceptionally high proportions of oil [10].

Pools in which *all* gas is dissolved in oil are termed "undersaturated," whereas those in which some (associated) gas exists as a free gas cap are said to be "saturated."

Migration

As overburdens thicken from ~1500 to 4000 m, and geothermal heat raises source-rock temperatures from ~60 to ~130°C, progressively more intense kerogen degradation alters the available pore volumes, gradually converts increasing proportions of the labile kerogen to more refractory matter, transiently increases hydrocarbon saturation at the expense of H_2O, and by these changes initiates a migration of hydrocarbons through the pore system. But as organic material and pore sizes in a source rock are always very inhomogeneously distributed, *primary* migration to points from which hydrocarbons can move into a more open-structured carrier—usually a sandstone or porous limestone—is still not well understood [11].

A major role is intuitively assigned to progressive compaction of the source rock. This is thought to trigger migration by expelling water, and so makes it reasonable to associate the extent of movement with lithostatic gradients and source-rock permeability. However, at depths at which hydrocarbon liquids are intensively generated, all but the most strongly sorbed water (mainly sited on smectites, illites, and other clay minerals prone to pronounced swelling) will already have been expelled [12], and new water, which might play a role in hydrocarbon movement, could only be provided by transformation of montmorillonite into illite. It must therefore be supposed that the principal migration mode entails transfer of discrete hydrocarbon phases—that is, liquids *per se*, liquids variously saturated with gas (mainly C_1–C_4), gas variously saturated with liquids, and gas *per se* (Hunt, 1979; Leythaeuser *et al.*, 1984; Mackenzie *et al.*, 1987).

An important contributor to such migration is sometimes the entrainment and dissolution of hydrocarbon liquids in gas that moves upward through microfractures [13]. Where catagenesis has reached a stage at which most of the water has been expelled from the rock, this mode is, in fact, a necessary outcome of rock compaction—if only because a greatly reduced porosity would make capillary forces and fluid potential gradients propel organic matter from small pores, which it filled, into coarser-grained rock where capillary pressures are much lower [14].

In earlier stages of catagenesis, when rates of hydrocarbon generation are still relatively low and appreciable amounts of H_2O remain in the pore space, some primary migration could conceivably result from hydrocarbon dissolution in water (Table 3.1.7). But that is only likely to involve the lightest and most soluble hydrocarbons, and would therefore contribute relatively little to primary migration. Although solubility of CH_4 in H_2O increases rapidly with pressure from ~25 ppm at STP to ~2500 and 7500 ppm at 2500 and 6000 m, respectively, the solubilities of two oils, measured at autogenic pressure between 25 and 125°C, rose in one case only from ~5 to 20 ppm, and in the other from ~20 to 125 ppm (Price, 1976).

Other modes—in particular, movement of hydrocarbon liquids in micellar form (Cordell, 1972; Bray and Foster, 1980)—cannot be ruled out, but are deemed unlikely. Aside from indications that the mean diameter of ionic micelles (~6.5 nm) would equal or exceed the diameter of the pores (5–10

TABLE 3.1.7 Solubility of Selected Hydrocarbons in H_2O[a]

Compound class	Compound	Solubility at 25°C (ppm)
	methane	24.4 ± 1
	ethane	60.4 ± 1.3
	propane	62.4 ± 2.1
	n-butane	61.4 ± 2.1
	isobutane	48.9 ± 2.1
n-Paraffins	*n*-pentane	38.5 ± 2.0
	n-hexane	9.5 ± 1.2
	n-heptane	2.9 ± 0.2
Isoparaffins	2-methylbutane	13.0 ± 0.2
	isopentane	48.0 ± 1.0
	2,3-dimethylpentane	5.2 ± 0.02
	2,2,4-Me_3pentane	2.1 ± 0.1
	2-methylhexane	2.5 ± 0.02
	3-methylheptane	8.0 ± 0.03
	4-methyloctane	0.1 ± 0.01
Cycloparaffins	cyclopentane	158.0 ± 3.0
	methylcyclopentane	42.0 ± 1.5
	n-propylcyclopentane	2.0 ± 0.1
	cyclohexane	61.0 ± 5.0
	methylcyclohexane	15.0 ± 1.0
Aromatics	benzene	1760.0 ± 20.0
	toluene	535.0 ± 20.0
	o-xylene	170.0 ± 5.0
	m-xylene	134.0 ± 2.0
	p-xylene	157.0 ± 1.0
	ethylbenzene	142.0 ± 9.0
	isopropylbenzene	50.0 ± 3.0
	isobutylbenzene	10.0 ± 0.4
Hetero compounds	thiophene	3015 ± 34
	2,7-Me_2quinoline	1795 ± 130
	indole	3558 ± 170
	indoline	10800 ± 700

[a] McAuliffe (1966); Price (1976); *Handbook of Physics and Chemistry*.

nm) through which they would have to move (Baker, 1963), micellar migration is discounted by laboratory investigations. For example:

1. Shales possess semipermeable membrane characteristics that retard ions significantly smaller than ionic micelles: soap solutions made to flow through silty clays were thus found to be retained by sediments with permeabilities < 10^{-3} mD (Hunt, 1979) [15].

2. Whereas the amount of soap required for micelle formation ranges, depending on water temperature, from 600 to 6000 ppm, only 2–30 ppm

of potential solubilizers are usually available in formation waters (Zhuze *et al.*, 1971).

3. Rates of micelle formation peak at ~70°C, and fall off dramatically toward lower and higher temperatures (Trofimuk and Kantorovich, 1977).

The mechanisms that drive hydrocarbon transport during primary migration are also uncertain, but at least three hypothesized processes seem to be consistent with the evidence at hand and may be alternatives that come into play in particular circumstances.

One concept envisages accumulation of hydrocarbons until an oil-wetted pore network or a network of (still-heavy) organic matter offers escape paths (Yariv, 1976; McAuliffe, 1980; Stainforth and Reinders, 1990). Scanning electron micrographs of source-rock material (freed of mineral matter) even showed three-dimensional organic carbon networks in shales with <6% organic matter (McAuliffe, 1978), and two-dimensional carbon networks have been seen in bedding planes (Momper, 1978a). Hydrocarbon transport along

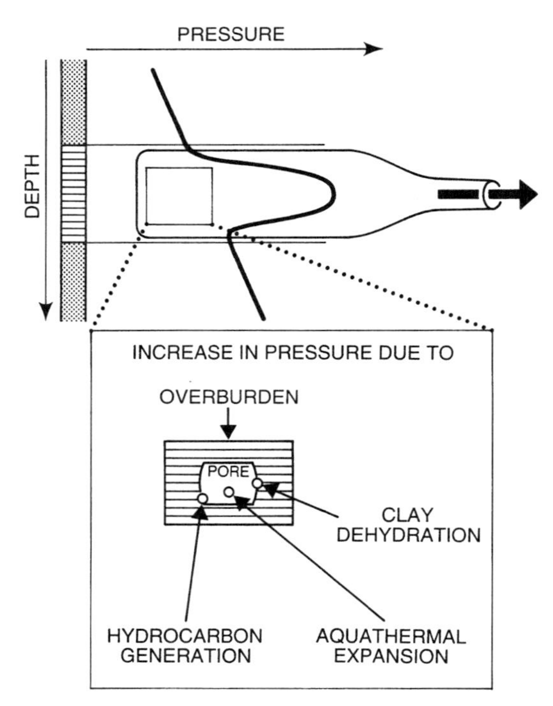

FIGURE 3.1.1 Contributors to the "buried bottle" model (after Mann, 1990). (Reproduced with permission)

TABLE 3.1.8 Effective Molecular Diameters of Some Molecular Species and Hydrocarbon Types[a]

Molecule	Diameter (nm)
Helium	0.20
Hydrogen	0.23
Water	0.32
Carbon dioxide	0.33
Methane	0.38
Benzene	0.47
n-Alkanes	0.48
Cyclohexane	0.54
Polynuclear aromatics	1–3
Asphaltenes	5–10

[a] After Tissot and Welte (1978).

such carbon "wicks" would be initiated by differential pressures due to compaction, hydrocarbon generation, and thermal expansion and would not require more than 3–10% oil saturation of the organic matter.

A second concept (Momper, 1978b; Jones, 1980; Honda and Magara, 1982) assumes that migration proceeds mostly through the largest pore throats (which are sometimes termed *atypical* pore throats) and echoes a "buried bottle" model (Fig. 3.1.1), in which pressure increases are ascribed to volume changes resulting from hydrocarbon generation [16] and dehydration of clays, and to partial transfer of overburden pressures to pore liquids.

A third concept hypothesizes that hydrocarbons accumulate until they establish an internal pore pressure that exceeds the mechanical strength of the source rock, and thereby creates a fracture network through which they can freely escape (Tissot and Pelet, 1971).

But whatever the migration modes and mechanisms, the hydrocarbon phases always undergo some partitioning of molecular species en route to the reservoir. To some extent, this arises from molecular-sieve properties of constricted pore systems and from consequent retardation, if not actual retention, of large molecules [17]. The structure of hydrocarbon molecules can, however, also—much as in chromatographic separation—cause retardation through allowing transient attachment of molecules to pore walls. Both cases are implicitly illustrated in Table 3.1.8, which lists effective molecular diameters of some relevant hydrocarbons [18], and in Table 3.1.9, which shows a form of partitioning that clearly reveals primary migration of discrete hydrocarbon phases favoring light over heavy compounds and saturates over aromatics.

Because separation of components is also a response to greater salinity, lower pressures, and lower temperatures, similar (and sometimes drastic) com-

TABLE 3.1.9 Partitioning of Some Light Hydrocarbons between Oil and Its Source Rock (Vol %)[a]

	Source rock	Crude oil
n- and isopentane	2.9	14.4
n- and isohexane	3.6	18.4
n- and isoheptane	5.6	25.9
Cyclopentane	0.2	0.9
Methylcyclopentane	1.5	6.7
Cyclohexane	1.6	3.8
Benzene	10.0	0.2
Toluene	58.0	1.2

[a] Martin *et al.* (1963).

positional changes can occur during migration of solutions, whether of hydrocarbons in water or of water in hydrocarbons. These changes are commonly associated with secondary migration through more porous rock, begin to manifest themselves as soon as the solutions leave the source rock, and are critically dependent on the type of water associated with the hydrocarbons. Hydrocarbon solubilities in saline waters fall by a factor of 10–15 as salt contents approach 3×10^5 ppm (Price, 1976); but, although that would cause one to anticipate connate waters promoting more far-reaching exosolution than meteoric waters, measurements of toluene solubility in chloride–Ca^{2+} connate waters and bicarbonate–Na^+ meteoric waters at >140°C indicate that the reverse more likely to occur.

"Secondary" migration through porous carrier beds toward, and within, a

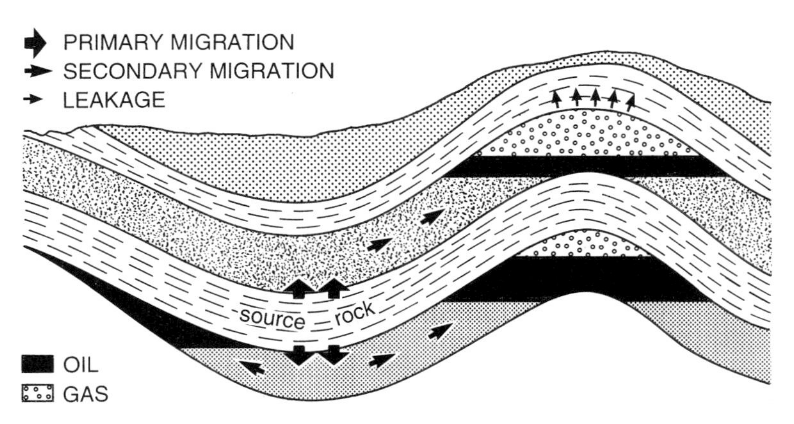

FIGURE 3.1.2 Development of an oil/gas pool: schematic of oil migration into a trap.

reservoir rock or "trap" can thus facilitate the formation of separate oil and gas phases, and either or both can therefore gradually move vertically upward to form a pool wherever further upward movement is prevented by low rock permeability (Fig. 3.1.2).

2. THE HOST ROCKS OF HEAVY HYDROCARBONS

The ubiquitous, and often massive, occurrence of heavy liquid and semisolid hydrocarbons poses questions about aspects of their origin that are still unresolved.

Concentrations of such hydrocarbons in sediments with petroleum *source-rock* characteristics are substantial (Gallup, 1974). They run typically to 30–49 mg g^{-1}; mirror geothermal degradation of kerogen by increasing with depth to 150–180 mg g^{-1} before declining in concert with falling rates of oil generation; and are thus consistent with the view that they have been retained by the molecular-sieve-like properties of the source rock. Accumulations of asphaltic substances largely composed of asphaltene and resin molecules (see Chapter 7) in host strata made up of unconsolidated sandstones and the like are therefore broadly ascribed, as noted in Chapter 2, to biodegradation and water-washing of oils—with the latter assumed to have removed light ($<C_{15}$) hydrocarbons. There is, however, little persuasive evidence for such alteration as a *general* case, and it is equally possible to suppose that heavy hydrocarbons retained in source rocks represent moieties left behind during primary migration from less compacted, and therefore more porous sediments (see Fig. 2.5.2). It is not unlikely that both mechanisms have operated, albeit in different geological environments. But as matters stand, choices between them in particular instances are virtually impossible—partly because conditions for heavy-hydrocarbon migration from source rocks have not yet been seriously investigated.

Bituminous Substances

Heavy oils occur in strata and reservoir structures much like those that accommodate conventional light crudes, and need little further comment here. But a few geologically explored accumulations of bituminous matter merit description because they illustrate similarities as well as differences between them.

Bitumen and bitumen-like materials are widely distributed, with substantial deposits reported from the Americas as well as from Asiatic Russia, continental

FIGURE 3.2.1 The oil sands and heavy oil deposits of Alberta.

Europe, and Africa. Detailed information about such deposits—variously referred to as bituminous sands, oil sands, or tar sands—is, however, only available for the massive occurrences in Alberta (Canada), Utah (USA), and Venezuela. Smaller accumulations elsewhere have so far only been briefly noted and largely remain "virgin territory."

The Alberta Oil Sands

Alberta's oil sands are of Lower Cretaceous (Mississippian) age and form a discontinuous trend that stretches from the Peace River area eastward to the Wabasca and Athabasca deposits before turning south to the Cold Lake deposit and some scattered heavy-oil (16–24 API) pools in the Lloydminster area (Fig. 3.2.1). Depositional histories of the sands have been discussed by Jardine (1974), and the more detailed geological features of the Athabasca, Wabasca, and Cold Lake deposits have been described by, among others, Gallup (1974), Kramers (1974), and Minken (1974), as well as in a series of Alberta Research

TABLE 3.2.1 Reservoir Characteristics, Alberta Oil Sands[a]

	Athabasca[b]	Athabasca[c]	Cold Lake	Wabasca	Peace R.
Formation	McMurray[b]	McMurray[c]	Upper Manville	Wabasca/ Grand Rapids	Bluesky/ Bullhead
Depth (m)	0–75	75–600	300–600	75–750	300–750
Area (M · ha)	310	2020	1280	705	475
Av. pay (m)	32.0	21.3	12.8	9.1	2.8
Bitumen[d]	19.1	80.3	26.1	8.4	7.9
Porosity (%)	30–35	30–35	37	35	23
Ave. permeability (mD)	5,000	~5,000	~3,000	~3,500	~1,000
°API	6–8	8–10	10–12	10–13	8–23

[a] After Jardine (1974).
[b] Surface mining area.
[c] Subsurface.
[d] In place (10^9 m^3).

Council monographs, Reservoir characteristics of the four adjoining deposits are summarized in Table 3.2.1.

The Peace River and Athabasca deposits lie in nonmarine L. Mannville sediments that developed in predominantly fluvial environments from material that originated in the Canadian Shield. The Wabasca and Cold Lake accumulations, which occur in U. Mannville nonmarine clastics, formed from similar material. The thickness of the sands is in all cases largely determined by the topography of a pre-Cretaceous unconformity that underlies them, but fluvial and deltaic sedimentation patterns have resulted in complex reservoir distributions that are particularly pronounced in the Athabasca and Cold Lake deposits, in which removal of Devonian salt has caused collapse of the strata.

The migratory pathways are still uncertain, but there are indications that the bitumen may have resulted from progressive biodegradation, water-washing, and, possibly, some abiotic oxidation of L. Cretaceous oil as it moved updip and encountered water that invaded the basin from the edge of the Canadian shield (Deroo and Powell, 1978; Deroo *et al.*, 1974). Figure 3.2.2 illustrates such movement. Jha *et al.* (1978) have indeed suggested that this alteration of bitumen still proceeds: gas extracted from the bitumen was found to contain *neopentane* and *acetaldehyde*—the former the *only* detectable C_5 hydrocarbon [19], and the latter an unusual component that is too reactive to persist in oil sands, and therefore believed to be a relatively recent reaction by-product.

The Utah Tar Sands

Of the several U.S. bitumen deposits, variously referred to as *oil* or *tar* sands (Hosterman and Meyer, 1988), by far the largest lie in Utah and Alaska [20]

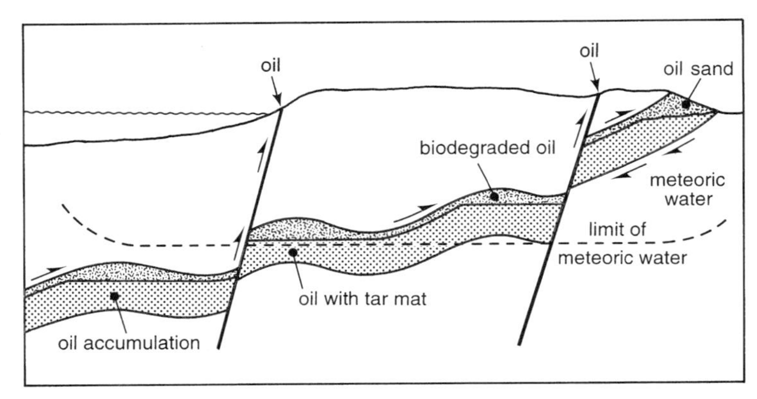

FIGURE 3.2.2 Hypothesized development of a heavy oil and oil sand deposit: the Athabasca oil sands deposit may have formed in this manner.

(Table 3.2.2). However, thus far only the Utah deposits seem to have attracted serious interest (Werner, 1984; Trulle, 1986), and of the 54 separate occurrences recognized in the state, only the PR Spring and NW Asphalt deposits, in NE Utah's Uinta Basin, and the Tar Sand Triangle, between the Dirty Devil and Colorado Rivers in SE Utah, have been subjects of fairly detailed exploration.

The PR Spring deposit, located on the southeastern margin of the Uinta Basin, is made up of at least five bitumen-bearing sandstone (arkose) zones with a combined thickness of up to 120 m (400 ft). All are of lacustrine origin and composed of sandstones, siltstones, shales, and oolitic algal and ostracodal limestones, which were deposited near the shoreline of ancient Lake Uinta. There is close association with the principal oil shales of the Green River Formation (Oliver and Sinks, 1986). Identification of authigenic pyrite in the sediments has been advanced as evidence of biodegradation by sulfate-reducing

TABLE 3.2.2 U.S. Bitumen (Tar Sand) Deposits (IOCC, 1984)

	Estimated bitumen in place	
	10^9 bbl	10^9 m^3
Utah	19.4	3.08
Alaska	10.0	1.59
Alabama	6.4	1.02
Texas	4.8	0.76
California	4.5	0.71
Kentucky	3.4	0.54

bacteria resembling *Desulfovibrio* and *Desulfococcus* (Mason and Kirchner, 1992).

The neighboring NW Asphalt Ridge deposit is a 25-km-long northwest-trending hogback made up of clastic quartz, chert, clay minerals, and feldspars.

The Tar Sand Triangle consists of Paleozoic and Mesozoic strata in which the major oil-bearing unit, the White Rim Sandstone, is represented by a friable, light gray/reddish brown sandstone. The up-dip pinchout immediately west of the Colorado R., with potential for petroleum accumulation, is erosionally dissected, and consequently allowed oil seepages as well as devolatilization of hydrocarbons.

Table 3.2.3 summarizes the more important characteristics of the three oil (tar) sand occurrences.

The Orinoco Tar Belt

Venezuela's tar sand deposits lie in a 55-km-wide band that parallels the northern banks of the Orinoco River over a distance of some 600 km (Ayaleto *et al.*, 1974). This region holds 9–12 API oil believed to have originated in the Oficina trend, which also contains more conventional lighter oils immediately north of the poorly defined northern limits of the tar sands. However, resource data are incomplete and the information at hand relates primarily to the Zuata area of the tar belt's south-central zone (Galavis and Velarde, 1967). Here, the net pay zone is ~53.5 m (175 ft) thick, shows an average 70% oil saturation and 25% porosity, and contains an estimated 4×10^9 m^3 oil in place. The oil is assumed to have migrated, but where it originated and how it reached its current host rocks—a structural–stratigraphic trap formed by folding, faulting, and unconformity—remain matters for speculation.

Oil Shales

Oil shales are impervious marine shales in which organic matter came to be occluded while the sediments accumulated. Migration of organic material was therefore confined to transport of some exogenous kerogen precursors and kerogen particles into the developing sediments, and because hydrocarbons can only be generated by pyrolyzing or extracting the shales at temperatures above 400–450°C (see Chapter 9), questions of source-rock maturity are irrelevant. However, different depositional environments are indicated by varying contents of *labile* kerogen and/or by the quality of oil produced from this organic matter.

Shales that developed in large lacustrine basins—especially in basins that resulted from crustal warping associated with orogeny—will, as a rule, deliver >170 liters of oil per tonne (>40 U.S. gal/ton). Typical of such shales are the Eocene-age Green R. formations of Colorado, Utah, and Wyoming, the

TABLE 3.2.3 Characteristics of the Major Utah Tar-Sand Deposits[a,b]

P.R. Spring	
Age	Eocene
Lithology	sandstone, siltstone
Depth	0–300 ft (0–91.5 m)
Pay zone thickness	40 ft (12 m) av.
porosity	18–30%
permeability	700 mD
Bitumen saturation	52% av.
Bitumen gravity	9–9.5 °API
Reserves/resources	4.3×10^9 bbl (0.68×10^9 m^3)
Asphalt Ridge	
Age	Cretaceous
Lithology	sandstone
Depth	20–600 ft (6.1–183 m)
Pay zone thickness	40 ft (12 m) av.
porosity	17–22%
permeability	>1,000 mD
Bitumen saturation	48% av.
Bitumen gravity	10.4 °API
Reserves/resources	1.1×10^9 bbl (0.17×10^9 m^3)
Tar Sand Triangle	
Age	Permian
Lithology	sandstone, conglomerate
Depth	200–1500 ft (61–460 m)
Pay zone thickness	75 ft (23 m) av.
porosity	27% av.
permeability	268 mD av.
Bitumen saturation	40% av.
Bitumen gravity	4.3 °API
Reserves/resources	12.5–16 $\times 10^9$ bbl (2–2.5 $\times 10^9$ m^3)

[a] Crysdale and Schenk (1988).
[b] Approximate metric equivalents in parentheses.

Mississippian-age Albert shales of New Brunswick, and the Congo's Triassic Stanleyville Basin (Duncan, 1976).

Similar volumes of oil are also delivered by shales that formed by alternating marine and fluvial sedimentation in small lakes, lagoons, and bogs, and that are therefore often associated with cannel or boghead coals.

However, contrary to what might perhaps be expected, shales deposited in shallow marine seas on continental platforms and shelves will rarely yield >125 liters/tonne (30 gal/ton). Examples are the Cambrian black shales of northern Europe and Siberia, the Devonian black shales of the eastern and midwestern United States, the Permian black shales of southern Brazil and

Argentina, and a number of Jurassic black shale formations in Europe, Alaska, and east Asia (Lee, 1991).

Extensive hydrocarbon source beds—i.e., deposits that cover several hundred km^2 or more—are rarely attributable to a single cause. They commonly result from complex interplay of several, not always fully understood, factors that are unique in time and place. The Devonian and early Mississippian black shales of the east-central United States are a case in point. Some 700,000 km^2 of these shales, which contain 7% organic matter equivalent to 45×10^6 m^3 oil in place, are believed to be economically exploitable (Ettensohn, 1992).

It has been argued that oil shales could be considered to be immature oil source rocks (Kinghorn, 1983), but in that connection it should be noted that unlike petroleum source rocks, which need only contain >0.5% organic matter in order to produce a significant volume of reservoired hydrocarbons, oil shales must contain substantially more (by some estimates 3–5%) to cover the additional mining and retorting costs.

Coals

Other than sapropels, which formed in dysoxic lacustrine and shallow marine environments, coals developed from accumulated terrestrial plant debris in (a) shallow paralic basins fronting the approaches to folded mountains; (b) limnic basins that developed in intermontane areas; and (c) coastal environments that resembled a contemporary subtropical peat bog or brackish marine swamp. "Source-rock" equivalents are therefore the primordial, slowly maturing coal masses whose carbon contents or rank are the formal equivalents of source-rock maturity; "host rocks" derive from sediments that were deposited by (sometimes cyclical) incursions of the sea and progressively buried the precursor vegetal debris. These sediments are mostly fine-grained clastics, sometimes interbedded by thin marine strata. The predominant clastics are mudstones, siltstones, and shales, which are collectively far more common than sandstones and occasionally altered by magmatic intrusions manifest as dykes and sills (Duff and Walton, 1962). The transformation of kaolinite into pyrophyllite is one example of such alteration.

Detailed descriptions of coal-bearing rock sequences and the associated geochemistry are, however, frequently bedevilled by traditional regional nomenclatures and terms that connote *location* rather than composition or structure. Examples are *seat-earths, underclays,* and *fireclays*—all terms that refer to a group of rocks which, although varying in composition and structure, share many features and are, in fact, characteristic components of a sequence of cyclothemic deposits in a paralic basin. Like tonstein, an argillaceous kaolinite-rich rock whose mineralogy and chemistry reflect extensive weathering at its source prior to transport (Moore, 1964), most are of detrital origin, and some may represent fossil soils (Huddle and Patterson, 1961).

There are indications that "coalification," primarily driven by progressively greater overburden pressures and increasing geothermal heat, was here and there also promoted by land subsidence, and rates of such subsidence can be inferred from the vitrinite : inertinite ratios of coal [21]. An example presents itself in characteristic differences between NW Europe's Carboniferous and South Africa's Permian-age (Gondwana) coal deposits. The former are associated with relatively rapid development of an extensive synclinal structure and are vitrinite-rich, whereas the latter formed in a stable flat continental trough in which the granitic pre-Cambrian massif was overlain by conglomerates from a Paleozoic glacial invasion of the area, and which therefore contain much higher proportions of inertinites (Mackowsky, 1968).

Transformation of plant debris into coal usually proceeded at the growth sites of precursor vegetation, and transport (≡ "migration") of organic matter to other (allochthonous) sites was atypical. Where it did occur, however, it profoundly affected the kind of coal that developed. Carried into dysoxic lacustrine or shallow offshore waters, vegetal debris—mainly composed of terrestrial fungal spores and pollen—developed into H-rich sapropelic matter that resembles liptinitic kerogens; transported by floodwaters that redeposited it on land, the much more inclusive debris came to incorporate finely disseminated clays that contributed to often very much higher-than-usual mineral matter contents in the coals that eventually developed from it.

NOTES

[1] Such features, which comprise well-preserved pollen, fungal spores, leaf cuticles, and woody fragments, can commonly be seen and identified when thin sections or polished surfaces of immature (low-rank) coals are examined under a microscope.

[2] The oxygen index may only be reliable for rocks for which TOC $> 2\%$, HI > 300, and OI < 50.

[3] Strata with poor thermal conductivities are heat flow barriers, and therefore cause higher temperatures below them than would be expected from the average geothermal gradient.

[4] The procedure has been described in detail by, among others, Stach (1963, 1985) and Mackowsky (1971, 1975). A summary can be found in Berkowitz (1994).

[5] Such differentiation implies that *in-situ* kerogen decomposition is a pyrolytic carbon rejection process—but bearing in mind the chemistry of decomposition, that inference can hardly be surprising.

[6] In the latter case, the t/T diagrams, which depict oil genesis at specified wells, are calculated from field data.

[7] Extreme cases are exemplified by lava and pumice stone, both rocks characterized by high (vesicular) void volumes, but very low effective porosities for storage of fluids.

[8] Porosity enhancement is exemplified by dolomitization, a process which results when pore

water or percolating mineral waters substitute Mg^{2+} for Ca^{2+} in limestone-rich carbonate rocks. The greater porosity is then caused by the smaller ionic volume of Mg^{++}.

[9] Produced crude oil is variously reported in tonnes, m^3, barrels (bbl), or bbl/acre-ft of reservoir rock. One acre-foot equals 1233.48 m^3.

[10] In Italy's Malossa field, reservoir depth, temperature and pressure are, respectively, 6000 m, 155°C, and 1050 bar (Neglia, 1979), and under these conditions, hydrocarbons up to C_{18} could dissolve in the gas.

[11] Primary migration is sometimes divided into early and late phases—the former identified with initial compaction and expulsion of water, and the latter with intensive generation of hydrocarbons at 3000–4000 m (at which most of the water has already been forced out). But how long the onset of migration is delayed after hydrocarbon generation begins depends on the rate of hydrocarbon generation; that, in turn, is determined by the concentration of labile organic matter in the source rock. A minimum of 850 ppm hydrocarbon material is considered necessary before migration can begin (Momper, 1978), and this is one reason for setting lower limits of 0.3% and 0.5% for evaporites and clastic rocks before regarding them as source rocks.

[12] Smectites form a class of monoclinic silicate clay minerals with the general composition $X_{0.3}Y_{2-3}Z_4O_{10} \cdot nH_2O$, where

$$X = Ca/2,\ Li\ or\ Na$$
$$Y = Al,\ Cr,\ Cu,\ Fe^{2+},\ Fe^{3+},\ Li,\ Mg,\ Ni\ or\ Zn$$
$$Z = Al\ or\ Si.$$

They are noted for high cation-exchange and swelling capabilities, and H_2O sorbed on such clays differs markedly from bulk water. Its viscosity is almost 4× greater, and its solvent power zero; in Na^+ and K^+ smectites such "structured" water can be several molecular layers thick. Forcing it through the pores of smectites would therefore require far higher pressures than needed to transport bulk H_2O through small pores.

[13] Such fractures would develop as reaction to high pore pressures set up by formation of hydrocarbons in fine-grained source rocks.

[14] Parenthetically, it might be noted here that the ratio of bulk H_2O to structured H_2O (see [12] above) falls with decreasing porosity. Since oil could only move with or in bulk water, sorbed "structured" water limits bulk space and thus also forces oil expulsion.

[15] This includes almost *all* shales.

[16] This is estimated to increase the volume by 15 ± 5%.

[17] The data in Table 3.1.1 indicate a 15–17% source-rock porosity, equivalent to average pore diameters of ~10 nm, at the *onset* of catagenesis. High-molecular-weight moieties generated in the early stages of kerogen degradation could therefore quite conceivably move out with H_2O well before the source rock is further compacted.

[18] For *n*-alkanes, the effective molecular diameter shown in Table 3.1.8 is the chain diameter, and passage through slightly larger pores is only possible if the molecules are aligned in the direction of laminar flow. In turbulent flow, when they would randomly rotate, effective diameters would be functions of chain length.

[19] Neopentane, which commonly occurs in crude oils in amounts ranging from 0.1 to 0.7% of the total C_5 hydrocarbon fraction (Martin and Winters, 1969), is never found alone.

[20] This is the Kuparuk deposit on Alaska's North Slope. It has more recently been reclassified as a heavy oil occurrence estimated to contain as much as 40×10^9 bbl (U.S. Geol. Surv. Bull. 1784, 1988).

[21] The use of this ratio is based on the fact that vitrinites formed in strongly anoxic conditions, whereas precursor materials, which developed into macerals of the inertinite group, were altered by limited or episodal exposure to atmospheric oxygen.

REFERENCES

Albrecht, P. *Memoirs du Service Carte Geol. d'Alsace Lorraine No. 32*, p. 119 (1970).
Ayaleto, M. E., and L. W. Louder. In *Memoir 3* (L. V. Hills, ed.), p. 1, 1974. Calgary: Can. Soc. Pet. Geol.
Baker, E. G. *Fundamental Aspects of Petroleum Geochemistry* (B. Nagy and U. Columbo, eds.), p. 299, 1963. Amsterdam: Elsevier.
Berkowitz, N. *An Introduction to Coal Technology*, 2nd ed., 1994. New York: Academic Press.
Bray, E. E., and W. R. Foster. *Amer. Assoc. Petrol. Geol. Bull.* **64**, 107 (1980).
Cooper, B. S. *Practical Petroleum Geochemistry*, 1990. London: Robertson Sci. Publ.
Cordell, R. J. *Amer. Assoc. Petrol. Geol. Bull.* **57**, 1618 (1972).
Crysdale, B. L., and C. J. Schenk. Bitumen-bearing deposits of the United States, *U.S. Geol. Surv. Bull.*, 1784 (1988).
Deroo, G., B. Tissot, R. G. McCrossan, and F. Der. In *Memoir 3* (L. V. Hills, ed.), p. 148, 1974. Calgary: Can. Soc. Pet. Geol.
Deroo, G., and T. G. Powell. *Oil Sand and Oil Shale Chemistry* (O. P. Strausz and E. M. Lown, eds.), p. 11. New York: Verlag Chemie.
Doebl, F., D. Heling, W. Karweil, M. Teichmüller and D. Welte. *Inter-Union Commission on Geodynamics, Sci. Rept. 8, Stuttgart* (1974).
Duff, P. McL. D., and E. K. Walton. *Sedimentology* **1**, 235 (1962).
Duncan, D. C. In *Oil Shale* (T. F. Yen and G. V. Chilingarian, eds.), Chapter 2, 1976. Amsterdam: Elsevier.
Espitalié, J., J. L. Laporte, M. Madec, F. Marquis, P. Leplat, and T. Paulet. *Rev. Inst. Fr. Pet.* **32**, 23 (1977).
Ettensohn, F. R. *Fuel* **71**, 1487 (1992).
Galavis, J. A., and H. Velarde. In *Proc. 7th World Petroleum Congr., 1967.*
Gallup, W. B. In *Memoir 3* (L. V. Hills, ed.), p. 100, 1974. Calgary: Can. Soc. Pet. Geol.
Honda, H. and K. Magara. *J. Petrol Geol.* **4**, 407 (1982).
Hunt, J. M. and R. N. Meinert. U.S. Pat. No. 2,854,396 (1954).
Hunt, J. M. *Geokhimi Khimico-Fizi Voprosam Razvedki i Dobychi Neft i Gaza* **1**, 219 (1972).
Hunt, J. M. *Petroleum Geochemistry and Geology*, p. 207, 1979. San Francisco: Freeman & Co.
Interstate Oil Compact Commission, U.S.A. *Major Tar Sand and Heavy Oil Deposits of the United States* (V. A. Kuuskraa and E. C. Hammershaib, eds.), 1984. Oklahoma City.
Jardine, D. In *Memoir 3* (L. V. Hills, ed.), p. 50, 1974. Calgary: Can. Soc. Pet. Geol.
Jha, K. N., D. S. Montgomery, and O. P. Strausz. In *Oil Sand and Oil Shale Chemistry* (O. P. Strausz and E. M. Lown, eds.), p. 33, 1978. New York: Verlag Chemie.
Jones, R. W. *Amer. Assoc. Pet. Geol., Studies in Geol.* **10**, 476 (1980).
Jüntgen, H., and J. Klein. *Erdöl Kohle, Suppl. Vol.* **1**, 52, 74 (1975).
Kinghorn, R. R. F. *Introduction to the Physics and Chemistry of Petroleum*, 1983. New York: Wiley & Sons.
Kramers, J. W. In *Memoir 3* (L. V. Hills, ed.), p. 68, 1974. Calgary: Can. Soc. Pet. Geol.
Larter, S. *Mar. Pet. Geol.* **5**, 195 (1988).
Lee, H. K., and S. Uyeda. *Terrestrial Heat Flow, Geophys. Monogr.* **8**, 87 (1965).
Lee, S. *Oil Shale Technology*, 1991. Boca Raton, FL: CRC Press.
Le Tran K. *Advances in Organic Geochemistry*, p. 717, 1972. New York: Pergamon Press.
Levorsen, A. I. *Geology of Petroleum*, 2nd ed., 1967. San Francisco: Freeman & Co.
Leythaeuser, D., A. S. Mackenzie, R. G. Schaefer, and M. Bjoroy. *Amer. Assoc. Petrol. Geol. Bull.* **68**, 196 (1984).
McAuliffe, C. D. *J. Phys. Chem.* **70**, 1267 (1966).

McAuliffe, C. D. *AAPG Short Course, Vol. 2, Natl. Mtg. Amer. Assoc. Petrol. Geol., Oklahoma City* (1978).
McAuliffe, C. D. *AAPG Studies in Geology* **10**, 89 (1980).
Mackenzie, A. S., and T. M. Quigley. *Amer. Assoc. Petr. Geol. Bull.* **72**, 399 (1988).
Mackenzie, A. S., I. Price, D. Leythaeuser, P. Mueller, H. Radke and R. G. Schaefer. In *Petroleum Geology and NW Europe* (J. Brooks and K. Glennie, eds.), 1987. London: Graham and Trotman.
Mackowsky, M.-Th. In *Coal and Coal-Bearing Strata* (D. Murchison and T. S. Westoll, eds.), Chapter 14, 1968. New York: Elsevier.
Mackowsky, M.-Th. *Fortschr. Geol. Rheinl. Westfalen* **19**, 173 (1971); *Microsc. Acta* 77(2), 114 (1975).
Mann, U. In *Sediments and Environmental Geochemistry* (D. Heling, P. Rothe, U. Foerstner & P. Stoffers, eds.), 1990. Berlin: Springer.
Martin, R. L., J. C. Winters, and J. A. Williams. *6th World Pet. Congr., Sec. V, Paper #13* (1963).
Martin, R. L., and J. C. Winters. *Anal. Chem.* **31**, 1954 (1969).
Mason, G. M., and G. Kirchner. *Fuel* 71, 1403 (1992).
Minken, D. F. In *Memoir 3* (L. V. Hills, eds.), p. 84, 1974. Calgary: Can. Soc. Pet. Geol.
Momper, J. A. *AAPG Short Course, Vol. 1, AAPG Natl. Mtg., Oklahoma City* (1978a).
Momper, J. A. *AAPG Course Note* **8**, B1-60 (1978b).
Monin, J. C., D. Barth, M. Perrut, J. Espitalié, and B. Durand. *Org. Geochem.* **13**, 1079 (1988).
Moore, L. R. *Proc. Yorks. Geol. Soc.* **34**, 235 (1964).
Neglia, S. *AAPG Bull.* **63**(4), 573 (1979).
Oliver, R. L. and D. J. Sinks. *Proc. Tar Sand Symposium, Jackson, WY (July 1986), paper #2-6.*
Philippi, G. T. *20th Internatl. Geol. Congr., Mexico City,* **Sec. 3**, 25 (1956).
Price, L. C. *Amer. Assoc. Petrol. Geol. Bull.* **60**, 213 (1976).
Price, L. C., and C. E. Barker. *J. Petr. Geol.* **8**(1), 59 (1985).
Pusey, W. C. *World Oil* **176**, 71 (1973).
Quigley, T. M., A. S. Mackenzie, and J. R. Gray. In *Migration of Hydrocarbons in Sedimentary Basins* (B. Doligez, ed.), 1987. Paris: Editions Technip.
Redmond, J. L., and R. P. Koesoemadinata. *Proc. 5th Mtg., Indonesian Pet. Assoc., Jakarta* (1976).
Stach, E. *Textbook of Coal Petrology*, 1985. Berlin: Borntraeger; *Internatl. Handbook of Coal Petrology*, 2nd ed., 1963. Paris.
Stainforth, J. G., and J. E. A. Reinders. *Org. Geochem.* **16**, 61 (1990).
Tissot, B. P., and J. Espitalié. *Rev. Inst. Fr. Petrol.* **30**, 743 (1975).
Tissot, B. P., and R. Pelet. *Proc. 8th World Petrol. Congr.* **2**, 35 (1971).
Tissot, B. P., and D. H. Welte. *Petroleum Formation and Occurrence*, 1978. New York: Springer.
Trofimuk, A. A., and A. E. Kantorovich. *8th Internatl. Congr. Org. Geochem. Moscow, Abstracts* **1**, 160 (1977).
Trulle, L. G. *DOE/Fe/60177-2250* (1986).
Ungerer, P., and R. Pelet. *Nature (London)* **327**, 6117 (1987).
Ungerer, P., F. Bessis, P. Y. Chenet, B. Durand, E. Nogaret, A. Chiarelli, J. L. Oudin, and J. F. Perrin. *Amer. Assoc. Petr. Geol. Mem.* **35**, 53 (1988).
Waples, D. W., In *Hydrocarbon Source Rocks of the Greater Rocky Mountain Region* (J. Woodward, F. F. Meissner, and J. L. Clayton, eds.), 1984. Denver: Rocky Mtn. Assoc. Geol.
Welte, D. H. *J. Geochem. Explor.* **1**, 117 (1972).
Werner, M. R. *Amer. Assoc. Petrol. Geol., Research Conf. Vol. 2, Santa Maria, CA* (1984).
Yariv, S. *Clay Sci.* **5**, 19 (1976).
Zhuze, T. P., V. I. Serge-evich, V. F. Burmistrova, and Y. A. Yesakov. *Dokl. Acad. Nauk SSSR* **198**(1), 206 (1971).

CHAPTER 4

Classification

Technological advances since the mid-1900s; greater flexibility of modern processing and refining practices; and an ability to generate information by instrumented analytical methods—all these have tended to marginalize comprehensive fossil hydrocarbon classifications by type and composition [1]. But by formally ordering an otherwise unmanageable welter of fossil hydrocarbons, such classifications have created data bases [2] that make it possible to characterize oils and coals by parameters that do not need to be specifically derived from prior chemical measurements.

1. CRUDE OILS

Because alteration of oil during migration and accumulation in a reservoir precludes any systematic variation of composition with age, as exists for coal, considerable importance attaches to detailed classifications that specify types and relative concentrations of molecular entities, and thereby define differences between different oils. Table 4.1.1, a classification proposed by Sachanen (1950), exemplifies such schemes. However, the application of this or other classifications of that genre to specified oils is contingent on sets of measurements that cannot be justified in industrial operations, and several simplified systems, often differing only in minutiae, have therefore been proposed to widen the utility of oil classifications based on chemical parameters. Tables 4.1.2 and 4.1.3 illustrate these.

The scheme advanced by Tissot and Welte (Table 4.1.2) stipulates concentration limits for aromatics, naphthenes, and alkanes in oil fractions boiling at >210°C/760 mm Hg, and uses these limits to define oils as:

1. *Paraffinic:* Usually very fluid light oils with specific gravities < 0.85, but more viscous if containing high proportions of wax; resins and asphaltics total <10%; sulfur contents are low

TABLE 4.1.1 Chemical Classification of Crude Oils[a]

Class	Composition	
Paraffinic	paraffinic side chains	>75%
Naphthenic	naphthenic hydrocarbons	>75%
Aromatic	aromatic hydrocarbons	>75%
Paraffinic/naphthenic	paraffin side chains	60–70%
	naphthenes	>20%
Asphaltic	resins and asphalts	>60%
Paraffinic–naphthenic–aromatic	paraffins ≈ naphthenes ≈ aromatics	
Naphthenic–aromatic	naphthenes, aromatics	each >35%
Naphthenic–aromatic–asphaltic	naphthenes, aromatics, asphalts	each >25%
Asphaltic–aromatic	asphalts, aromatics	each >35%

[a] Sachanen (1950).

TABLE 4.1.2 A Simplified Chemical Classification of Oils[a]

Group	Concentrations (wt%)	Oil type
	paraffins > naphthenes	paraffinic
	paraffins > 40%	
Sat. > 50%	paraffins > 40%	paraffinic–naphthenic
Arom. < 50%	naphthenes < 40%	
	naphthenes > paraffins	naphthenic
	naphthenes > 40%	
	paraffins > 10%	aromatic–intermediate
Sat. < 50%	paraffins ≥ 10%	aromatic–asphaltic
Arom. > 50%	naphthenes > 25%	
	paraffins ≥ 10%	aromatic–naphthenic
	naphthenes ≥ 25%	

[a] Tissot and Welte (1978).

TABLE 4.1.3 A Simplified Classification of Crude Oils[a] (Concentrations in wt%)

Class	Paraffinic	Naphthenic	Aromatic	Wax	Asphaltics
Paraffinic	46–61	22–32	12–25	1–10	0–6
Paraffinic/ naphthenic	42–45	38–39	16–20	1–51	0–6
Naphthenic	15–26	61–76	8–13	trace	0–6
Paraffinic/ naphthenic/ aromatic	27–35	36–47	26–33	0.05–1	0–10
Aromatic	0–8	57–78	20–25	0–0.5	0–20

[a] After Speight (1989).

2. *Paraffinic–naphthenic:* Heavier and more viscous than paraffinic oils; typically composed of >50% saturates, 25–40% aromatics, and 5–15% resins plus asphalts; [S] usually low and mostly existing as benzothiophenes
3. *Naphthenic:* Relatively rare; commonly biodegraded oils with >50% saturated cyclics; <20% *n*- and isoalkanes derived from paraffinic and paraffinic–naphthenic moieties
4. *Aromatic–intermediate:* Contain 40–70% aromatics, many derived from steroids; [S] mainly held in thiophenes that may represent >25% of total aromatics
5. *Aromatic–asphaltic:* Mostly biodegraded,[1] heavy aromatic-intermediate oils; typically with 30–60% asphaltenes and resins; [S] <1 to 9%
6. *Naphthenic–aromatic:* Biodegraded oils derived from paraffinic and/or paraffinic–naphthenic crudes; contains up to 25% resins

The formulation in Table 4.1.3, cited (uncredited) by Speight (1989), follows a very similar approach, but differs from Tissot and Welte's system in deriving proposed concentration limits for the different oil components from the composition of the 250–300°C fractions of crude oils.

Data from such simplified chemical classifications have been used to place different oil types into constructs that show compositions in a graphic format (see Fig. 4.1.1), define the broad compositional limits of conventional, intermediate, and heavy oils (Fig. 4.1.2.), and illustrate the broad effects of oxidation and biodegradation (Fig. 4.1.3).

However, for rapid "typing" of oils, chemical parameters have now been largely superseded by graphic devices and/or empirical correlations that make it possible to characterize crude oils and oil fractions in terms of one or two physical parameters. Data gained in this manner are inevitably less informative than specifications of composition *per se,* but evidently suffice to meet all immediate practical needs [2].

Graphic devices for quick assessment of oil types are illustrated by a correlation index (CI), which was developed from relationships between specific gravities and reciprocal boiling points of hydrocarbons (Smith, 1940, 1966, 1968). If the straight line for *n*-paraffins in this diagram (see Fig. 4.1.4) designates Cl = O, and a parallel line passing through a point for benzene denotes Cl = 100, the correlation index is defined by

$$\mathrm{Cl} = 473.7(1/d) - 456.8 + (48{,}640/\mathrm{b.p.}),$$

where $1/d$ is the reciprocal of the specific gravity of the oil and b.p. its average

[1] This characteristic is *assumed.* Cf. Chapter 3, Section 2.

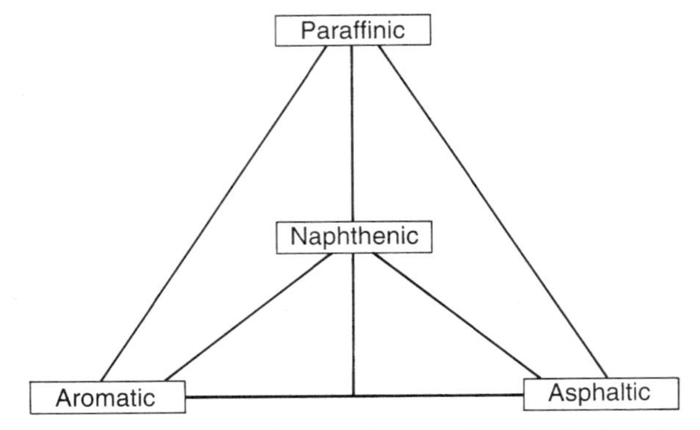

FIGURE 4.1.1 Base diagram for classification of crude oils (see Fig. 4.1.2).

boiling point (determined by the USBM standard distillation method). An index of <15 then indicates a predominantly paraffinic oil, Cl = 15–50 a mainly naphthenic oil or an ill-defined mix of paraffins, naphthenes, and aromatics, and Cl >50 an essentially aromatic oil or oil fraction.

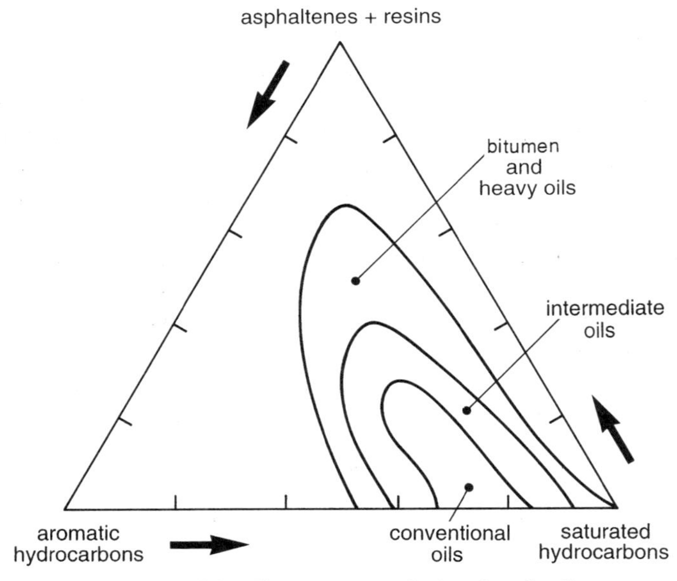

FIGURE 4.1.2 The composition limits of crude oils.

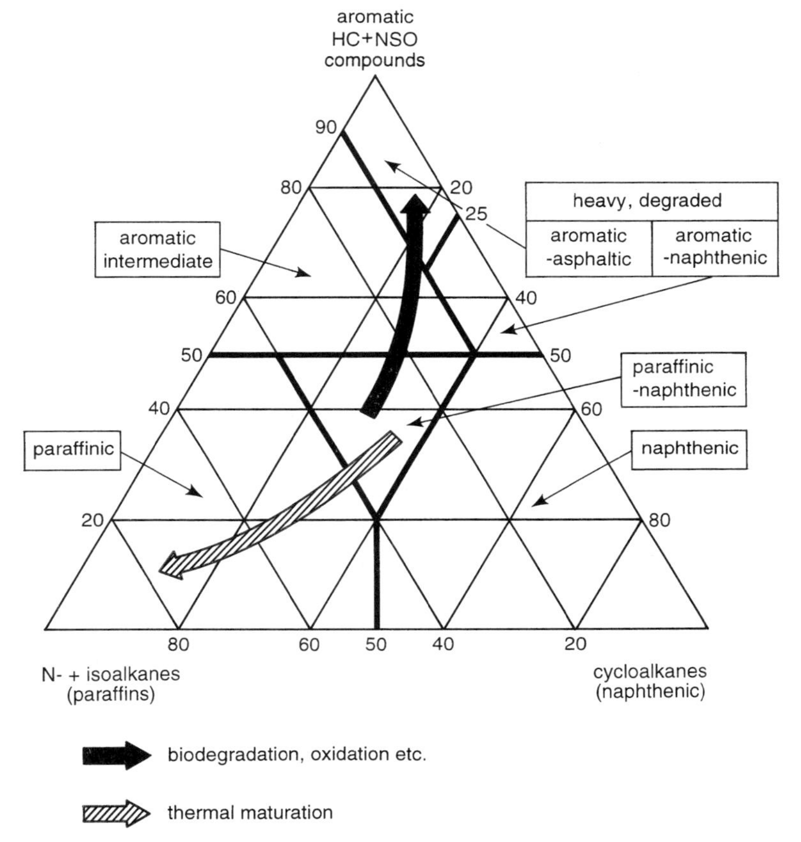

FIGURE 4.1.3 The class compositions of crude oils in a triangular format: the effects of maturation, oxidation, and biodegradation.

Other, more direct empirical correlations that permit quick "typing" of oils are exemplified by a system that makes use of the dependence of API gravities on composition and derives its required basic data from two "key" fractions deemed to characterize oil types. Of these fractions, both obtained by the USBM Hempel distillation, one accrues at 250–275°C/1 atm (in the kerosine range) and the other at 275°C/40 mm Hg (in the lubricating oil range). Tables 4.1.4 and 4.1.5 define them and the nine oil classes based on them.

An alternative version of this system (Tables 4.1.6a and b) makes use of "light" and "heavy" distillate fractions obtained at 250–275°C/1 atm and

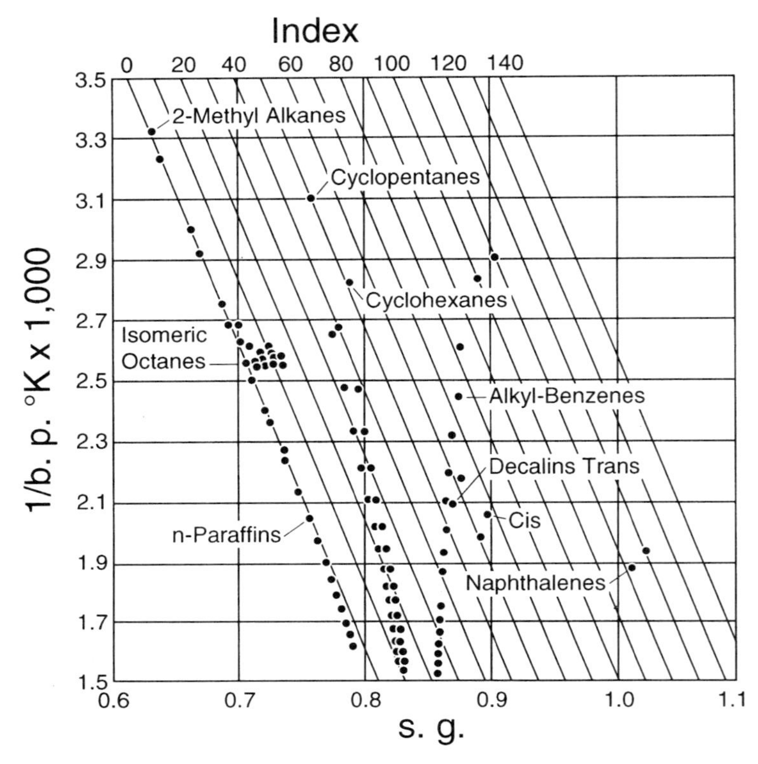

FIGURE 4.1.4 Reference data for the correlation index.

275–300°C/40 mm Hg, respectively, and another, more problematic system endeavors to specify the distribution of carbon among paraffinic, naphthenic, and aromatic structures from refractive indices, densities, and molecular weights.

Parenthetically, it might be noted that crude oils can also be broadly "typed" by the UOP (or so-called Watson) characterization factor:

$$Z = (T_a)^{1/3}/\text{s.g.},$$

where T_a is the average boiling point of the fraction in °Rankine (= °F + 460) and s.g. its specific gravity at 15°C. Since Z is additive on a weight basis, values between 10.5 and 12.5 would indicate a predominantly naphthenic oil, whereas

TABLE 4.1.4 Classification of "Key" Fractions from Hempel Distillation

Key fraction	API gravity	Classification
#1	40 and higher	paraffin
	30.1–39.9	intermediate
	<30	naphthene
#2	30 and higher	paraffin
	20.1–29.9	intermediate
	<20	naphthene

TABLE 4.1.5 Oil Classes Defined by Key Fractions from Hempel Distillation

Class	Key fraction #1	Key fraction #2
1. Paraffin	paraffin	paraffin
2. Paraffin–intermediate	paraffin	intermediate
3. Intermediate–paraffin	intermediate	paraffin
4. Intermediate	intermediate	intermediate
5. Intermediate–naphthene	intermediate	naphthene
6. Naphthene–intermediate	naphthene	intermediate
7. Naphthene	naphthene	naphthene
8. [a]Paraffin–naphthene	paraffin	naphthene
9. [a]Naphthene–paraffin	naphthene	paraffin

[a]Not recognized in any known crude oil.

TABLE 4.1.6a Classification Based on API Gravity of Light (250–275°C) Fractions

Class	°API	s.g.	Type
Paraffin	>40.0	<0.8251	paraffinic
Paraffin–intermediate	>40.0	<0.8251	paraffinic
Paraffin–naphthene	>40.0	<0.8251	paraffinic
Intermediate–paraffin	33.1–39.9	0.8256–0.8597	intermediate
Intermediate	33.1–39.9	0.8256–0.8597	intermediate
Intermediate–naphthene	33.1–39.9	0.8256–0.8597	intermediate
Naphthene–intermediate	<33.0	>0.8602	naphthenic
Naphthene	<33.0	>0.8602	naphthenic
Naphthene–paraffin	>33.0	>0.8602	naphthenic

TABLE 4.1.6b Classification Based on API Gravity of Heavy (275–300°C) Fractions

Class	°API	s.g.	Type
Paraffins	>30.0	<0.8762	paraffinic
Intermediate–paraffin	>30.0	<0.8762	paraffinic
Naphthene–paraffin	>30.0	<0.9340	paraffinic
Paraffin–intermediate	20.1–29.9	0.8767–0.9334	intermediate
Intermediate	20.1–29.9	0.8767–0.9334	intermediate
Naphthene–intermediate	20.1–29.9	0.8767–0.9334	intermediate
Intermediate–naphthene	<20.0	>0.9340	naphthenic
Naphthene	<20.0	>0.9340	naphthenic
Paraffin–naphthene	<20.0	>0.9340	naphthenic

12.5–13.0 would suggest a paraffinic one. However, this method provides no better characterization than the correlation index CI, and is *a priori* less satisfactory.

2. HEAVY HYDROCARBONS

Bituminous Substances

Between heavy oils (see Fig. 4.1.2) and coals stands a profusion of semisolid and solid bituminous substances that at present, with very few exceptions, command little commercial interest, and that are therefore not classified *in sensu stricto* [4]. However, most are qualitatively defined by their descriptions:

1. *Asphalts:* Semisolid or solid brown-to-black aggregates of high-boiling hydrocarbons, which occur naturally and as residua from crude oil refining. They are characterized by relatively high proportions of sulfur, oxygen, and nitrogen, as well as by small amounts of inert matter; when filling pores and crevices in limestones and sandstones, they are often referred to as rock asphalts.
2. *Asphaltites:* Naturally occurring dark-brown or black nonvolatile solid bituminous substances that superficially resemble bitumen. They contain high proportions of asphaltenes, and consequently soften over a wide temperature range (~115–330°C). Recognized varieties of asphaltite are:
   (a) *Gilsonite,* which usually occurs in near-vertical vein forms, but contains very little occluded mineral matter; is characterized by a bright luster, a conchoidal fracture, and 10–20% fixed carbon
   (b) *Grahamite,* which resembles gilsonite, but is differentiated from it by a higher fusion temperature, higher "fixed carbon" content (35–

55%), and a tendency to swell (or "intumesce") when heated to the fusion temperature

(c) *Glance pitch* (German "glanz" = shiny), a variety of Gilsonite characterized by higher luster and some properties of Grahamite

3. *Asphaltoids:* Naturally occurring brown-to-black solid bituminous substances, which differ from asphaltites in being infusible and only slightly soluble in CS_2; are sometimes also termed asphaltic pyrobitumen.

4. *Bituminous sands:* Frequently termed oil sands or tar sands; form more or less unconsolidated sandy strata, sandstones, or limestones that contain, as pore-filling or interstitial matter, semisolid or soft solid bitumens characterized by >95% benzene solubility.

5. *Mineral wax:* Yellow to dark-brown, solid paraffinic matter, which occurs in vein form or in interstitial spaces of rocks; fuses between ~60 and 95°C; and is in older literature often referred to as ozokerite.

Oil Shale Kerogens

Widely distributed, oil shales contain organic matter (kerogen) that usually amounts to less than 15–20%, but may on occasion, as in the some parts of Colorado's Eocene-age Green River shales, approach 35–40 wt%.

Analogous to the categorization of bituminous substances, kerogens are broadly classified by their precursor source materials and/or appearance under a microscope (cf. Chapter 2). Occasional references to "coaly" kerogen relate to extensively fragmented and dispersed coallike material [5], but in view of the definition of "kerogen," this is not synonymous with coal *per se.*

The following are separately specified as forms of kerogen:

1. *Tasmanites,* massively accumulated organic matter derived from marine algae (*Tasmanaceae*); exemplified by very extensive Permian-age deposits in Australia and by Jurassic/Cretaceous occurences in Alaska
2. *Torbanites,* mainly derived from fresh and/or brackish water algae such as *Botryococcus,* and exemplified by Carboniferous accumulations in Scotland as well as by Permian formations in Australia and South Africa
3. *Coorongites,* like tobanites, mostly derived from *Botryococcus* and found in some Australian Quaternary deposits

Coals

Since publication of classifications that made use of simple "rank" indicators (such as carbon, hydrogen, and/or volatile matter contents) to define coal classes and explore relationships between rank and coal properties (Seyler, 1899, 1900; Grout, 1907; Ralston, 1915; Rose, 1930), classification schemes have proliferated through formulation of several national and international specifications, and adopted tabular formats which employ numerical codes for expressing the rank-dependence of properties deemed to affect end-uses. Unlike characterization of oil, which became simpler but less explicit, coal classifications have, in short, become more explicit and much more complex.

Sapropels

Because of the unique properties of coallike solids that developed in dysoxic or suboxic environments and that, except for their higher O/C ratios, closely resemble amorphous (sapropelic) kerogens, a distinction is made between two types of coal:

1. *Sapropelic* coals (or sapropels), which formed from marine biota and accumulations of allochthonous terrestrial fungal spores and pollen in lacustrine and open offshore waters
2. *Humic* coals, which developed mainly from terrestrial plant debris that was transiently exposed to atmospheric oxygen, and therefore passed through a distinctive peat stage [6]

The differentiation between these two types rests primarily on the exceptionally high H-contents of sapropels—a feature reflecting liptinitic matter that often represents 60% of their total organic mass and, except for their relative rarity [7], makes these coals more generous sources of liquid hydrocarbons than many oil shales.

Of the two sapropel classes, *cannel* coal derives from the biodegraded debris of contemporaneous peat swamps that was, here and there, augmented by algal matter and wind-borne spores (Moore, 1968). The proportions of these components vary and define the character of the coal as algal or spore cannels. H-contents run to ~8%.

The other class, *boghead* coal, formed exclusively from algal masses in large dysoxic basins to which transport of exogenous organic matter was limited. The common contributors to the source materials were *Botryococcus*-type algae, which are often also seen in algal cannels. Average H-contents run to ~12%.

Characteristic properties of both classes include a dull appearance of freshly broken surfaces, a uniformly fine-grained texture, conchoidal fracture patterns, and an absence of the banding that is characteristic of mature humic coals (see Chapter 5). Associated mineral matter consists for the most part of almost colloidally dispersed clay minerals.

Humic Coals

Like chemically explicit oil classifications, early classifications of humic coals were mainly designed to facilitate development of coal systematics. The framework was therefore a rectangular coordinate system that displayed the covariance of primary rank indicators (usually [C], [H], [O], and/or volatile matter contents). To allow direct comparison, these were always expressed on dry ash-free (daf.) or dry mineral matter-free (dmmf) coal, and the band delineated in such diagram was then graphically amplified by data that showed the covariation of other properties. Figure 4.4.1 exemplifies "development lines" (White, 1933) that illustrate semiquantitatively how [C], [H], and [O] change relatively as the coal matures from the lignite to the anthracite stage [8], and that compare these trends with the corresponding development of sapropels. Figure 4.4.2 shows a correlation diagram based on an [H]/[C] plot. Both illustrate compositional and other changes that characterize metamorphic development and assist exploration projects when decisions about such projects must be based on properties of weathered outcrops (Berkowitz, 1994).

However, because chemical maturation of coal does not always parallel corresponding changes in physical properties (Berkowitz, 1950, 1952), intuitive delineation of coal classes in a continuous coal band is not sufficiently precise to meet the needs of industry—and this recognition has led to replacement of graphic schemes by classification in tabular formats.

A long-standing example of such format is the ASTM classification, which defines four classes, each subdivided into two or more groups (Table 4.4.1). This formulation has established itself as the definitive North American classification and has also gained wide acceptance elsewhere. It is, however, not much more explicit than its graphic forerunners, offers no information about coal quality, and because it is, strictly speaking, only valid for the usually vitrinite-rich Northern Hemisphere coals, it effectively discounts the massive Gondwanaland deposits of Australia, China, and South Africa [10].

An *a priori* more complete characterization presents itself in a rank-based classification, which the British National Coal Board (NCB, 1946) developed for commercial use. This employs *numerical* codes for specifying the category, class, and subclass to which a coal must be assigned on the basis of its elemental composition and properties, and designates heat-altered and weathered coals

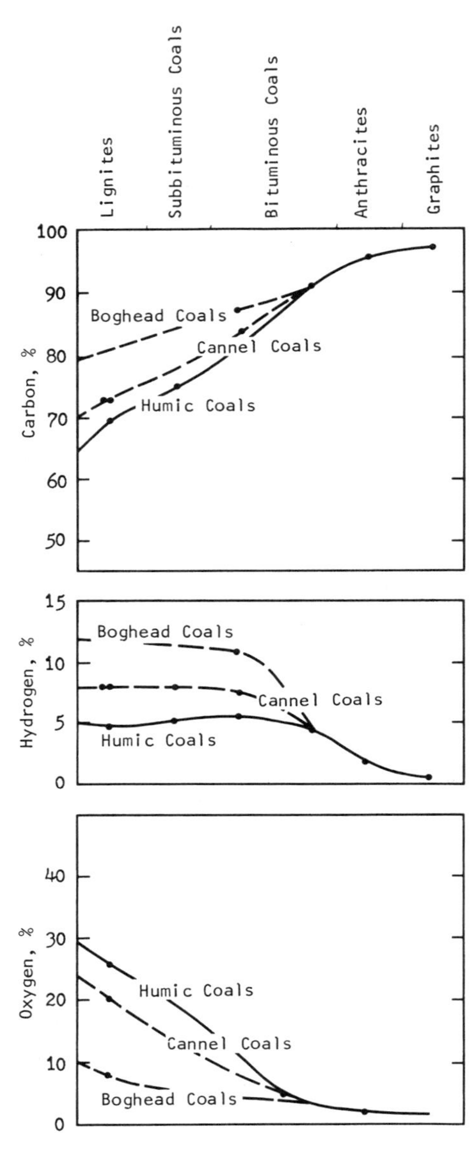

FIGURE 4.4.1 "Development lines" showing the effects of progressive maturation on the composition of sapropelic and humic coals (after White, 1933).

by respectively adding H or W to the rank code (Table 4.4.2). But this scheme, too, is restricted: it is only applicable to Carboniferous bituminous coals and anthracites, and its usefulness as means for classifying mature *post*-Carboniferious coals—or indeed, coals other than those of England, Scotland, and Wales—remains untested.

As matters stand, then, the most detailed and potentially most useful system is an international rank-based classification of mature (or hard) coal (UN, 1988), which in its latest version (Table 4.4.3, Fig. 4.4.3) uses a 14-digit code to define eight parameters. Procedures for measuring these are specified in the relevant ISO standards [11], but a major drawback of the scheme is the need for very careful decoding.

The hard-coal classification is augmented by a scheme that, albeit in preliminary form [12], characterizes the "soft" coals—broadly, the brown coals, lignites, and subbituminous B and C coals of the ASTM classification—in terms of moisture contents and the tar yields they would be expected to furnish when pyrolyzed (Table 4.4.4).

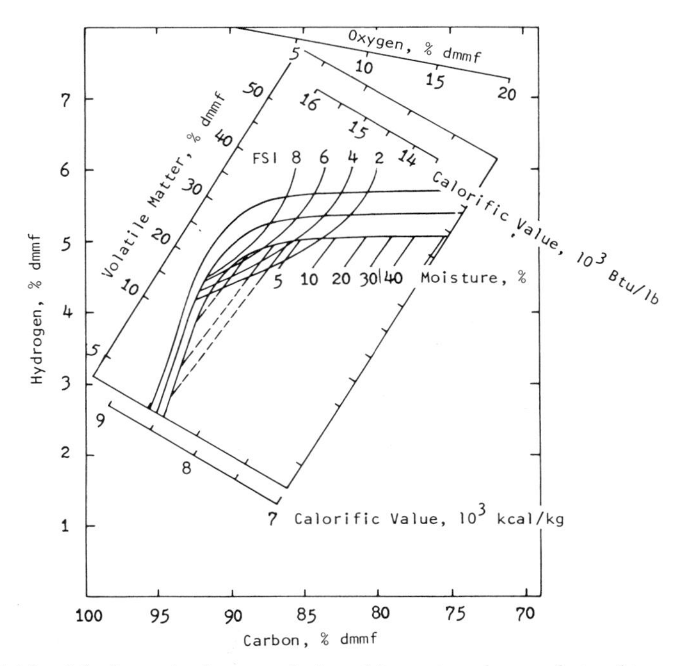

FIGURE 4.4.2 A hydrogen/carbon correlation: this version shows relationships among [C], [H], [O], [VM], [H_2O], and caking propensities as reflected in free swelling indeces (FSI). (Adapted from Seyler's coal chart 47B.)

TABLE 4.4.1 ASTM Classification of Coal by Rank[a]

Class	Group	Abbreviation	Fixed carbon (% dmmf) ≥	Fixed carbon (% dmmf) <	Volatile matter (% dmmf) ≥	Volatile matter (% dmmf) <	Calorific value[b] (MJ/kg mmmf) ≥	Calorific value[b] (MJ/kg mmmf) <	Agglomerating character
I. Anthracite	1. Metaanthracite	ma	98			2			Nonagglomerating
	2. Anthracite	an	92	98	2	8			Nonagglomerating
	3. Semianthracite[c]	sa	86	92	8	14			Nonagglomerating
II. Bituminous	1. Low volatile bituminous	lvb	78	86	14	22			Commonly agglomerating
	2. Medium volatile bituminous	mvb	69	78	22	31			Commonly agglomerating
	3. High volatile A bituminous	hvAb		69	31		32.56[d] [14,000]		Commonly agglomerating
	4. High volatile B bituminous	hvBb					30.24 [13,000	32.56 14,000]	Commonly agglomerating
	5. High volatile C bituminous	hvCb					26.75 [11,500	30.24 13,000]	Commonly agglomerating
							24.42 [10,500	26.75 11,500]	Agglomerating
III. Subbituminous	1. Subbituminous A	subA					24.42 [10,500	26.75 11,500]	Nonagglomerating
	2. Subbituminous B	subB					22.10 [9,500	24.42 10,500]	Nonagglomerating
	3. Subbituminous C	subC					19.31 [8,300	22.10 9,500]	Nonagglomerating
IV. Lignitic	1. Lignite A	ligA					14.65 [6,300	19.31 8,300]	Nonagglomerating
	2. Lignite B	ligB						14.65 [6,300]	Nonagglomerating

[a] ASTM (1984).
[b] Bracketed data show calorific values in Btu/lb. (mmmf = mineral matter and moisture-free).
[c] If agglomerating, classify as lvb coal.
[d] Coals with more than 69% fixed shall be classified according to fixed carbon, regardless of calorific value.

TABLE 4.4.2 NCB (Britain) Coal Classification System

Group	Class	Volatile matter (% dmmf)	Gray–King coke type	Description
100	101	6.1[a]	A	Anthracite
	102	6.1–9.0[a]	A	Anthracite
200	201	9.1–13.5	A–G	Dry steam coals
	201a	9.1–11.5	A–B	
	201b	11.6–13.5	B–C	
	202	13.6–15.0	B–G	Coking steam coals
	203	15.1–17.0	B–G4	
	204	17.1–19.5	G1–G8	
	206	9.1–19.5	A–B[b] A–D[c]	Heat-altered low-volatile bituminous coals
300	301	19.6–32.0	G4	Prime coking coals
	301a	19.6–27.5		
	301b	27.6–32.0		
	305	19.6–32.0	G–G3	Heat-altered medium-volatile bituminous coals
	306	19.6–32.0	A–B	
400	401	32.1–36.0	G9	Very strongly caking coals
	402	>36.0		
500	501	32.1–36.0	G5–G8	Strongly caking coals
	502	>36.0		
600	601	32.1–36.0	G1–G4	Medium caking coals
	602	>36.0		
700	701	32.1–36.0	B–G	Weakly caking coals
	702	>36.0		
800	801	32.1–36.0	C–D	Very weakly caking coals
	802	>36.0		
900	901	32.1–36.0	A–B	Noncaking coals
	902	>36.0		

[a] To distinguish between classes 101 and 102, it is sometimes more convenient to use a hydrogen content of 3.35% instead of 6.1% volatile matter.
[b] For volatile matter contents between 9.1 and 15.0%.
[c] For volatile matter contents between 15.1 and 19.5%.

A parallel comprehensive classification, specifically designed for Australian (and presumably other Souther Hemisphere) hard coals, has also been published by the Standards Association of Australia (1987). This scheme (Table 4.4.5) resembles the international system, but goes beyond it by distinguishing between high- and low-rank coals, and thereby implicitly differentiating between mature coals, which pose no significant hazard from autogenous heating during long-distance transport or prolonged storage, and immature coals, which may pose such hazards.

TABLE 4.4.3 International Classification of Hard Coals[a]

Vitrinite reflectance		Reflectogram characteristics				Maceral group composition (mmf)				Crucible swelling no.	
						Inertinite[b]		Liptinite			
Code	$R_{rand'}$ (%)	Code	Standard deviation		Type	Code	Vol%	Code	Vol%	Code	Number
0.2	0.2–0.29	0	≤0.1	No gap	Seam coal	0	0–<10	1	0–<5	0	0–0.5
03	0.3–0.39	1	≥0.1 ≤ 0.2	No gap	Simple blend	1	10–<20	2	5–<10	1	1–1.5
0.4	0.4–0.49	2	≥0.2	No gap	Complex blend	2	20–<30	3	10–<15	2	2–2.5
⋮	⋮	3		One gap	Blend with one gap	⋮		⋮		⋮	
48	4.8–4.89	4		Two gaps	Blend with two gaps	7	70–<80	7	30–<35	7	7–7.5
49	4.9–4.99	5		Two gaps	Blend with two gaps	8	80–<90	8	35–<40	8	8–8.5
50	≥5.0					9	≥90	9	≥40	9	9–9.5

Volatile matter (daf)		Ash, dry		Total sulfur, dry		Gross calorific value (daf)	
Code	mass%	Code	mass%	Code	mass%	Code	MJ/kg
48	≥48	00	0–<1	00	0–<0.1	21	>22
46	46–<48	01	1–<2	01	0.1–<0.2	22	22–<23
44	44–<46	02	2–<3	02	0.2–<0.3	23	23–<24
⋮							
12	12–<14	20	20–<21	29	2.9–<3.0	37	37–<38
10	10–<12			30	3.0–<3.1	38	38–<39
⋮						39	≥39
02	2–<3						
01	1–<2						

[a] "Higher rank coals" are coals with gross calorific values (daf) > 24 MJ/kg or <24 MJ/kg if the mean random reflectance exceeds 0.6%.
[b] Some inertinites may be reactive..
[c] Where ash contents exceed 10%, they must be reduced before analysis to <10% by dense medium separation.

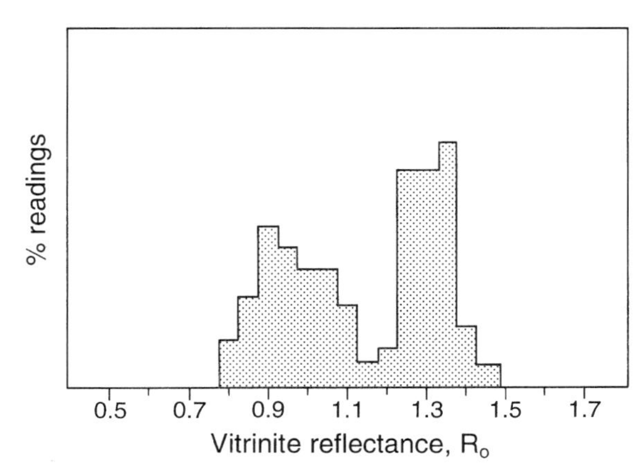

FIGURE 4.4.3 An illustrative "reflectogram"—a parameter used in the International Coal Classification.

TABLE 4.4.4 International Classification of Soft Coals

Group parameter—tar yield (% daf)	Group number	Code number					
>25	4	14	24	34	44	54	64
20–25	3	13	23	33	43	53	63
15–20	2	12	22	32	42	52	62
10–15	1	11	21	31	41	51	61
<10	0	10	20	30	40	50	60
Class number		1	2	3	4	5	6
Class parameter[b]		<20	20–30	30–40	40–50	50–60	60–70

[a] ISO (1974).
[b] Total moisture content of ash-free run-of-mine coal.

NOTES

[1] Some extremely explicit coal classifications were, nevertheless, published in the late 1980s in order to assist international coal trade. These classifications are noted under "COALS".

[2] Commonly referred to as *systematics*, such data quantity relationships among different properties of a substance and thereby enable its probable composition or behavior under specified conditions to be predicted with some confidence.

[3] When needed, additional information is presumably obtained by in-house analyses.

[4] The compositions of these substances (see Chapter 5) are usually so variable that the isolated data now at hand have little meaning.

[5] The origin of such coaly material is uncertain. One possibility is that it derived from, or

TABLE 4.4.5 Classification and Coding of Australian Coals[a]

Vitrinite reflectance, R_{max}—higher rank coals		Vitrinite reflectance, R_{max}—lower rank coals		Gross calorific value (daf)		Volatile matter (daf)	
Code	%	Code	%	Code	MJ/kg	Code	%
03	0.3–0.39			15	15–15.98	08	8–8.9
04	0.4–0.49	0	0.0–0.09	16	16–16.98	09	9–9.9
05	0.5–0.59	1	0.1–0.19	17	17–17.98	10	10–10.9
·		2	0.2–0.29	·		·	
·		3	0.3–0.39	·		·	
·		4	0.4–0.49	·		·	
19	1.9–1.99	5	0.5–0.59	35	35–35.98	49	49–49.9
20	2.0–2.09			36	36–36.98	50	50–50.9

Crucible Swelling No.—higher rank coals		Bed moisture (as sampled)—lower rank coals		Ash, dry		Total sulfur, dry	
Code	No.	Code	%	Code	%	Code	%
0	0 or 0.5	20	20–20.9	00	0–0.9	00	0–0.09
1	1 or 1.5	21	21–21.9	01	1–1.9	01	0.1–0.19
2	2 or 2.5	·		·		·	
·		·		·		·	
·		·		·		·	
8	8 or 8.5	64	64–64.9	29	29–29.9	31	3.1–3.19
9	9 or 9+	65	65–65.9	30	30–30.9	32	3.2–3.29

Note. Higher rank coals are coals with gross calorific values of ≥21 MJ/kg (moist, ash-free) or ≥27 MJ/kg (daf); lower rank coals are coals with gross calorific values of ≤21 MJ/kg (moist, ash-free) or ≤27 MJ/kg (daf).

[a] Standards Association of Australia (1987).

was, *sea coal*—apparently a coallike substance that was, as recorded in late-12th-century English records, washed up on beaches and collected there.

[6] Some authors refer to humic coals as "normal" coals, but this term is inappropriate and can prove misleading because it is more commonly used in reference to coals that were subject to normal rather than regional metamorphism.

[7] Unfortunately, most coal classifications pay scant, if any, regard to them.

[8] Although the continuity of the coal band might suggest progressive maturation from the youngest to the most mature coal, it by no means proves such development (see Fuchs, 1952; Dryden, 1956). Fuchs does, in fact, offer an imporant alternative to a kinetic approach based on continuous progressive maturation (see Karweil *et al.*, 1955, 1966).

[9] Details of this diagram (chart 47B), which was repeatedly modified since it was first published in 1899, are, strictly speaking, only valid for British Carboniferous-age bituminous coals.

Reliable data for coals from North America and regions in the Southern Hemisphere, which later came to hand, enabled more encompassing compilations (see Mott, 1948).

[10] *Gondwanaland* denotes the Southern Hemisphere's landmass, which did not begin to break up until the late Mesozic. Coal first appeared there in Permo-Carboniferous times, the earliest deposits being those of present-day South Africa, Australia, northern China, northern India, and Australia. Later, during the Cretaceous, coal accumulated in Queensland (Australia), Assam (northern India), Hokkaido (northern Japan), and New Zealand, and formed the vast deposits of China and South Africa that, collectively, rival those of the Americas and Europe.

[11] Two digits are used to define the vitrinite reflectance (R_o); one for a reflectogram characteristic (see Figure 4.4.3); one each for the inertinite and exinite (liptinite) content; one for the FSI (a measure of caking propensities); two for volatile matter contents; two for ash content; two for sulfur; and two for the gross heat value. (For a definition of these parameters, see, among others, Berkowitz, 1994.)

[12] The parameters of this scheme reflect some, but not all, current conventional uses of immature coals, and will undoubtedly in due course have to be supplemented by others that relate to, e.g., suitability for briquetting, gasification, or liquefaction.

REFERENCES

American Society for Testing Materials D-338, 1966; revised 1984.

Berkowitz, N. *Introduction to Coal Technology*, 2nd ed., 1994, pp. 109–110. New York: Academic Press.

Berkowitz, N. *Fuel* **29**, 138 (1950); *Fuel* **31**, 19 (1952).

Dryden, I. G. C. *Coke and Gas*, April/May (1956).

Fuchs, W. *Chemiker-Ztg.* **61** (1952).

Grout, F. F. *Econ. Geol.* **2**, 225 (1907).

Karweil, J. *Z. Dtsch. Geol. Ges.* **107**, 132 (1956); *Brennst. Chem.* **47**, 161 (1966).

Mott, R. A.; *J. Inst. Fuel (London)* **22,** 2 (1948).

Moore, L. R. In *Coal and Coal-Bearing Strata* (D. Murchison and T. S. Westoll, eds.), Chapter 2, 1968. Elsevier Publ. Co., NY 1968.

National Coal Board (U.K.). DSIR Survey Paper No. 58, HMSO London (1946).

Ralston,, O. C. *U.S. Bur. Mines Tech. Paper No. 93* (1915).

Rose, H. J. *Trans. Amer. Inst. Min. Metall. Eng.* **88**, 541 (1930).

Sachanen, A. N. In *Science of Petroleum* (B. T. Brooks and A. E. Dunstan, eds.), 1950. Oxford Univ. Press.

Seyler, C. A. *Proc. S. Wales Inst. Engrs.* **21**, 483 (1899); **22**, 122 (1900); **53**, 254, 396, (1938).

Speight, J. G. *The Chemistry and Technology of Petroleum,* 2nd ed., 1989. New York: Dekker.

Smith, H. M. *U.S. Bur. Mines Tech. Paper No. 610* (1940); *U.S. Bur. Mines Info. Circ. 8286* (1966); *U.S. Bur. Mines Bull.* 642 (1968).

Standards Association of Australia. AS 2096 (1987).

Tissot, B. P., and D. H. Welte. *Petroleum Formation and Occurrence,* 1978. New York: Springer-Verlag.

United Nations Publication #1957 II/E/Min.20, Geneva (1957); International Standards Organization (ISO) (1974).

United Nations Publication ECE/Coal/115 (1988).

White, D. *Econ. Geol.* **28**, 556 (1933).

CHAPTER 5

Composition and Chemical Properties

1. CHEMICAL SYSTEMATICS

Chemically, fossil hydrocarbons become progressively more complex as one proceeds from natural gas and crude oils to the heavier members of the family, and it may therefore be helpful to preface a discussion of chemical compositions with a summary of the rules that govern the designations of organic compounds and atomic configurations [1].[1]

The simplest of the disparate entities that make up fossil hydrocarbons are *n*-paraffins or *alkanes*—a homologous series of compounds that begins with CH_4, is characterized by the empirical formula C_nH_{2n+2}, and for all practical purposes ends with 40 or so $—CH_2—$ bridges between the terminal $—CH_3$ groups. However, except for the first three members of this series, *n*-paraffins present themselves in isomeric forms whose number rises quickly with molecular weight. A C_4 alkane can thus exist in two forms, as *n*-butane or isobutane:

$$CH_3—CH_2—CH_2—CH_3 \quad \text{or} \quad CH_3—\underset{\displaystyle CH_3}{\underset{|}{CH}}—CH_3.$$

Pentane (C_5H_{12}) can assume three forms, viz., n-pentane, isopentane, and tetramethylmethane [2]:

$$CH_3—CH_2—CH_2—CH_2—CH_3 \quad \text{or}$$

$$CH_3—\overset{\displaystyle CH_3}{\overset{|}{CH}}—CH_2—CH_3 \quad \text{or} \quad CH_3—\underset{\displaystyle CH_3}{\underset{|}{\overset{\displaystyle CH_3}{\overset{|}{C}}}}—CH_3.$$

Hexane can appear in six, and so on. By IUPAC rules, which count C atoms

[1] This may also be of assistance for Chapter 9, which reviews processing, and for chapter 10, which summarizes chemical as well as technological aspects of conversion.

TABLE 5.1.1 Some Properties of Hexane Isomers[a]

		m.p.	b.p.	s.g.
n-hexane	$CH_3(CH_2)_4CH_3$	−95.3	104.3	0.664
2-Methylpentane	$CH_3\underset{\substack{\vert \\ CH_3}}{C}HCH_2CH_2CH_3$	−153.6	95.8	0.658
3-Methylpentane	$CH_3CH_2\underset{\substack{\vert \\ CH_3}}{C}HCH_2CH_3$	−118.0	98.8	0.669
2,3-Dimethylbutane	$CH_3\underset{\substack{\vert \\ H_3C}}{C}H\underset{\substack{\vert \\ CH_3}}{C}HCH_3$	−128.5	93.5	0.666
2,2-Dimethylbutane	$CH_3\overset{\substack{CH_3 \\ \vert}}{\underset{\substack{\vert \\ CH_3}}{C}}CH_2CH_3$	−99.8	85.3	0.654

[a] m.p. and b.p. in °C, s.g in g/cm³ at 15°C.

in a chain from the *left*, and designate alkanes by the number of C-atoms in the "ancestral" straight chain, a C_7 alkane can consequently present itself, among other forms, as *n*-heptane,

$$\begin{array}{ccccccccccccc} CH_3 & — & CH_2 & — & CH_2 & — & CH_2 & — & CH_2 & — & CH_2 & — & CH_3, \\ 1 & & 2 & & 3 & & 4 & & 5 & & 6 & & 7 \end{array}$$

or be structured as

$$\begin{array}{ccccccccccc} & & CH_3 & & & & & & & & \\ & & \vert & & & & & & & & \\ CH_3 & — & CH & — & CH_2 & — & CH_2 & — & CH_2 & — & CH_3, \\ 1 & & 2 & & 3 & & 4 & & 5 & & 6 \end{array}$$

which is formally a heptane isomer, but is considered to be a *hexane* derivative and designated as 2-methylhexane; or appear as

$$\begin{array}{ccccccccc} & & H_3C & & CH_3 & & & & \\ & & \vert & & \vert & & & & \\ CH_3 & — & CH & — & CH & — & CH_2 & — & CH_3, \\ 1 & & 2 & & 3 & & 4 & & 5 \end{array}$$

which is viewed as a pentane derivative and termed 2,3-dimethylpentane.

The significance of isomeric form lies in its effect on the properties of the compound. In straight-chain alkanes, C—C bonds are energetically equivalent, and as their molecular weights increase, each additional CH_2 will therefore contribute a nearly constant increment to the specific gravity and boiling point [3]. But the different isomer forms of any one compound possess appreciably different properties. Table 5.1.1 exemplifies this with C_6 alkanes.

Much the same applies to two other series of aliphatic compounds, known as *alkenes* and *alkynes.*

Alkenes, commonly referred to as *olefins,* are partially unsaturated straight-chain hydrocarbons characterized by a $—C{=}C—$ moiety in their molecules, and are consequently defined by the empirical formula C_2H_2. The first two members of the series, which correspond to the C_2 and C_3 alkanes, are ethylene (or ethene: $CH_2{=}CH_2$), and propylene (or propene: $CH_3—CH{=}CH_3$).

But beginning with the third, butylene (or butene), two special aspects of alkenes come into play. By IUPAC rules, the position of the double bond is identified by the first C-atom nearest to that bond when numbered from the *nearest chain end,* so that

$$\underset{4}{CH_3}—\underset{3}{CH_2}—\underset{2}{CH}{=}\underset{1}{CH_2} \quad \text{and} \quad \underset{4}{CH_3}—\underset{3}{CH}{=}\underset{2}{CH}—\underset{1}{CH_3}$$

are designated as 1-butylene and 2-butylene, respectively. Starting with C_4, alkenes also exist in two *geometrically* isomeric forms that require differentiation between *cis-* and *trans-*configurations [4], as in

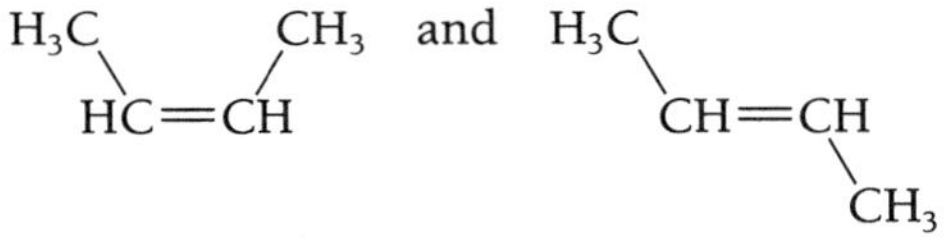

```
H3C         CH3   and  H3C
   \       /              \
    HC=CH                  CH=CH
                                \
                                 CH3
cis-2-butylene         trans-2-butylene
```

The physical properties of alkenes resemble those of the corresponding alkanes; but since alkenes are unsaturated, they are more reactive and generally seek to eliminate the $—C{=}C—$ double bond by hydrogen additions.

Alkene derivatives with two or three double bonds in their molecules are termed (alka)dienes or (alka)trienes, but are often simply referred to as dienes or trienes, and are as reactive as the corresponding alkenes. Examples of dienes and trienes are

	1,2-butadiene	$CH_2{=}C{=}CH—CH_3$
	1,3-butadiene	$CH_2{=}CH—CH{=}CH_2$
and	1,2,3-pentatriene	$CH_2{=}C{=}C{=}CH—CH_3$

The third series of aliphatic compounds, the alkynes, are characterized by the same empirical formula as (alka)dienes, i.e., by C_nH_{2n-2}, but have a $—C{\equiv}C—$ moiety in their molecules. Nomenclature rules are the same as for alkenes, so that a C_6 alkyne isomer structured as

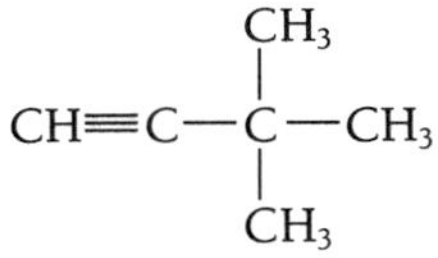

```
             CH3
             |
   CH≡C—C—CH3
             |
             CH3
```

FIGURE 5.1.1 The *trans* and *cis* isomeric forms of cyclohexane.

is identified as 3,3-trimethyl-1-butylene, and analogous to reactions of alkenes, most alkyne reactions entail sequential H-addition, as in

$$-C{\equiv}C- \rightarrow -HC{=}CH- \rightarrow -CH_2-CH_2-$$

Cyclic alkanes, more commonly known as *naphthenes* or *cycloparaffins*, are saturated ring structures and represented by unadorned rings—a triangle for cyclopropane, a square for cyclobutane, a pentagon for cyclopentane, or a hexagon for cyclohexane. But the reactivities of cyclic alkanes depend on their C—C bond angles and vary. In cyclohexane, which can assume *cis* and *trans* configurations (Fig. 5.1.1), C—C bonds form 109.5° angles, and the two isomers are therefore as stable and inert as *n*-alkanes. The same is true of cyclopentane, in which C—C bond angles are only slightly smaller (108°). However, in cyclopropane and cyclobutane the angles are substantially smaller, and both are therefore much more reactive [5].

Condensed naphthenic rings, which are fully saturated counterparts of polynuclear aromatics, are exemplified by bicyclooctane (comprised of two fused cyclopentane rings) and decalin (composed of two fused cyclohexane rings).

Of aromatic compounds (or *arenes*), the simplest are benzene (C_6H_6), naphthalene ($C_{10}H_8$), and anthracene ($C_{14}H_{10}$). For these and other aromatics—for example, phenanthrene, in which one of the three rings is angled with respect to the other two—no systematic nomenclature exists, and most are assigned "trivial" names that do not reflect their structure. Methylbenzene ($C_6H_5CH_3$) is thus generally referred to as toluene, and dimethylbenzene $C_6H_4(CH_3)_2$ as xylene; the three isomers of xylene are usually termed *o(rtho)*-, *m(eta)*-, or *p(ara)*-xylene, rather than 1,2-, 1,3-, or 1,4-dimethylbenzene.

Cycloalkanoaromatics such as tetrahydronaphthalene (tetralin), which are also routinely designated by trivial names, are condensed ring systems that contain cycloparaffinic and aromatic structure elements.

Hydrocarbon *derivatives* are designated by their hydrocarbon precursors and class (Table 5.1.2), with the latter defining the substituent function that affects chemical reactivity and physical properties, and polyfunctional com-

TABLE 5.1.2 The Nomenclature of Hydrocarbon Derivatives

Class	Function
Alcohol[a]	—OH
Aldehyde	—CHO
Amide	$—NH_2$
Carboxy acid	—COOH
Ether	—O—
Ketone	>C=O
Nitrile	—C≡N
Organometallic	≡C—M or =C=M
Phenol[b]	—OH
Thioether	—S—
Thiol	—SH

[a] —OH on aliphatics.
[b] —OH on aromatics.

pounds, which contain more than one function, follow the same nomenclature rules. Examples are

ethylene glycol $HO—CH_2—CH_2—O—CH_2—CH_2—CH_3$

ethanolamine $HO—CH_2—CH_2—NH_2$

acetamide $CH_3—C(=O)—NH_2$

and ethyl acetate $CH_3—C(=O)—O—CH_2CH_3$

Replacement of C atoms in naphthenic and aromatic rings by hetero atoms (most often N, O, or S) furnishes compounds such as thiophene (C_4H_4S), pyridine (C_4H_5N), or quinoline (C_9H_7N) that possess chemical and physical properties profoundly different from those of the pure hydrocarbons from which they derive. That is to some extent also true when a metal ion replaces H in a molecule to form an organometallic compound. In crude oils, in which such ion is commonly nickel or vanadium, the resultant organometallic compound usually takes the form of a porphyrin whose structure (see Fig. 5.1.2)

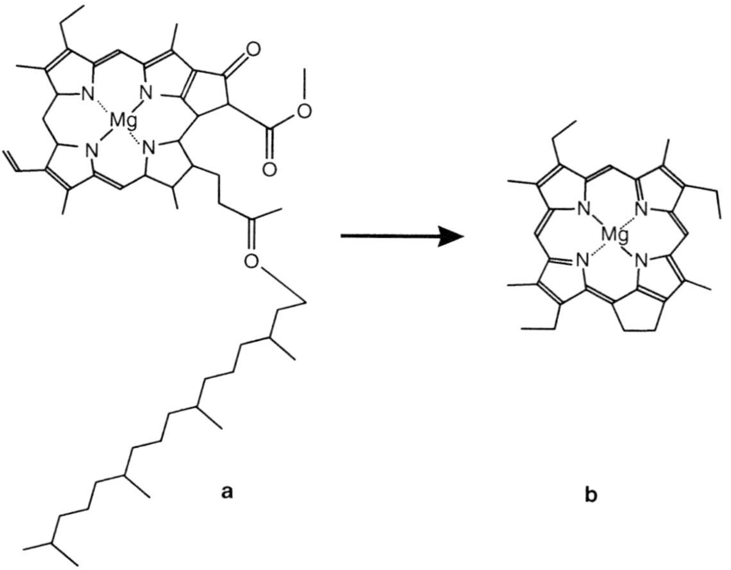

FIGURE 5.1.2 Chlorophyll (a) and a derived porphyrin (b).

betrays its origin in the chlorophyll of living plants; in coals, however, metal ions reside, as a rule, only in substituent functions, i.e., in structures such as

$$\begin{array}{l} =\!\mathrm{C}\!-\!\mathrm{COOM} \\ \quad | \\ =\!\mathrm{CH} \end{array} \quad \text{or} \quad \begin{array}{l} =\!\mathrm{C}\!-\!\mathrm{O} \\ \quad | \qquad\quad \mathrm{M}, \\ =\!\mathrm{C}\!-\!\mathrm{O} \end{array}$$

in which the "carrier" is normally a —COOH or —OH function, and M a mono- or bivalent metal.

Compounds that form when a nonmetallic such as chlorine or phosphorus replaces H in —COOH or phenolic —OH are organic *salts*.

2. COMPOSITIONS

Natural Gas

The simplest of the lighter petroleum hydrocarbons [6] is natural gas, which is often identified with CH_4, but in its raw form actually contains significant proportions of C_2–C_6 alkanes, H_2S, and several other inorganic contaminants

TABLE 5.2.1 Composition of Raw (Dry) Natural Gas

Hydrocarbons (%)	
Methane	70–98
Ethane	1–10
Propane	trace–5
Butanes	trace–2
Pentanes	trace–1
Hexanes	trace–0.5
Heptanes	0–trace
Inorganics (%)	
Hydrogen	trace–15
Carbon dioxide	trace–1
Hydrogen sulfide	trace–5
Helium	trace–8

[7]. Tables 5.2.1 and 5.2.2 list composition ranges and boiling points of the hydrocarbon components; Table 5.2.3 identifies some parameters that "classify" natural gas; and Table 5.2.4 presents some atypical cases of sour gas in which H_2S constitutes 8–98% of the total gas volume.

The hydrocarbon components, as well as most H_2S, CO_2, and N_2, are products of kerogen degradation. However, some CH_4, H_2S, and N_2 was apparently also generated by microbial activity in marine near-surface sediments by

$$\text{SO}_4\text{ reduction:}\quad SO_4^{2-} + 6H_2 \rightarrow H_2S + 4H_2O$$
$$\text{CO}_2\text{ reduction:}\quad CO_2 + 4H_2O \rightarrow CH_4 + 2H_2O$$

and

$$\text{NO}_3\text{ reduction:}\quad NO_3^- + 6H_2 \rightarrow N_2 + 6H_2O$$

TABLE 5.2.2 Boiling Points of Lower Alkanes (°C)

Methane	−162
Ethane	−89
Propane	−42
Isobutane	−10
n-Butane	−1
Isopentane	28
n-Pentane	36

TABLE 5.2.3 Natural Gas "Classifiers"

"Associated"	occurring with crude oil
"Nonassociated"	occurring alone
"Wet"	>0.040 liter condensible hydrocarbons/m^3
"Dry"	<0.013 liter condensible hydrocarbons/m^3
"Sweet"	<0.55% H_2S
"Sour"	>2.5% H_2S

Appreciable additional volumes of H_2S are also thought to have been produced during catagenesis by desulfurization, as in

$$R{-}CH_2{-}CH_2{-}SH \rightarrow R{-}CH{=}CH_2 + H_2S,$$

which would have proceeded more readily than dehydroxylation because the C—S bond is substantially weaker than the C—O bond, and in some environments, N_2 may additionally have been generated by oxidation of NH_3 (Getz, 1977).

Helium accumulated in natural gas by diffusion from granites in which it was produced by decay of radioactive elements such as thorium and uranium; when present in >2% concentrations, it is extracted [8].

Analogous processing of "wet" gas by condensing it and fractionating the condensate under pressure to prevent unacceptable vaporization (Schobert, 1990) delivers ethane and propane—the latter then being the main component of liquefied petroleum gas (LPG).

TABLE 5.2.4 Some Occurrences of "Sour" Natural Gas[a]

Location	Reservoir age	Lithology	Depth (m)	H_2S (%)
South Texas, USA	Late Cret.	limestone	3350	8
Weser-Ems, Germany	Permian	dolomite	3800	10
Alberta, Canada	Mississippian	limestone	3500	13
Lacq, France	U. Jurassic/ L. Cretaceous	dolomite/ limestone	3600–4500	15
Asmari-Bandar, Iran	Jurassic	limestone	3600–4800	26
Irkutsk, Russia	Late Cambrian	dolomite	2540	42
Wyoming, USA	Permian	limestone	3050	42
Mississippi, USA	U. Jurassic	limestone	3760	78
Alberta, Canada	Devonian	limestone	3800	87
South Texas, USA	U. Jurassic	limestone	5790–6100	98

[a] After Hunt (1979).

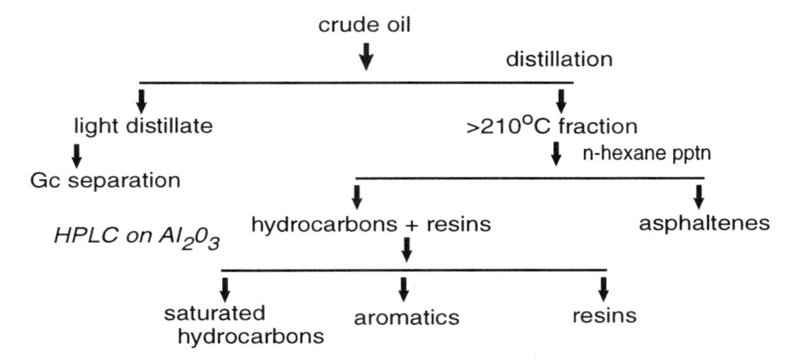

FIGURE 5.2.1 An illustrative crude-oil fractionation scheme.

Crude Oils

Unlike coal, for which elemental compositions provide some guide to properties and behavior, and in which chemical complexity manifests itself mainly during thermal decomposition, crude oils are complex mixtures *per se,* and gross compositions expressed by elemental analyses are generally of little technical interest [9]. Importance centers instead on the relative proportions of *compound types* that might affect downstream processing, and these can be identified by appropriately modified distillation and chromatographic separation techniques or, sometimes, by measurements of density and viscosity [10].

Figure 5.2.1 exemplifies the procedures by which a crude can be separated into fractions that can be more easily characterized. The saturated hydrocarbon fraction can subsequently be resolved by gas chromatography (GC) or urea adduction into *n*-alkanes and a mix of iso- and cycloalkanes, and aromatics can be separated into mono-, di-, and polyaromatics by high-pressure liquid chromatography (HPLC) on Al_2O_3, or by thin layer chromatography (TLC).

Table 5.2.5 shows the average composition of >210°C aromatic fractions from some 120 crude oils determined in this manner.

TABLE 5.2.5 Average Composition of >210°C Fractions of Crude Oils (wt%)[a]

Monoaromatics	33.0
Diaromatics	23.4
Triaromatics	12.9
Tetra- and higher aromatics	7.3
Thiophene derivatives	23.4

[a] Tissot and Welte (1978).

Over the years, such analyses of oil compositions have brought together a vast body of data about crude oils, and demonstrated the presence in them of almost all types of aliphatic, naphthenic, and aromatic hydrocarbons.

On average, *n*-alkanes, positively identified to $>C_{40}$, represent 15–20% of a crude (with its cloud point mainly due to $>C_{20}$ *n*-alkanes). Also present are C_4–C_{10} isoalkanes (which are generally accompanied by isolated higher members of that series, in particular squalene), and isoprenoids, which have been identified to C_{25}. Alkenes are rare, but minute amounts have been reported in some Pennsylvania crudes (Tissot and Welte, 1978).

Of the cycloalkanes, by far the most prominent are cyclopentane and cyclohexane, but polycyclic members of this class—in particular, two-, three-, and four-ring compounds such as decahydronaphthalene, tricyclodecane, and cholestane—also often occur in small concentrations.

Among aromatics—which are, as a rule, encountered as alkyl derivatives with one, two, or three CH_3 groups [11]—the most abundant are benzene, naphthalene, and phenanthrene, but anthracene-, chrysene-, and benzanthracene derivatives are frequent minor components [12]. As well, most crudes contain substantial amounts of alkylated naphtheno-aromatics. In immature (shallow) oils and in oil fractions with relatively high boiling points, such compounds—mostly found as methyl derivatives of indane, tetrahydronaphthalene, and tetrahydrophenanthrene—are, in fact, more abundant than aromatics [13].

However, in addition to the hydrocarbons that constitute their bulk, crude oils also contain appreciable proportions of organometallic and heteroatomic compounds, as well as minute amounts of exogenous organic and inorganic matter. Organometallic exist mainly as porphyrins complexed with Ni or V (see Chapter 2, [17]) and are usually associated with resin fractions. But hetero-atomics present themselves in a much wider spectrum of compounds.

Least known are forms of nitrogen. In the United States, the highest N contents—~0.89% as against a weighted average of 0.15% for all U.S. crude oils—have been reported for some California oils and are thought to be partly due to dissolved natural gas with relatively high N contents. However, basic heterocycles—mainly pyridines and quinolines—have also been encountered in some distillates.

Sulfur, which occurs in concentrations between 0.1 and 5.5% in almost all crudes and oil fractions, is usually present as elemental S and H_2S; in thiols characterized by —SH; in sulfides such as

$$R—CH_2—S—CH_2—R'$$

and steranes, in which S replaces a C atom in a five-membered ring; in disulfides,

$$R—CH_2—S—S—CH_2—R';$$

and in sulfoxides,

$$R—CH_2—S(O)—CH_2—R'$$

which develop by oxidation of sulfides. As a rule, sulfur contents vary inversely with the API gravity [14]; in crudes of similar gravity, they tend to be highest in paraffinic and lowest in naphthenic oils.

Sulfur-bearing hydrocarbons are polar compounds that affect the interfacial tension at solid/liquid, liquid/liquid, and gas/liquid interfaces, and that tend to be particularly common in cracked distillates, in which their concentrations may be enhanced by high-temperature distillation.

Oxygen contents range from 0.1 to 4%, but usually run to less than 2% and are accounted for by free O_2 and [O] in phenolic -OH, fatty acids and their alkyl derivatives, carboxy acids of naphthenes and resins, and asphaltics believed to have formed by oxidative polymerization of hydrocarbons in oils. An example, cited by Levorson (1967), is a nonwaxy crude from Grozny (Chechnya), in which resins with an empirical formula $C_{41}H_{57}O$ and an apparent molecular weight of 589 represent 8.2 wt% of the crude.

Exogenous organic matter is usually made up of microscopic fragments of coal, petrified wood, spores, leaf cuticles, resins, and algal material that withstood total decay, and exogenous inorganics are typically ashes of indeterminate origin. Analysis of 110 West Virginia (U.S.) pools show ash ranging from 0.04 to 400 ppm, and in South American, Mexican and Middle Eastern crudes, it has been reported as varying between 0.003 and 0.72%.

Oil Fractions

Since boiling points increase with molecular weights, specific oil cuts produced by fractionation of a crude show pronounced composition shifts:

1. *Liquified petroleum gases* (LPGs) are made up of propane (C_3H_8) and butane (C_4H_{10})—i.e., light hydrocarbons that can be liquefied by compression without cooling
2. *Ligroin,* which boils up to 70°C, but is sometimes taken to 100°C and then referred to as a "light gasoline," is composed of *n*- and branched isomers of pentane (C_5H_{12}) and hexane (C_6H_{14})
3. *Gasoline* is defined as a fraction boiling between 70 and 180°C and characterized by a C_7–C_{11} hydrocarbon range; the dominant paraffins are *n*-isomers; among branched compounds, 2-methyl isomers are the

most abundant; and dominant cycloalkanes and aromatics are CH_3-substituted

4. *Kerosine* boils in the 180–270°C interval, consists of C_{11}–C_{15} hydrocarbons (of which many are produced by thermal cracking of heavier feeds), and normally contains an abundance of aromatics (10–40%), condensed naphtheno-aromatics, and multiringed cycloparaffins
5. *Gas oils* boil between 270 and 400°C and are formed by (mainly cyclic) C_{15}–C_{25} hydrocarbons, in particular tricycloparaffins, derivatives of anthracene and phenanthrene, biphenyls, and tetrahydronaphthalenes
6. *Lubricating oils* are composed of C_{26}–C_{40} hydrocarbons, a range that may sometimes be extended to C_{50} and contain straight-chain and branched paraffins, one-, two-, and three-ring cycloparaffins, and up to three-ring aromatics, the last mainly phenanthrene- and naphtheno-aromatic derivatives
7. *Residua,* colloquially referred to as "resids," are repositories of resins and asphaltenes, and when obtained from highly paraffinic crudes may also contain very high-molecular-weight alkanes. The resins and asphaltenes (see Chapter 7) hold most of the [N], [S], and [O] of the parental crude.

Microbiological Alteration of Oils

Microbial action on crude oils, and in particular the possibility that such activity might accelerate detachment of oil from the sand grains to which it adheres and thereby enhance oil recovery, was demonstrated by ZoBell (1947), and later studied in order to identify the reservoir properties that affect such activity (Donaldson and Clark, 1983; Zajic and Donaldson, 1985). As now understood, these properties include reservoir depth, temperature, and pressure, rock porosity and permeability, oil gravity (API), and the pH and salinity of associated formation fluids.

Organisms of special interest for microbially enhanced oil recovery (MEOR) are thermophiles thriving in anaerobic environments, notably

1. thermophilic species of *Clostridia,* spore-forming obligate anaerobes that ferment sugars and celluloses;
2. CH_4-producing organisms, exemplified by *Methanobacterium thermoautotrophicum; and*
3. sulfate-reducing biota such as *Desulfovibrio thermophilus* (which can at times cause problems by forming water-insoluble sulfides that induce corrosion and nonselective plugging).

However, these organisms operate primarily by generating surfactants [15] and do not appreciably alter oil *compositions*. How and by what (presumably aerobic) organisms more appreciably alter oil *compositions*. How and by what (presumably aerobic) organisms more destructive effects—i.e., biodegradation and attendant degeneration of light crudes into heavy oils and bituminous substances (see Chapters 2 and 3)—are caused is still largely speculative.

The Heavy Hydrocarbons

Bituminous Substances

Heavy fossil hydrocarbons nominally as diverse as bitumens, asphalts, tars, and residua are collectively referred to as bituminous substances—a term that may by some definitions even embrace extra-heavy oils. But because of the profusion of seemingly different bituminous substances [16], the applicable terminology is as much confused as confusing [17]. Uncertainties about the previous histories and sampling of analytical specimens make data on composition ambiguous and, at best, illustrative rather than representative. And widely held, but not necessarily correct, opinions about formative aspects of bituminous substances—for example, the rarely challenged assertion that they *always* represent microbially degraded light crude oils—further complicate interpretation of analytical data.

In general terms, characteristic chemical compositions of bituminous substances show free alkanes (which form the bulk of light crudes) present in only low concentrations; naphthenes are mostly derivatives of cyclohexane, decalin, and sterane; hydroaromatics are predominantly derivatives of 9,10-dihydronaphthalene and 1,2,3,4-tetrahydronaphthalene (tetralin); and aromatics are mainly substituted benzenes, naphthalenes, and phenanthrenes that often carry relatively long (up to C_{10}) paraffinic side chains and are interconnected by S or O bridges [18]. Heteroatomic species are the same as those in lighter crude oils, but much more abundant.

Overall, as indicated by ^{13}C NMR spectra of seven residua boiling between 510 and 538°C, aromatic carbon contents (C_{ar}) of bituminous substances can thus range to >35% (Beret and Reynolds, 1990).

But more detailed information has been specifically adduced for *bitumens* that can, in the context of this survey, be appropriately taken as representative of bituminous substances.

A 1962 API definition, which parallels an earlier ASTM formulation, broadly identifies *bitumen* as a mixture of liquid, semisolid, and solid natural and/or pyrogenous hydrocarbons that are readily soluble in CS_2 and contain substantial

TABLE 5.2.6 **Composition of Two Athabasca Oil Sand Samples**[a]

			Bitumen composition[b]			
Preasphaltenes	Bitumen	Moisture	1	2	3	4
0.45 ± 0.12	12.92 ± 0.28	0.55 ± 0.01	8	31	32	29
0.34 ± 0.08	15.76 ± 0.23	0.50 ± 0.01	36	11	37	16

[a] Berkowitz and Calderon (1990).
[b] 1, aliphatics; 2, aromatics; 3, polar hydrocarbons (resins); 4, asphaltenes.

proportions of hetero compounds that influence their properties. However, much more definitive information has been obtained from the response of bitumens to fractionation, and that response can be conveniently assessed by a class separation scheme, which combines selective solvent extraction with sorption on clays. Commonly referred to by its acronym SARA, which denotes the main fractions (saturates, aromatics, resins, and asphaltenes) it separates, this delivers

1. *n*-Pentane-soluble matter (or maltenes) that can be resolved by column chromatography into
   (a) resins, which are sorbed from solution by silica gel or clay, and
   (b) oils, which are not sorbed, but can be further separated into

TABLE 5.2.7 **Simulated Distillation of Four Bitumens**[a,b]

	Cumulative wt% distilled			
Cut point (°C)	1	2	3	4
200	3.0	2.3	0.7	1.7
250	6.5	4.4	2.4	4.4
300	14.0	7.5	4.9	8.4
350	18.1	11.7	8.0	15.2
400	26.2	16.8	12.5	22.4
450	33.1	23.7	20.0	28.9
500	40.0	34.0	25.0	35.1
535	44.3	44.0	28.0	40.0
535+	55.7	56.0	72.0	60.0

[a] Speight (1991).
[b] Bitumens sources: 1, Athabasca; 2, NW Asphalt Ridge; 3, P.R. Springs; 4, Tar Sand Triangle.

TABLE 5.2.8 Compositions of Resins

Carbon (wt%) (23 samples):	82.1–87.8
Arithmetic average	85.29
Geometric mean	84.95
Atomic H/C ratio:	1.36–1.69
Arithmetic average	1.66
Geometric mean	1.52

(i) aromatics, sorbed from solution in *n*-pentane by SiO_2/Al_2O_3, and
(ii) saturates, which are not sorbed and therefore remain dissolved in *n*-pentane

2. Asphaltenes, which are insoluble in *n*-pentane but soluble in benzene (or toluene) and can be further separated into soluble carbenes and insoluble carboids by extraction with CS_2
3. Preasphaltenes, which are insoluble in benzene or toluene, but completely soluble in tetrahydrofuran (THF)
4. "Coke," a misleadingly designated mix of high-molecular-weight polynuclear aromatics

In this context, saturates are aliphatic entities; aromatics form a class of compounds with variously long saturated substituent functions; resins are characterized by higher aromaticities, molecular weights, and heteroatom contents than the aromatics; and the asphaltene fractions contain the bulk of high-molecular-weight polar entities.

Assessed by such separation, bitumen compositions reportedly average 19–20% asphaltenes, 30–32% resins, and 48–50% oils (Speight, 1991). However, averages obscure the fact that bitumens, like other *generically* designated fossil hydrocarbons, form a class of substances whose compositional variability can manifest itself even when they occur in close geographic proximity (Tables 5.2.6 and 5.2.7); and it is therefore not surprising that resins and asphaltenes (Tables 5.2.8 and 5.2.9) should show similar diversity.

From a study of asphaltenes and carbenes in a high-boiling oil fraction generated by mild hydropyrolysis of maltenes, Blazek and Sebor (1993) have concluded (i) that asphaltenes form primarily by condensation of resins, but can under relatively severe conditions also come about by resin dehydrogenation and dealkylation, and (ii) that carbenes are produced from highly polar asphaltenes by dehydrogenation, dealkylation, and concurrent cracking of heterocyclic moieties in them. But such generalizations must be viewed with caution, if only because of the very limited information about asphaltene compositions *per se*. As matters stand, the most detailed data at hand relate

TABLE 5.2.9 Compositions of Asphaltenes

Canada (25 samples):	
Carbon (wt%)	79.0–88.7
Geometric mean	83.85
Atomic H/C ratio	0.98–1.56
Geometric mean	1.27
Middle East (13 samples):	
Carbon (wt%)	78.3–83.8
Geometric mean	81.8
Atomic H/C ratio	0.98–1.56
Geometric mean	1.14
USA (8 samples):	
Carbon (wt%)	83.2–88.6
Geometric mean	85.9
Atomic H/C ratio	1.00–1.29
Geometric mean	1.14
Venezuela (6 samples):	
Carbon (wt%)	81.1–84.7
Geometric mean	82.9
Atomic H/C ratio	1.13–1.19
Geometric mean	1.16

to asphaltenes precipitated from benzene or CH_2Cl_2 solutions of Athabasca bitumen by addition of *n*-pentane, and then separated by gel permeation chromatography into five molecular-weight fractions (Hepler and Chu Hsi, 1989)—and these (see Table 5.2.10) betray the heterogeneity of asphaltenes by their variable contents of aromatic carbon (C_{ar}).

TABLE 5.2.10 Elemental Composition of Fractionated Asphaltenes from Athabasca Bitumen[a]

Fraction no.	1	2	3	4	5
Yield (wt%)	22.5	30.5	13.6	11.1	20.8
Molecular wt	16,900	13,700	7,100	3,400	1,200
Elem. comp. (wt%)					
Carbon	79.8	79.8	80.1	80.0	79.8
Hydrogen	8.1	8.2	8.3	8.6	8.2
Nitrogen	1.2	1.2	1.1	1.1	1.1
Sulfur	7.9	8.0	7.8	7.9	7.0
Oxygen	3.0	2.8	2.6	2.6	3.9
Atomic H/C	1.22	1.23	1.24	1.24	1.23
C_{ar} (%)	35	34	39	41	48

[a] From Strausz, cited in Hepler and Hsi (1989).

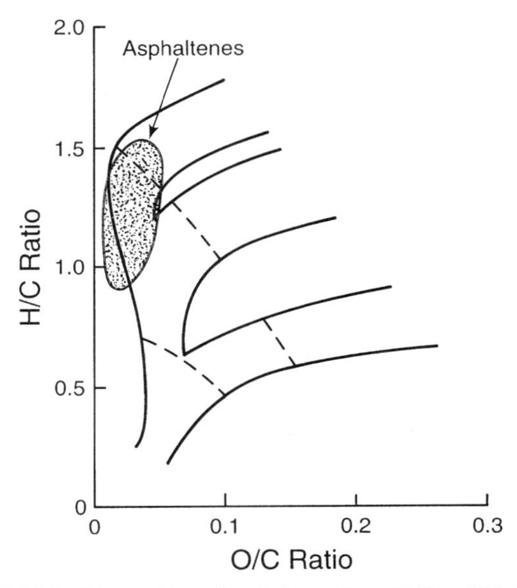

FIGURE 5.2.2 The position of asphaltenes in an H/C vs O/C diagram.

Heterogeneity is also indicated by the location of asphaltenes in an H/C vs O/C diagram (see Fig. 5.2.2), and by the dependence of asphaltene yields on the type and volume of the precipitating solvent [19]. Branched-chain paraffins and terminal olefins will, as a rule, precipitate substantially less material than the corresponding *n*-paraffins, and some attention has therefore been given to characterizing suitable solvents by a solubility parameter δ, defined by

$$\delta_1 = \sigma v^{-1/3}$$

or

$$\delta_2 = (\Delta E_v/v)^{1/2} = (\Delta H_v - RT)/v]^{1/2},$$

where σ denotes the surface tension of the solvent, v its molar volume, ΔE_v its energy of vaporization, and ΔH_v its enthalpy of vaporization; R and T are the gas constant and absolute temperature. Formulated in this manner, δ_1 equals the internal pressure p_i of the solvent and would be expected to be applicable to polar fluids, whereas δ_2 would be more appropriate for nonpolar ones. However, the validity of this approach has not been established, and past experience [18] does, in fact, make it questionable.

TABLE 5.2.11 Compositions and Oil Yields of Western and Eastern U.S. Oil Shales[a]

	Eastern	Western
Elemental composition (Dry basis) wt%)		
Carbon	13.7	13.6
Hydrogen	1.6	2.1
Sulfur	4.7	0.5
CO_2	0.5	15.9
Fischer assay		
Oil yield (wt%)	4.6	11.4
Liters/tonne	43.3	125.2
Hydropyrolysis		
Oil yield (wt%)	9.1	11.4
Liters/tonne	97.4	125.2

[a] Office of Technology Assessment (1980).

Oil-Shale Kerogens

Other than the generally small (<10%) proportion of bitumen it contains, organic matter in oil shales—that is, the kerogens derived from marine algae (*Tasmanaceae*), fresh and brackish water algae (such as *Botryococcus*), and/or other sapropelic matter—are virtually insoluble in common organic solvents, but yield fairly light oils when pyrolyzed [21].

In lacustrine deposits such as the exceptionally rich Tertiary Green R. oil shales of the western United States, the inorganic component is a $CaCO_3$- and $MgCO_3$-rich marlstone that contains up to 40% organic matter with an (apparent) average molecular weight near 3,000. High-resolution ^{13}C NMR spectra indicate that 18–20% of the total C_{org} of these shales resides in aromatics, 36–38% in alicyclics, ~26% in long-chain *n*-alkanes, ~5% in $—CH_3$ and $—CH_2—$ directly bonded to aromatic rings, and 8–12% in linkages to oxygen in ether-, ester-, and —COOH configurations (Abraham, 1960).

Representative analytical data for these and other major U.S. deposits—which include the Mesozoic marine deposits of Alaska, the Permian-age deposits of Montana's Phosphoria formation, and the Devonian/Mississippian oil shales of the Eastern and Midwestern states—are summarized in Table 5.2.11. Significantly, these data show that whereas Western shales will yield 2–3 times more oil than their Eastern counterparts when pyrolyzed *per se,* this difference is substantially reduced when retorting is conducted under H_2 or in the presence of an H donor [22].

More generally, elemental compositions of oil-shale kerogens vary as widely as those of bitumens and coals. Lee (1991) thus cites, expressed in wt%,

TABLE 5.2.12 Compositions of Three Kerogens of Different Origin[a,b]

	Torbanite	Tasmanite	Colorado oil shale
Carbon	83.6	78.1	79.2
Hydrogen	11.3	10.2	10.5
Oxygen	3.5	6.0	6.5
Nitrogen	0.6	0.6	2.6
Sulfur	1.0	5.1	1.2

[a] Lee (1991).
[b] wt% dry, ash-free.

$$
\begin{aligned}
[\mathrm{C}] &= 64.0\text{–}89.0 \\
[\mathrm{H}] &= 7.1\text{–}12.8 \\
[\mathrm{O}] &= 0.8\text{–}24.8 \\
[\mathrm{N}] &= 0.1\text{–}3.1 \\
\text{and}\quad [\mathrm{S}] &= 0.1\text{–}8.7,
\end{aligned}
$$

but cautions that the scatter of these values deprives generalizations of plausibility.

Some significant systematic differences appear to exist between immature and mature kerogens: for the former, defined as still in their diagenetic phase of development, Tissot and Welte (1978) report

$$
\begin{aligned}
[\mathrm{C}] &= 68.6\text{–}77.5\ \text{wt\%} \\
[\mathrm{H}] &= 6.0\text{–}10.8 \\
[\mathrm{O}] &= 8.4\text{–}18.0,
\end{aligned}
$$

and for the latter, variously altered by catagenesis, the corresponding ranges are

$$
\begin{aligned}
[\mathrm{C}] &= 80.6\text{–}85.4 \\
[\mathrm{H}] &= 5.9\text{–}9.9 \\
[\mathrm{O}] &= 4.4\text{–}9.5.
\end{aligned}
$$

There are, however, no clearly discernible differences between supposedly different types of kerogen (see Chapter 2) and, as illustrated by Table 5.2.12, very few between kerogens of different origin. In a modified Ralston diagram, kerogens would thus occupy an area between fatty acids—which reflect contributions from algal families (*Clorophyceae, Cyanophyceae,* etc.) characterized by high proportions of saturated as well as unsaturated C_{14}–C_{22} fatty acids—and the hydrocarbon mixtures that comprise crude oils.

Nevertheless, some important information on chemical compositions and, by inference, on the molecular structure of kerogens has accrued from studies of kerogen *concentrates* prepared by acid digestion of the shale envelope with HCI/HF at 20–65°C, gravity separation, and/or selective (differential) wetting.

Solvent extraction of such concentrates has shown the presence of paraffins, 2-methyl- and 3-methylalkanes, cycloalkanes, steranes, and one-, two-, and three-ring aromatics, as well as of alcohols, ketones, and *N*-heterocycles such as pyridines and pyrroles—all considered to be structure elements of, or in, the kerogen matrix (Forsman and Hunt, 1958; Robinson, 1969; Saxby, 1976; Durand and Nicase, 1980). This has been amplified by two more recent investigations.

In one, ruthenium tetroxide was used to selectively degrade a kerogen under mild conditions (Boucher *et al.*, 1991) and yielded CH_2Cl_2-solubles in which α,ω-dicarboxylic, branched mono- and dicarboxy, isoprenoid, and cyclic acids could be identified, and the dominance of polymethylene chains in the precursor was established by the preponderance of straight-chain carboxy acids. In the other, bitumen-free kerogen from Morocco's Timahdit Y-layer was isolated by multistage oxidation with alkaline $KMnO_4$ and studied by FTIR and ^{13}C NMR spectroscopy (Ambles *et al.*, 1994). IR absorptions of varying intensities were recorded at

<3000 and 1390–1460 cm^{-1} (—OH),
>3000 and 1390–1450 cm^{-1} (—CH_3 and —CH_2—)
1720 cm^{-1} (=CO), and near
1150 and 1200 cm^{-1} (indicative of —OH, esters, and ethers).

In ^{13}C NMR spectra, a strong peak at 30 ppm could be unequivocally assigned to *n*-alkyl chains, whereas another between 90 and 160 ppm established the presence of aromatic carbon. The acids yielded by $KMnO_4$ oxidation were consequently identified as

1. Aliphatic	α,ω-dicarboxy	C_4–C_{16}
	branched diacids	C_4–C_{16}
	n-monocarboxy	C_6–C_{16}
	branched monocarboxy	C_6–C_{10}, C_{13}, C_{14}, C_{17}
2. Alkanes	tricarboxy	C_8–C_{14}
	tetracarboxy	C_7–C_{10}, C_{12}, C_{13}
3. Cycloalkanoic	monocyclic monocarboxy	C_8–C_{10}
4. Aromatic	monocarboxy	C_6H_5, $C_6H_4CH_3$ $C_6H_4C_6H_{13}$, $C_6H_4C_7H_{15}$ $C_{10}H_6CH_3$, etc
	dicarboxy	C_6H_4, $C_6H_3CH_3$, $C_6H_3C_4H_9$
	tricarboxy	C_6H_4

Aliphatics represented ~58 wt% of this melange, alkanes ~11 wt%, cycloalkanes ~6.5 wt%, and aromatics ~24.5 wt%.

That the spectral details of this study stand in good agreement with FTIR data from other sources (see Ganz and Kalkreuth, 1987; Benalioulhaj and

Trichet, 1990) underscores the view that the molecular structure of oil-shale kerogens, like those of bituminous substances, forges a direct link between the lighter petroleum hydrocarbons and coal (see Chapter 8).

Coals

Structural complexities—which are in bituminous substances masked by substantially unimpeded molecular mobility, and in kerogens superficially hidden by dispersal in inorganic matrices—manifest themselves in coal in its *petrographic* make-up as well as in its molecular architecture.

Petrographic Composition

A prominent feature of hand specimens of bituminous coals is a striated appearance caused by four irregularly alternating "banded components," or *lithotypes,* which make coals the organic analogs of conglomerates [23]. Stopes (1919) definitively described these lithotypes:

1. *Vitrain* typically forms narrow glossy or vitreous black bands, appears structureless to the naked eye, and fractures conchoidally or into small cubes
2. *Clarain* occurs as black, often laterally striated layered matter, possesses a silky luster, and displays a glossy surface when freshly broken
3. *Durain* forms dull, grey-black layers or lenticular masses and is characterized by tight granular texture and, when broken, by a fine-grained or matte surface
4. *Fusain* resembles charcoal, usually forms small lenses parallel to bedding planes of the coal, but is inherently soft and friable, and easily breaks into fibrous strands or a powder.

Later work has shown these lithotypes to be variously composed of microscopic entities that reflect formative environments [24], and to exist, although macroscopically less well expressed, in immature as well as mature coals. This has led to recognition of a petrographic organization that parallels mineralogical organization in rocks (Stach, 1985; Mackowsky, 1971, 1975). Each lithotype is thus made up of several *micro*lithotypes, the counterparts of mineral phases, and each microlithotype is composed of a number of *macerals*—the organic

TABLE 5.2.13 Relationships among Lithotypes, Microlithotypes, and Maceral Groups[a]

Lithotype	Microlithotype	Dominant maceral group(s)
Vitrain	vitrite (dominant)	vitrinite
	clarite	vitrinite + exinite (liptinite)
	vitrinertite	vitrinite + inertinite
	trimacerite	vitrinite + exinite + inertinite
Durain	clarite	vitrinite + exinite (liptinite)
	durite	inertinite + exinite (liptinite)
	vitrinertite	vitrinite + inertinite
	trimacerite	vitrinite + exinite + inertinite
Fusain	inertite (dominant)	inertinite

[a] Clarain, not listed, consists of variously composed vitrain/durain mixtures; liptinite is an alternative term for exinite and more expressive of its chemical characteristics, i.e., its relatively high H-content.

analogs of minerals, that collectively form three *maceral groups*—the counterparts of mineral classes. Tables 5.2.13 and 5.2.14 detail this organization [25].

Petrographic compositions bear importantly on coal behavior and are quantitatively assessed by analyses that center on measuring the reflectivity R—a parameter that varies systematically with maceral maturity (or "rank") and is defined by

$$R = [(n - n')^2 + n^2k^2]/[(n + n')^2 + n^2k^2],$$

where n and k are the refractive and absorptive indices of the maceral, and n' is the refractive index of the medium. Specimens for analysis are prepared by ASTM (or corresponding) standard procedures (ASTM, 1972; Stach, 1985); the incident light is a plane-polarized beam that is usually monochromatized at $\lambda = 546 \pm 5$ nm; and measurements of R are now almost invariably made with the microscope lens immersed in a standard oil (e.g., Cargille's Type A or B) [26]. For calibration, standard optical-glass prisms with known refractive indices and reflectivities or, on occasion, synthetic garnets or sapphires are used; the reported values of R are the averages of at least 100 individual "point" readings; and if the coal is optically anisotropic—which is the case when its carbon content exceeds ~87–88%—these averages are calculated from *maximum* readings of R.

In practice, petrographic compositions are reported as percentages of the three maceral groups; however, since the dominant component is most often the vitrinite group [27], vitrinite reflectivity is also consensually used to define the maturity of coals and petroleum source rocks (see Chapter 3). Figure 5.2.3 shows how this parameter tracks metamorphic development or maturity [28].

TABLE 5.2.14 **Maceral Groups, Macerals, and Maceral Source Materials**

Maceral group	Maceral	Source materials
Vitrinite (V)	collinite	humic gels
	telinite	wood, bark, cortical tissues
	vitrodetrinite[a]	
Exinite (E)	alginite	algal remains
	cutinite	leaf cuticles
	resinite	plant resin bodies and waxes
	sporinite	fungal and other spores
	liptodetrinite[a,b]	
Inertinite (I)	macrinite[c]	unspecified detrital matter
	micrinite	similar
	fusinite	"carbonized" woody tissues
	semifusinite	similar
	sclerotinite	fungal sclerotia and mycelia
	inertodetrinite[a]	

[a] These designations apply to small entities that must be assigned to the group, but that cannot be unequivocally identified with any maceral.
[b] If cited as *liptinite*, it is indexed as L.
[c] Macrinite: 10–100 μm entities; micrinite: <10 μm entities.

Chemical Composition

Despite episodically intensive studies (De Marsilly, 1862; Guignet, 1879; Bedson, 1902; Fischer and Gluud, 1916; Cochram and Wheeler, 1927, 1931; Dryden, 1951, 1952), fractionation by selective solvent extraction has not

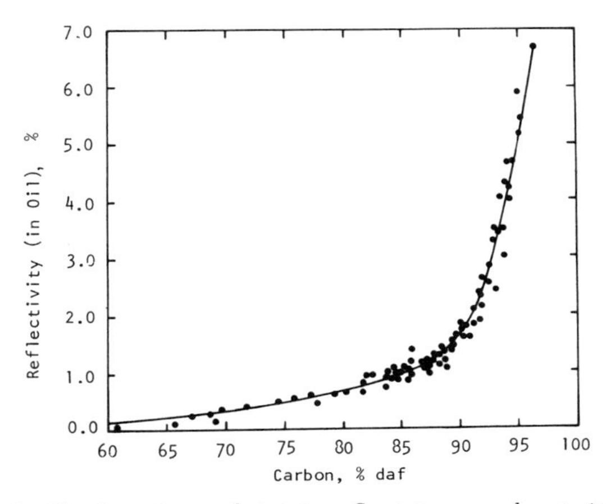

FIGURE 5.2.3 The dependence of vitrinite reflectivity on coal maturity (% carbon).

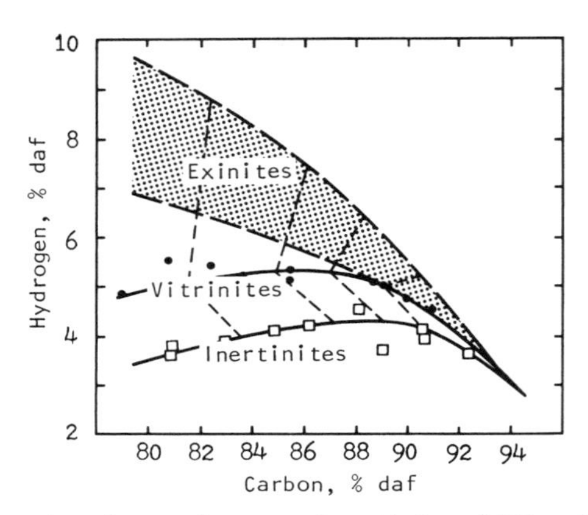

FIGURE 5.2.4 Maturation of maceral groups: the variation of [C] and [H] among exinites, vitrinites, and inertinites.

proved useful for characterizing coal, and immediate needs for data on chemical composition are met by two complementary analyses—*proximate* analysis, which assesses moisture, volatile matter and ash contents, and quantifies a residual "fixed" carbon content as

$$\text{F.C.} = 100 - ([H_2O] + [\text{vol. matter}] + [\text{ash}]);$$

and *elemental* analysis, which measures concentrations of carbon, hydrogen, nitrogen, and organic sulfur, and determines oxygen indirectly as

$$[O] = 100 - ([C] + [H] + [N] + [S_{org}].$$

Direct measurement of [O], for which several techniques have been described [29], is considered unnecessary or too cumbersome for routine use, but is occasionally reported in the context of structural studies (see Chapter 7).

Sample collection, preparation, and analysis are detailed in national and international specifications [30]. But, as noted in Chapter 4, meaningful comparison of different coals requires eliminating contributions from variable mineral matter contents by recalculating raw analytical data to a dry, *ash*-free (d.a.f.) or dry, *mineral* matter-free (d.mm.f) coal basis [31]–and even then, valid inferences about probable behavior of a coal may depend on information about petrographic composition. Figure 5.2.4 and Table 5.2.15 show that elemental compositions of immature maceral groups differ significantly and converge only slowly toward ~92% carbon; the effect of this on the elemental composition of "whole" coal is clearly seen when, as in Table 5.2.16, compositions of ostensibly similar coals are expressed in atom weight percent.

TABLE 5.2.15 Elemental Compositions of Maceral Groups at Different Stages of Metamorphic Development[a,b]

	Ash	Vm.	C	H	N	S	O	H/C
1. Exinite	0.5	68.8	85.5	7.3	0.5	0.9	5.8	1.03
Vitrinite	1.3	36.1	83.5	5.1	0.8	0.9	9.7	0.73
Inertinite	3.8	22.5	86.8	3.9	0.6	0.7	8.0	0.54
2. Exinite	0.6	59.8	87.4	6.7	0.6	0.5	4.8	0.93
Vitrinite	0.5	32.0	85.7	4.9	0.8	0.8	7.8	0.68
Inertinite	5.9	23.4	88.0	4.2	0.6	0.5	6.7	0.57
3. Exinite	0.1	37.1	89.1	6.0	0.7	0.5	3.7	0.80
Vitrinite	1.6	28.4	88.4	5.1	0.8	1.0	4.7	0.69
Inertinite	1.1	19.2	89.6	4.3	0.6	0.5	5.0	0.58
4. Exinite	1.9	22.6	89.3	4.9	1.5	0.6	3.7	0.66
Vitrinite	2.3	23.5	88.8	4.9	1.6	0.7	4.0	0.67
Inertinite	5.8	17.0	89.8	4.3	0.9	0.5	4.5	0.57

[a] Teichmüller (1975).
[b] Ash expressed on dry coal basis; other parameters are calculated on d.a.f. coal.

Aside from their bearing on coal preparation and use, compositional differences between maceral groups are, however, also significant in another context. As in the case of kerogens, contributors to elemental compositions of coal have been broadly identified by oxidative degradation. Thus:

TABLE 5.2.16 Elemental Compositions of Typical Subbituminous and Bituminous Coals; Expressed in Atom Weight Percent[a]

Carbon		Atoms per 100 atoms				
(% daf)	H/C	C	H	O	N	S
80.8	0.767	52.8	40.5	5.6	1.1	—
81.5	0.788	52.4	41.3	5.2	1.0	0.1
82.9	0.784	53.3	41.8	4.1	0.8	—
83.1	0.802	52.6	42.2	4.1	0.7	0.4
85.1	0.746	54.4	40.6	3.9	0.8	0.3
85.5	0.725	56.0	40.6	3.4	1.1	—
86.5	0.772	54.4	42.0	2.4	1.2	—
87.1	0.765	54.4	41.6	2.6	1.3	—
87.1	0.718	56.1	40.3	2.3	0.8	0.5
89.3	0.721	56.7	40.9	1.4	1.0	—
89.3	0.672	58.0	39.3	1.8	0.7	0.2
89.8	0.613	60.5	37.1	1.4	0.9	0.1

[a] Where [S] not shown, it is included with [N].

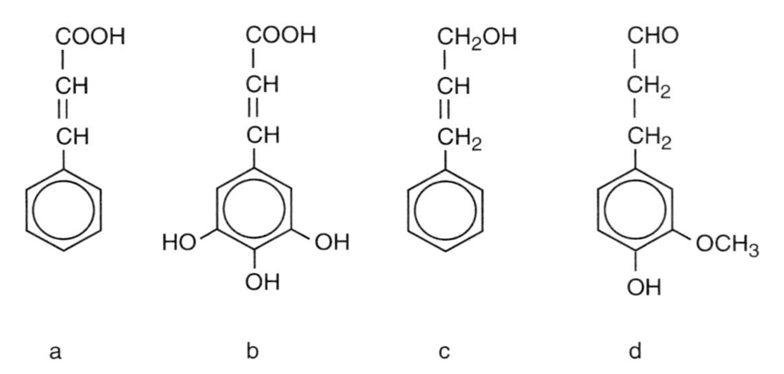

FIGURE 5.2.5 Some products of coal oxidation with performic acid (Raj, 1976; Bimer *et al.*, 1977, 1978). Four of eight compounds with 3-C chains are shown: (a) cinnamic acid, (b) 3,4,5-trihydroxy cinnamic acid, (c) phenyl 3-pyruvic alcohol, and (d) 3-methoxy, 4-hydroxyphenyl 3-propionaldehyde.

1. A significant presence of aliphatics, sometimes extending to C_{20}–C_{22}, is established by mono- and dicarboxy acids among the products of oxidation with $KMnO_4$ (Yohe and Harris, 1961; Kukharenko and Belikova, 1968), HNO_3 (Deno *et al.*, 1981), and $Na_2Cr_2O_7$ (Hayatsu *et al.*, 1981): yields of these acids varied, as expected, inversely with coal maturity and depended on reaction conditions, but ranged from as much as 38.3% of the total precursor carbon for sapropelic coal to 0.5–2.5% for a mature bituminous coal.

2. Oxidation with performic acid (HCO_3H) at 55–60°C (Raj, 1976; Bimer *et al.*, 1977, 1978) furnished OH- and OCH_3-substituted phenyl derivatives with a 3-C side chain (see Fig. 5.2.5), which betrayed close relationship to biochemical degradation products of lignin and certain flavonoids.

3. Aromatics have been defined *inter alia* by oxidation with $Na_2Cr_2O_7$ at 250°C/autogenic pressure (Hayatsu *et al.*, 1975, 1978; [32]): product slates from this reaction contained a variety of water- and alkali-soluble acids that, depending on coal maturity, accounted for 50–70% of the substrate carbon and appeared in 32 different forms—20 of them being N-, O-, and S-bearing hetero aromatics (see Fig. 5.2.6).

If these contributors are apportioned to exinites, vitrinites, and inertinites on the basis of their H/C ratios, the *maceral groups emerge as structurally related analogs of the asphaltene, preasphaltene, and coke fractions defined by solvent fractionation of bitumens:* they can, in other words, be viewed—as implicitly suggested in Chapter 1—as the most highly aromatized members of a fossil hydrocarbon continuum that begins with light oils.

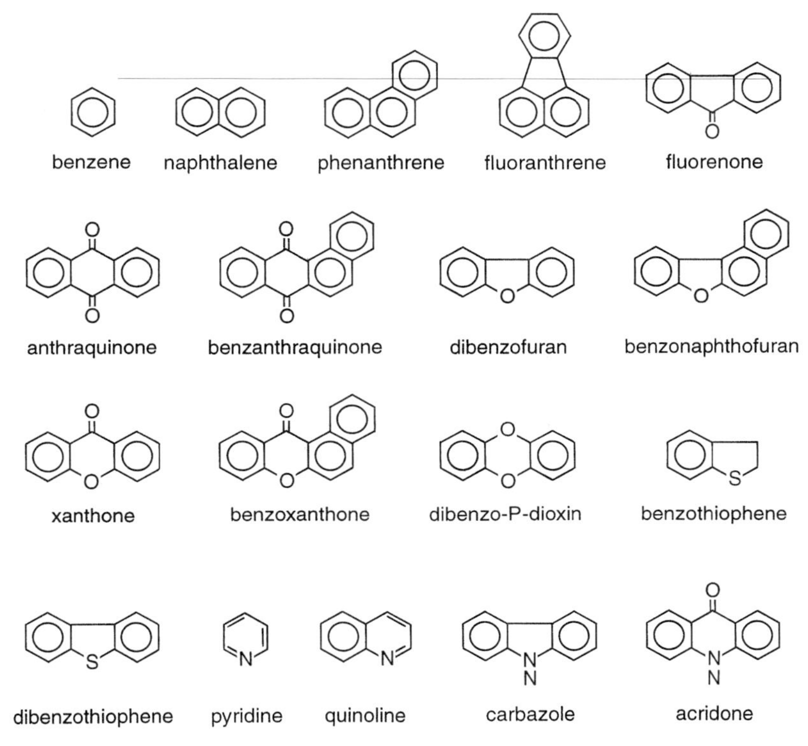

FIGURE 5.2.6 Aromatic moieties from coal oxidation with sodium dichromate (Hayatsu *et al.* 1975, 1978). These compounds have also been identified in coal tars.

Associated Inorganic Matter

Water: Moisture in coal can exist massively in fissures and large pores (and then possess the normal vapor pressure of water), or as physically sorbed H_2O in small pores (in which case its vapor pressure reflects its status as adsorbed matter).

Undisturbed *in situ* coal is normally water saturated and functions as an active acquifer. In that state, moisture contents range from 15–35% at 75% C (d.a.f.) to a low of 1–5% at 87–89% C before increasing again to 5–12% among anthracites with 93–95% C. However, the more-often-cited moisture contents relate to coal in equilibrium with air at ~20 ± 5°C and 30–60% relative humidity, and in such "air-dried coals," moisture runs typically to 45–65% of saturation.

Mineral matter: Inorganic material, regardless of type simply referred to as

TABLE 5.2.17 IR Absorption Bands of Common Minerals in Coal

Mineral	Absorption (cm^{-1})
Illite	3620, 1640, 1070, 1015, 920, 820, 750, 460
Kaolinite	3695, 3665, 3650, 3620, 1108, 910, 782, 749, 690, 530, 460, 422, 360, 340, 268
Na-montmorillonite	3625, 3400, 1640, 1110, 1025, 915, 835, 790, 515, 460
Calcite	1782, 1420, 871, 842, 710, 310
Dolomite	1435, 875, 730, 390, 355, 310
Pyrite	411, 391, 340, 284
Quartz	1160, 1065, 790, 770, 687, 500, 450, 388, 362, 256
Gypsum	3605, 3550, 1615, 1150, 1110, 1090, 1010, 660, 595, 450

mineral matter, was syngenetically incorporated into the primordial humic mass during diagenesis or entered it epigenetically during metamorphic development. It includes trace elements derived from parental vegetation or later inserted by ion exchange, but for the most part represents material carried into the exposed vegetal debris by wind and water, or deposited in cracks and bedding planes of the coal by percolating mineral waters. The dominant types of mineral matter are always:

Clay minerals, especially illite, kaolinite, montmorillonite, and mixed illite-montmorillonites, which commonly constitute 50–55% of the total
Silica, predominantly quartz, which often accounts for 15–20%
Carbonates, predominantly calcite ($CaCO_3$), dolomite ($CaCO_3 \cdot MgCO_3$), siderite ($FeCO_3$), and ankerite ($2CaCO_3 \cdot FeCO_3 \cdot MgCO_3$)
Sulfides, mainly FeS as pyrite and marcasite [33], but sometimes including smaller amounts of pyrrhotite (Fe_{1-x}) and galena (PbS).

Also frequently encountered are minor amounts of alkali feldspars (K,Na-$AlSi_3O_8$), plagioclase ($Na \cdot AlSi_3O_8$–$CaAl_2Si_2O_8$), hydrated Fe sulfates, jarosite (a mixed NaKFe sulfate), serpentine (a hydrated Mg silicate), chlorite (a hydrated MgAlFe aluminosilicate), and rutile (TiO_2).

How this melange—in different raw coals totaling <5% or exceeding 35%, and even in a single seam varying widely over relatively short distances—is disseminated in the organic matrix can be established radiographically (John, 1926; Kemp, 1929) or, better, by optical microscopy (Mackowsky, 1956). However, low-temperature ashing, a technique that destroys the organic material at <150°C without appreciably altering mineral matter (Gluskoter, 1965, 1967; Estep *et al.*, 1968), has enabled more detailed instrumental definition of mineral matter. Table 5.2.17, which summarizes data recorded by dispersive IR spectroscopy (Jenkins and Walker, 1979) and FTIR (Painter *et al.*, 1981) for the most common components, exemplifies this [34].

3. CHEMICAL PROPERTIES

In the context of this survey, chemical properties of fossil hydrocarbons are mainly of interest insofar as they bear on preparation and processing, and such properties are considered in Chapters 8 and 9. It is, however, appropriate to refer here briefly to two aspects that affect handling and storage.

OXIDATION

Atmospheric oxidation during storage usually causes progressive quality deterioration and can pose significant environmental hazards. But oxidation rates, mechanisms, and products are specific. They depend on the hydrocarbon as well as on the severity of the conditions under which it is oxidatively attacked, and few comments on the subject possess generally validity.

In *oils,* atmospheric oxidation initiates O_2-promoted polymerization which generates poorly defined sludges with increasingly high asphaltene contents. The process can proceed at room temperature, but is, like all chemical reactions, accelerated by higher temperatures and manifests itself visually in slow darkening and increasing viscosity. However, although superficially a facile reaction, the mechanisms of oxidation are speculative. There is some evidence that oil viscosities increase at faster rates than accumulation of asphaltenes (Rao and Serrano, 1986), that metal oxide contaminants accelerate oxidation (Mochida, 1986; Ruzicka and Ostvold, 1988), and that N as well as S must be assigned a major role in sludge formation. But experimental data on oil oxidation are ambiguous and sometimes contradictory [35], and efforts to model oxidation and sludge formation have so far not met with signal success (Ruzicka and Nordenson, 1990). Of the parameters *believed* to correlate with oxidative deterioration (η, asphaltene content, [N] and/or [S]), none does so reasonably unequivocally.

On the other hand, laboratory experiments in which petroleum asphalts and asphaltenes were variously oxidized (Ronvaux-Vankeerbergen and Thyrion, 1989) have indicated that oxidation may indeed involve condensation reactions [36]. If so, air-blowing tests (Moschopedis and Speight, 1976) suggest that some of the effects of such reactions are eventually reversed by prolonged oxidation: while O/C ratios continue to increase, the apparent molecular weights of asphaltenes gradually fall (Table 5.3.1).

This process—i.e., slow oxidative condensation followed by molecular degradation during extended oxidation—is precisely what occurs during prolonged atmospheric oxidation of coal (which has for various reasons been more closely investigated). Here, initial chemisorption of oxygen at active sites develops inherently unstable peroxides and hydroperoxides, as in

TABLE 5.3.1 Effects of Air Blowing on Asphaltenes

Temperature (°C)	Time (h)	Mol. wt.[a]	O/C
		4,920	0.016
76	4	4,620	0.026
132	8	4,420	0.036
	24	3,605	0.039
150	8	1,845	0.048
	24	1,645	0.044

[a] Determined by vpo in benzene.

$$\phi\text{—}\phi \rightarrow \phi\text{—O—O—}\phi$$
$$\phi\text{—CH}_3 \rightarrow \phi\text{—CH}_2\text{—O—OH},$$

which, when they decompose, initiate free radical reactions, as in

$$\phi\text{—CH}_2\text{—O—OH} \rightarrow \phi\text{—}^*,\ \phi\text{—C=O}^* \quad \text{and/or} \quad \phi\text{—CH}_2\text{—O}^*,$$

and thereby furnish H_2O, CO, and CO_2, and concurrent direct oxidative attack on peripheral alkyl substituents or bridge configurations introduces —COOH, phenolic —OH, and/or carbonyl moieties, as in

$$\phi\text{—CH}_3 \rightarrow \phi\text{—CH}_2\text{OH} \rightarrow \phi\text{—COOH} \quad \text{and/or} \quad \phi\text{—OH}$$
$$\phi\text{—CH}_2\text{—}\phi \rightarrow \phi\text{—CO—}\phi,$$

which can launch condensation reactions and in some circumstances directly degrade aromatic rings (Tronov, 1940; Berkowitz, 1989). In the latter case, the process creates alkali-soluble humic acids [37], which then slowly decompose to hymatomelanic and fulvic acids, and finally furnish simple benzenoids, CO, CO_2, and water.

A study in which the roles of time (1–80 h), temperature (100–300°C), and oxygen concentration (10–21%) were explored (Jensen *et al.*, 1966) suggests concurrent reactions that can be qualitatively represented by

$$\begin{aligned} &\rightarrow \#1 \rightarrow \text{CO, CO}_2\text{, H}_2\text{O} \\ \text{coal} &\rightarrow \#2 \rightarrow \text{humic acids} \rightarrow \#2\text{a} \rightarrow \text{I}_1 \rightarrow \text{I}_2 \ldots \rightarrow \text{CO, CO}_2\text{, H}_2\text{O} \\ &\rightarrow \#2\text{b} \rightarrow \text{“regenerated coal”} + \text{CO, CO}_2\text{, H}_2\text{O} \end{aligned}$$

In this scheme, #1 shows decarboxylation/dehydroxylation of coal, #2 the formation of humic acids, which are (in #2a) sequentially broken down to intermediates I_1, I_2, etc., and eventually yield CO, CO_2, and H_2O, and/or which (in #2b) at >250°C decarboxylate/dehydroxylate with consequent loss of alkali solubility (and in that manner regenerate "coallike" matter). Reaction rates depend on coal rank, oxygen availability, and temperature; but, although

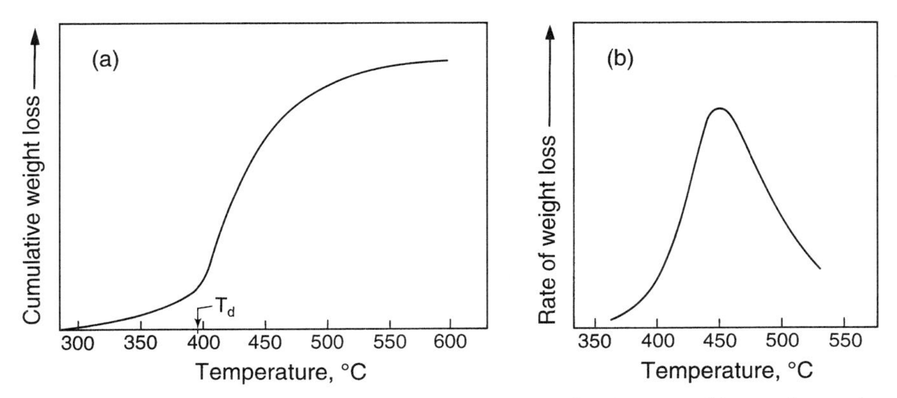

FIGURE 5.3.1 An illustrative weight-loss curve (a), and its first derivative (b). T_d indicates the onset of "active" thermal decomposition.

formation of humic acids (#2) dominates at <100°C, the exothermicity of oxidation can quickly accelerate #1 and 2b.

An illustration of such acceleration is afforded by *autogenous heating*, and eventual ignition, of low-rank coals during open storage. This begins when the coal, after drying out in hot, sunny weather, is wetted by heavy condensation or rain (Erdtmann and Stoltzenberg, 1908; Threllfall, 1909; Hoskins, 1928; Francis, 1938) and has been traced to heat releases associated with wetting (Berkowitz and Schein, 1951). In porous lignites and subbituminous coals, which accommodate more than 10–12% H_2O when *saturated*, heats of wetting can increase temperatures by as much as 25–30°C over ambient, and thereby raise oxidation rates to levels at which the process becomes a runaway reaction.

Thermolysis

Thermolytic behavior reflects the composition and chemical structure of a substance, and the responses of heavy fossil hydrocarbons therefore tend to be quite similar. At atmospheric pressure, all are thermally stable to ~350–375°C, and differences among bitumens, kerogens, and coals lie mainly in the proportions and compositions of liquids they yield when beginning to decompose near that temperature [38]. However, two or three aspects of thermolytic behavior merit more specific comment.

First, the onset of "active" decomposition, which is conventionally defined as the temperature at which the rate of weight loss increases sharply (see Fig. 5.3.1), does not imply that there is no limited decomposition (with little or

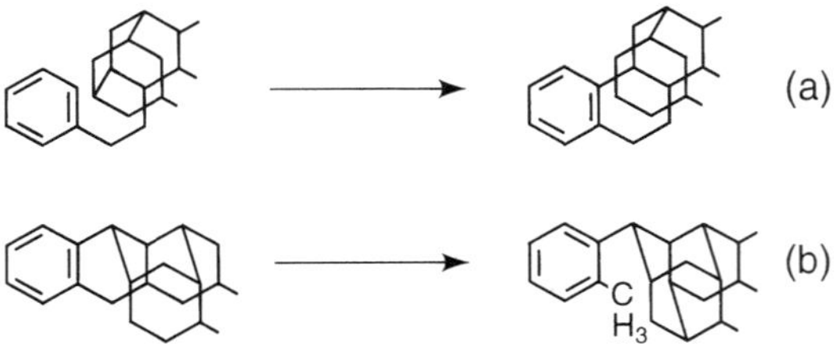

FIGURE 5.3.2 Two isomerization reactions at >200–225°C (Chakrabartty and Berkowitz, 1978): (a) benzylic carbon undergoes transannular formation; (b) transformation of benzylic carbon into methylphenyl derivatives.

no loss of volatile material) at much lower temperatures; aside from sorbed matter, heavy hydrocarbons do, in fact, commonly evolve some CO, CO_2, H_2O, and H_2S at temperatures as low as 175–200°C, mainly due to alteration or detachment of substituent functions such as

$$\phi\text{—COOH} \rightarrow \phi\text{—H} + CO_2$$
$$\phi\text{—OH} + \text{H—}\phi \rightarrow \phi\text{—}\phi + H_2O$$
$$\phi\text{—SH} + \text{H—}\phi \rightarrow \phi\text{—}\phi + H_2S$$

and $\phi\text{—COOH} \rightarrow [\phi\text{—C=O}^* + \text{OH}^* + \text{H—}\phi] \rightarrow \phi\text{—CO—}\phi + H_2O$,

where, as above, ϕ— denotes a phenyl ring, and the asterisk indicates a transient radical. At 200–225°C, some benzylic carbon is also thought to undergo transannular bond formation (see Fig. 5.3.2a), and to begin isomerizing to methylphenyl derivatives as in Fig. 5.3.2b (Chakrabartty and Berkowitz, 1978).

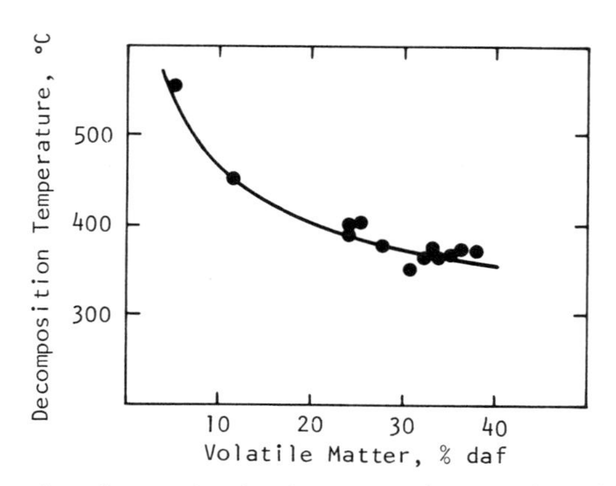

FIGURE 5.3.3 The influence of coal rank on onset of "active" thermal decomposition.

Second, unlike bitumens, which begin to actively decompose at much the same temperature (~375–400°C), coals cover a much wider range of C_{ar}/C (from <0.60 to >0.85), and consequently do so at temperatures that depend on, and increase with, rank (see Fig. 5.3.3).

Finally, liquids yields from heavy hydrocarbons increase with H/C ratios, and the composition (or "quality") of the liquids, expressed by C_{al}/C_{ar}, improves correspondingly. However, a history of prior weathering or oxidation affects yields as well as compositions—the former falling as a result of oxidative destruction of aliphatic/naphthenic moieties, and the latter therefore becoming increasingly aromatic and viscous.

NOTES

[1] These rules were developed and refined by the International Union of Pure and Applied Chemistry (IUPAC).

[2] This compound is sometimes also referred to as neopentane.

[3] Melting points also increase, but do so less regularly because they are influenced by crystallographic ordering as well as by molecular size.

[4] Less formally, these are sometimes also referred to as *boat* and *chair* configurations.

[5] This might explain the absence of these compounds in crude oils.

[6] This continuum can be legitimately defined as including conventional light and medium crude oils and, among oil fractions, as extending to gas oils.

[7] Biogenic gas, which contains >98% CH_4, is often encountered above an oil zone, but gives way suddenly to wet gas with C_2–C_6 hydrocarbons at the top of a mature zone at which the temperature has reached 55–60°C and catagenesis begins. Total gas and its wet gas component increase rapidly with depth until wet gas represents 50–80% of the total near the bottom of the oil zone (where $T \approx 140°C$), and it has therefore been suggested that C_2–C_4 concentrations might be taken as measures of oil maturity.

[8] Extraction of helium entails cooling the raw gas to temperatures at which the gas phase consists only of N_2 and He, and cooling that stream further to condense N_2. Thereafter, the residual He is purified by passage through activated carbon at −200°C.

[9] The much narrower ranges of elemental carbon, hydrogen, and oxygen contents underscore this point.

[10] The latter measurements can serve to define the distribution of molecular types in fractions characterized by similar melting or boiling points (see Chapter 4).

[11] The frequency of $—C_2H_5$ and longer functions decreases very rapidly with increasing number of C atoms.

[12] Significantly, these compounds are also major structural elements of the hypothesized coal macromolecules and components of coal tars (see Chapters 7 and 9).

[13] This underscores the view that thermal degradation of kerogen to a light oil involves a form of carbon rejection: it gradually increases the aromaticity (C_{ar}/C) of what is left behind in the source rock, and so mirrors thermal decompositon of coal.

[14] High sulfur contents are thus commonly encountered in asphalts, bitumens, tars, and oil-shale kerogens. Mexican heavy oil seepages, or *chapotes,* which contain 6.5–10.7% S, are examples.

[15] Surfactants can be classed as anionic (containing —COOH groups), cationic (amines, heterocyclics), amphoteric (amino acids, peptides), and nonionic (e.g., esters).

[16] The listing of asphaltic matter forms in Chapter 4 may serve to illustrate this profusion of "*seemingly* different bituminous substances."

[17] Thus, designations such as *bituminous* sands, *oil* sands, and *tar* sands are used interchangeably to mean almost identical accumulations of bituminous substances; bitumens, asphalts, tars and residua from crude oil processing are often collectively referred to as "asphaltics"; and there is generally little overt discrimination between natural tars (such as in "tar mats" associated with heavily oxidized bitumen accumulations) and the tars generated by pyrolysis of bituminous substances or coal.

[18] It is of some interest that many of these moieties also are, or closely resemble, components of the coal macromolecule.

[19] Significantly, yields of asphaltenes always increase toward a final constant value with increasing total volume of the precipitant.

[20] Several earlier attempts to link solvent (or, more correctly, extractant) potency for coal with the internal pressure of the fluid proved fruitless (Dryden, 1950), and later efforts to define potency by a solubility parameter ratio $\partial_x/\partial_{solvent}$, where ∂_x refers to the substance to be extracted (or precipitated?), remain to be validated: although $\partial_{solvent}$ can be experimentally obtained from $(\Delta H_{vap}/v_m)^{1/2}$, where v_m is the molar volume, ∂_2 can only be indirectly estimated from thermodynamic correlations of free-energy changes that accompany dissolution of structurally known polymers (van Krevelen, 1961, 1965).

[21] Shale oil was in fact commercially produced in the 1800s and early 1900s, but these operations were abandoned when natural gas, crude oil, and fuel oil (from oil refining) became freely available.

[22] That yields and (presumably) the compositions of shale oil are governed by the conditions under which the precursor shale is retorted suggests that classification of kerogens on the basis of their response to pyrolysis in inert atmospheres at 500°C—in essence, a classic destructive distillation—neglects processing options offered by contemporary technology.

[23] In low-rank (lignitic and subbituminous) coals, banding tends to be poorly expressed, perhaps because such coals, which are almost always of post-Jurassic age, formed mainly from angiosperms that decayed more completely and generated a more uniform humus than the gymnosperms, which were dominant source materials of older coals (Francis, 1961).

[24] The irregularly layered arrangements of the lithotypes have been ascribed to fluctuating conditions in the formative bog, to differential settling of heavier plant components and/or to syneresis, which entails exudation of liquid from a quiescent gel and is exemplified by the behavior of silicic acid gels and agar gels. In the heterogeneous humic gels from which most humic coals are thought to have developed, partial separation of vitrain precursors could quite easily have been caused by syneresis. In this connection, however, it should be noted that fusain probably represents remains of ancient forest fires.

[25] A helpful feature of the petrographic nomenclature is its uniformity: all lithotype designations end in *-ain,* microlithotypes in *-ite,* and macerals or maceral groups in *-inite.* As well, maceral and maceral group designations indicate source materials (e.g., cutinite, resinite, sporinite) or appearance (e.g., vitrinite, micrinite). An exception is inertinite, a term borrowed from carbonization practice and intended to convey that macerals of this group are relatively unreactive. In passing it might be noted here that Bustin *et al.* (1985) have proposed modifying the ICCP's nomenclature by using a set of "huminite" macerals (see table) to describe the vitrinites of low-rank coals.

Maceral subgroup	Maceral	Sources	Equivalents in bituminous coal
Humotellinite	textinite ulminite }	woody tissues	tellinite
Humocollinite	gelinite	colloidal humic gels	collinite
	corpohuminite	tannin condensation products	collinite
Humodetrinite	attrinite densinite }	finely comminuted humic detritus	vitrodetrinite

[26] Measured under oil, reflectivity, then indexed as R_o, is lower than when determined in air, but more accurate because of enhanced contrast.

[27] In Northern Hemisphere coals, vitrinites often account for more than 70% of the organic coal material, but they are less dominant in Southern Hemisphere (Gondwanaland) coals, in which they are commonly subordinate to durains.

[28] Although R_o continues to be universally used to determine coal rank, some caution is needed here: R_o is reportedly affected by paleoenvironmental factors (Jones *et al.*, 1984), and differently composed vitrinites can sometimes furnish identical R_o values (Neavel, 1986).

[29] One such direct measurement makes use of neutron activation (Bate 1963; Weaver, 1978)

$$^{18}O + n \rightarrow {}^{16}N + p$$

and involves counting the β- or γ-radiation from ^{16}N, and comparing the result with the count for a specimen of known oxygen content. The method is fast and accurate, but requires careful prior demineralization of the test sample if *organic* oxygen is to be assessed, and because the half-life of ^{16}N is ~7.3 s, the irradiated sample must be quickly transferred to the counting cell.
Other procedures determine [O] by pyrolysis in nitrogen at 1,100°C (Oita and Conway, 1962; Abernethy and Gibson, 1966), but seem to have evoked rather less interest.

[30] Such specifications are set out in standards issued by, among others, the American Society for Testing Materials (ASTM), the British Standards Institution (BS), the German Normenausschuss (DIN), the Polish Standards Committee (PN), and the International Standards Organization (ISO).

[31] Since ash is measured as a residue of high-temperature combustion (during which $CaCO_3 \rightarrow CaO + CO_2$, $FeS \rightarrow FeO, Fe_2O_3 + SO_2$, etc.) and is consequently not synonymous with mineral matter, there is an appreciable difference between dry *ash*-free and dry *mineral* matter-free bases. Very accurate conversion to the d.mm.f. base requires detailed information about the composition of mineral matter and is problematical. But for practical purposes, a close approximation is afforded by multiplying the raw analytical data by

$$100/\{100 - ([H_2O] + 1.1\ [ash] + 0.1\ [S])$$

or

$$100/\{100 - ([H_2O] + 1.08\ [ash] + 0.55\ [S])$$

Both expressions are due to Parr (1931) and have, in fact, been specified for use with the ASTM classification (see Chapter 4).

[32] This reaction destroys aliphatic and alicyclic substituent groups on aromatic rings, but substantially preserves the "stripped" rings.

[33] These are dimorphs of FeS that differ only in crystal habit, pyrite being cubic and marcasite orthorhombic.

[34] For an authoritative discussion of the spectrum of *trace* elements in coal, the reader is referred to Swaine (1990).

[35] For example, the observation that initially fast absorption of O_2 by asphalts and heavy oils slows quickly with increasing oxidation (Ronvaux-Vankeerbergen and Thyrion 1989) contradicts results that show no such dependence (Tadema and Weijdema, 1970; Darbous and Foulton, 1979).

[36] Conceptually, these might be preceded by formation of carboxy acids and phenols that could initiate condensation via their —COOH and —OH groups.

[37] How closely these substances structurally resemble humins and/or humic acids involved in the formation of coal (Chapter 2) and the similarly named extractable matter in modern soils is uncertain.

[38] That *carbonization, coking, destructive distillation, pyrolysis,* etc., all signifying thermolysis, are usually conducted at somewhat higher temperatures is merely designed to speed the process.

REFERENCES

Abernethy, R. F., and F. H. Gibson. *U.S. Bur. Mines Rept. Invest. 6753* (1966).

Abraham, H. *Asphalts and Allied Substances,* 6th ed., 1960. New York: Van Nostrand.

Ambles, A., M. Halim, J. C. Jacquesy, D. Vitorovic, and M. Ziyad. *Fuel* **73**, 17 (1994).

ASTM (American Society for Testing Materials). *ASTM D 2797-72* (1972).

Bate, L. C. *Nucleonics* **21**, 72 (1963).

Bedson, P. P. *J. Soc. Chem. Ind. (London)* **21**, 241 (1902).

Benalioulhaj, S., and J. Trichet. *J. Org. Geochem.* **16**, 649 (1990).

Beret, S., and J. G. Reynolds. *Fuel Sci. Technol. Int.* **8**, 191 (1990).

Berkowitz, N. *The Chemistry of Coal,* 1985. Amsterdam: Elsevier.

Berkowitz, N. In *Sample Selection, Aging and Reactivity of Coal* (R. Klein and R. Wellek, eds.), Chapter 5, 1989. New York: Wiley & Sons.

Berkowitz, N., and J. Calderon. *J. Fuel Process. Technol.* **25**, 33 (1990).

Berkowitz, N., and H. G. Schein. *Fuel* **30**, 94 (1951).

Bimer, J., P. H. Given, and S. Raj. *ACS Div. Fuel Chem., Prepr.* **22**(5), 169 (1977); *ACS Symp. Ser.* **71**, 86 (1978).

Blazek, J., and G. Sebor. *Fuel* **72**, 199 (1993).

Boucher, R. J., G. Standen, and G. Eglinton. *Fuel* **70**, 695 (1991).

Broome, D. C. In *Bituminous Materials: Asphalts, Tars and Pitches* (A. J. Hoiberg, ed.), 1965. Interscience Publ.

Bustin, R. M., A. R. Cameron, D. A. Grieve, and W. D. Kalkreuth. *Coal Petrology,* 1985. St. John's, Newfoundland: Geol. Assoc. Canada.

Chakrabartty, S. K., and N. Berkowitz. *ACS Symp. Ser.* **71**, 131 (1978).

Cochram, C., and R. V. Wheeler. *J. Chem. Soc.* 700 (1927); 854 (1931).

Darbous, M. K., and P. E. Foulton. *Soc. Pet. Eng. J.* **14**, 253 (1979).

De Marsilly, C. *Ann. Chim. Phys., Ser. 3* **66**, 167 (1862).

Deno, N. C., K. W. Curry, A. D. Jones, K. R. Keegan, W. R. Rakitsky, C. A. Richter, and R. D. Minard. *Fuel* **60**, 210 (1981).

Donaldson, E. C., and J. B. Clark (eds.). *Proc. Internatl. Conf. Microbial Enhanced Oil Recovery NTIS, Springfield, VA* (1983).

Dryden, I. G. C. *Fuel* **30**, 39 (1951); *Chem. Ind. (London)* **30**, 502 (1952).

Durand, B., and G. Nicase. In *Kerogen* (B. Durand, ed.), 1980. Paris: Editions Technip.

Erdtmann, E., and H. Stoltzenberg. *Braunkohle (Dusseldorf)* 7, 69 (1908).

Estep, P. A., J. J. Kovach, and C. Karr, Jr. *Anal. Chem.* **40**, 358 (1968).

Forsman, J. P., and J. M. Hunt. In *The Habitat of Oil* (L. G. Weeks, ed.), 1958. Oxford: Pergamon Press.

Francis, W. *Fuel* **17**, 363 (1938).

Francis, W. *Coal,* 1961. London: Arnold.

Ganz, H., and H. Kalkreuth. *Fuel* **66**, 708 (1987).

Getz, F. A. *Oil Gas J.* **75**, 220 (1977).

Gluskoter, H. J. *Fuel* **44**, 285 (1965); *J. Sediment. Petrol.* **37**, 205 (1967).

Guignet, E. *Comp. Rend. Acad. Sci. (Paris)* **88**, 590 (1879).

Hayatsu, R., R. G. Scott, L. P. Moore, and M. H. Studier. *Nature (London)* **257**, 378 (1975); **261**, 77 (1976).

Hayatsu, R., R. E. Winans, R. G. Scott, L. P. Moore, and M. H. Studier. *ACS Symp. Ser.* **71**, 108 (1978).

Hayatsu, R., R. E. Winans, R. G. Scott, R. L. McBeth, and L. P. Moore. *Fuel* **60**, 77 (1981).

Hepler, L. G., and Chu Hsi (eds.). *AOSTRA Technical Handbook 1 on Oil Sands, Bitumens and Heavy Oils, AOSTRA Tech. Publ. Ser. #6,* 1989 Edmonton, AB.

Hoskins, A. J. *Purdue Univ. Experimental Station Bull.,* No. 30 (1928).

Hunt, J. M. *Petroleum Geochemistry and Geology,* 1979. San Francisco: W. H. Freeman & Co.

Jenkins, R. G., and P. L. Walker, Jr. In *Analytical Methods for Coal and Coal Products* (C. Karr, Jr., ed.), Vol. 2, p. 265, 1979. New York: Academic Press.

Jensen, E. J., N. Melnyk, J. C. Wood, and N. Berkowitz. *ACS Adv. Chem. Ser.* **55**, 621 (1966).

Jones, J. M., A. Davis, A. C. Cook, D. G. Murchison, and E. Scott. *Internatl. J. Coal Geol.* **3**(4), 315 (1984).

Kemp, C. N. *Trans. Inst. Min. Engrs. (London)* **77**, 175 (1929).

Kukharenko, T. A., and V. I. Belikova. *Khim. Tverd. Topl.,* No. 13 (1968).

Lee, S. *Oil Shale Technology,* 1991. Boca Raton, FL: CRC Press.

Levorson, A. I. *Geology of Petroleum,* 2nd ed., 1967. San Francisco: W. H. Freeman & Co.

Mochida, I. *Oil Gas J.,* Nov. 1986, p. 58.

Mackowsky, M.-Th. *Proc. Internatl. Comm. Coal Petrogr.* **2**, 31 (1956).

Mackowsky, M.-Th. *Fortschr. Geol. Rheinl. Westfalen* **19**, 173 (1971); *Microsc. Acta* **77**, 114 (1975).

Moschopedis, S. E., and J. G. Speight. *Fuel* **55**, 228 (1976).

Neavel, R. C. *Fuel* **65**, 1632 (1986).

Office of Technology Assessment (USA). *An Assessment of Oil Shale Technologies, OTA-M-118,* 1980. Washington, D.C.: U.S. Congress.

Oita, I. J., and H. S. Conway. *Anal. Chem.* **189**, 91 (1962).

Painter, R. C., S. M. Rimmer, R. W. Snyder, and A. Davis. *Appl. Spectrosc.* **35**, 102 (1981).

Parr, S. W. *The Analysis of Fuel, Gas, Water and Lubricants,* 1932. New York: McGraw-Hill.

Raj, S. Ph.D. Thesis, Pennsylvania State University, 1976.

Rao, B. M. L., and J. E. Serrano. *Proc. Ntl. Mtg. ACS, Anaheim, CA, Sept. 1986,* p. 669.

Robinson, W. E. In *Organic Geochemistry* (G. Eglinton and M. T. J. Murphy, eds.), 1969. Berlin: Springer.

Ronvaux-Vankeerbergen, A., and F. C. Thyrion. *Fuel* **68**, 793 (1989).

Ruzicka, D. J., and G. Ostvold. *Mar. Eng. Rev.,* Sept. 1988, p. 27.

Ruzicka, D. J., and S. Nordenson. *Fuel* **69**, 710 (1990).

Saxby, J. D. In *Oil Shale,* (T. F. Yen and G. V. Chiligarian, eds.), 1976. Amsterdam: Elsevier.

Schobert, H. H. *The Chemistry of Hydrocarbon Fuels,* 1990. London: Butterworths.

Speight, J. G. *The Chemistry and Technology of Petroleum,* 2nd ed., 1991. New York: Marcel Dekker.

Stach, E. *Textbook of Coal Petrology,* 1985. Berlin: Borntraeger.

St. John, A. *Trans. AIME* **74**, 640 (1926).
Stopes, M. C. *Proc. R. Soc. London* **B90**, 470 (1919).
Swaine, D. J. *Trace Elements in Coal,* 1990. London: Butterworth.
Tadema, H. J., and X. Weijdema. *Oil Gas. J.* **68**, 77 (1970).
Teichmüller, M.; cited in M.-Th. Mackowsky. *Microsc. Acta* **77**, 114 (1975).
Threllfall, R. *J. Soc. Chem. Ind. (London)* **28**, 759 (1909).
Tissot, B. P., and D. H. Welte. *Petroleum Formation and Occurrence,* 1978. Berlin/Heidelberg: Springer.
Tronov, B. V. *J. Appl. Chem. USSR (Engl. Transl.)* **13**, 1053 (1940); *Chem. Abstr.* **35**, 1966 (1941).
van Krevelen, D. W. *Coal,* 1961. New York: Elsevier.
van Krevelen, D. W. *Fuel* **44**, 229 (1965).
Weaver, J. N. In *Analytical Methods for Coal and Coal Products* (C. Karr, Jr., ed.), Vol. 1, p. 377, 1978. New York: Academic Press.
Yohe, G. R., and J. M. Harris. *Fuel* **40**, 289 (1961).
Zajic, J. E., and E. C. Donaldson (eds.), *Microbially Enhanced Oil Recovery,* 1985. El Paso, TX: Bioresources Publ.
ZoBell, C. E. *World Oil* **126**, 36; **127**, 35 (1947).

CHAPTER 6

Physical Properties

With few exceptions, physical properties other than those bearing on production, marketing, and end use of fossil hydrocarbons have received relatively little attention [1], and what is known about them tends to be resource specific, fragmentary, and, as a rule, more illustrative than definitive. Resource specificity arises from different practical interests in the behavior of natural gas, crude oils, bituminous substances, kerogens, and coals, and from the consequent need to assess these materials by very different criteria. The second qualifier, i.e., that data at hand are mainly illustrative, is occasioned by close dependence of physical properties on the minutiae of composition, and by the fact that data on heavy fossil hydrocarbons other than coal are often questionable because most refer to single samples and/or samples of uncertain origin and/or history.

1. NATURAL GAS

Active interest in the physical properties of light (C_1–C_5) hydrocarbons is, and remains, largely confined to their densities, specific heats, and related parameters, and to so-called "gas hydrates," i.e., clathrate compounds that form in certain circumstances by interaction with water.

Densities

By definition, density (d) denotes the weight or mass of unit volume of a substance, and the commonly cited specific gravity (s.g.) relates this parameter to the density of air, oxygen, or water at the same temperature and pressure. In either case, measurements are normally made at 15.5°C and 760 mm Hg, and reported in g/cm^3, kg/m^3, or, sometimes, tonnes "oil equivalent" [2] (in which case 1000 m^3 ≡ 0.825 metric tonne). In the United States, gas densities and specific gravities are often also given in lb/ft^3.

Measured densities depend on gas compositions—and consequently on

TABLE 6.1.1 Latent Heats (Enthalpies) of Vaporization of *n*-Paraffins

	b.p. (°C)	ΔH_v (kJ mol^{-1})
CH_4	−161	8.20
C_3H_8	−42	18.79
C_7H_{16}	98	32.72
$C_{10}H_{22}$	174	39.30
$C_{14}H_{30}$	253	47.75
$C_{20}H_{42}$	343	57.75

whether they refer to raw "as-produced" gas, or to processed gas that has been stripped of $>C_1$ hydrocarbons, H_2S, and other contaminants, and which may therefore contain >99% CH_4. Specific gravities relative to air = 1.0 can thus run from 0.616 to 0.965 or, based on H_2O = 1.0, vary from 0.73 to 0.933×10^{-3}.

Specific Heats

The specific heat of a substance, measured at constant pressure (C_p) or constant volume (C_v), is the thermal energy required to raise the temperature of a unit mass through 1°C, and can be expressed in J/kg or J/m^3, or cited in Btu/lb. But because of the importance of this parameter in refinery practice, measurements must be carried out in strict compliance with specifications detailed in the applicable national standards (see, e.g., ASTM D-2766).

Regression analyses of the voluminous data for hydrocarbons have yielded an expression of the form

$$C_p = \text{s.g.}\ (0.388 + 0.00045\ T),$$

where C_p is the specific heat of the gas at a temperature T (in °F), and s.g. its specific gravity at 60°F (15.5°C). This equation is reportedly valid for most oils as well as for gases, and appears to be sufficiently accurate to obviate the need for separate direct measurements.

Enthalpy of Vaporization

The enthalpy (or heat) of vaporization (ΔH_v) quantifies the heat change that accompanies the transition from a liquid to a gaseous state at the normal boiling point of a (volatile) substance, but depends intimately on chemical composition. Values for methane and some higher *n*-paraffins are listed in Table 6.1.1.

TABLE 6.1.2 Structural Features of Gas Hydrates

	I	II
H_2O molecules/unit cell	46	136
Small cages/unit cell	2	16
Large cages/unit cell	6	8
Cage diameter (nm)		
Small cavity	7.95	7.82
Large cavity	8.60	9.46

Dew Points

The dew point defines the temperature at which, at any given pressure, condensation begins to manifest itself in formation of small droplets in the gas. That temperature does, of course, depend on gas composition, and is drastically lowered by the presence of condensible ($>C_3$) components.

Gas Hydrates

At low temperatures, pressurized wet gas will generate icelike crystalline gas hydrates in which a lattice of water molecules can form two sets of crystallographically defined unit cells capable of accommodating light hydrocarbons, H_2S, CO_2, or N_2. The smaller cell is made up of 46 H_2O molecules and forms cages that can each hold up to eight CH_4 molecules, whereas the larger cell comprises 136 H_2O molecules and has cages sufficiently large to accept isobutane. One cubic meter of hydrate in which all cages are filled by CH_4 molecules could thus hold ~172 m^3 CH_4 at STP. Table 6.1.2 lists some specifics for the two cells, and Table 6.1.3 shows which hydrocarbons or natural gas components could be accommodated in them.

Since gas hydrates, whose formation is favored by high pressure and low temperatures, tend to promote corrosion as well as impede pipeline flow by coating pipe interiors, operating conditions are, wherever possible, selected with a view to inhibiting hydrate formation.

2. CRUDE OILS

The almost limitless variety of crude oils that command industrial interest has, not surprisingly, focused attention on a substantially wider set of physical properties than are, in practice, required to be known for natural gas; and

TABLE 6.1.3 Accommodation in Gas Hydrate Cells[a,b]

	Unit cell I		Unit cell II	
	Small	Large	Small	Large
CH_4	+	+	+	+
C_2H_6		+		+
C_3H_8				+
C_4H_{10}				+
i-C_4H_{10}			+	
N_2	+	+	+	+
CO_2	+	+	+	+
H_2S	+	+	+	+

[a] Pederson *et al.* (1989).
[b] *Small* and *large* refer to cage size.

this, in turn, has generated data that permit correlation with aspects of chemical composition. Such linkages, usually cast in the form of *trend diagrams,* characterize the oils, allow classification, and assist selection of optimal preparation and/or processing options. Parameters for trend diagrams are properties—mainly melting points, distillation ranges, densities, and viscosities—that vary systematically with average molecular weights (which reflect transitions from light to heavy oils). However, also often evaluated when characterizing crudes are certain "quality" indicators (color, odors, flash points, etc.) and, on occasion, specifics that bear on reservoir mechanics (notably thermal expansion coefficients, pressure-dependent oil volumes, surface tensions, etc.).

Melting Points

The melting points of *components* of an oil increase with their molecular weight (or number of C atoms; Table 6.2.1) and are, in fact, frequently used to identify

TABLE 6.2.1 Melting Points of *n*-Paraffins

C atoms/molecule	m.p. (°C)
1	−182
5	−130
10	−30
15	10
20	36
30	66
40	82
50	92

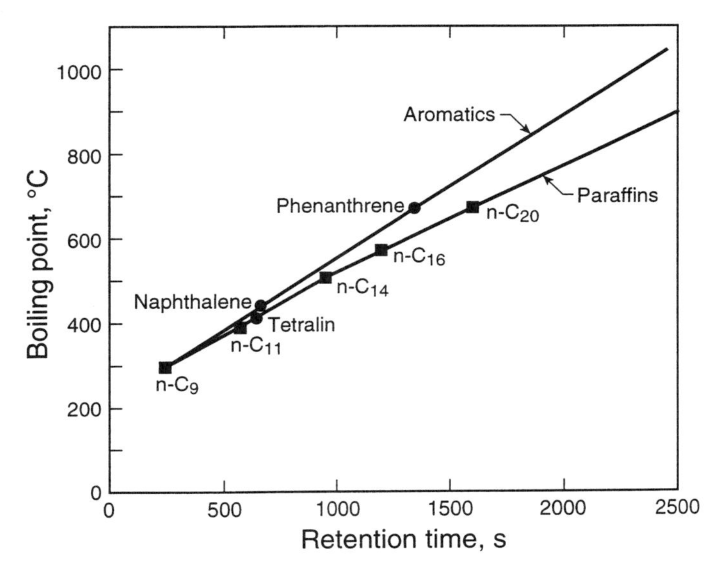

FIGURE 6.2.1 Hydrocarbon boiling points, determined from GC retention times.

the waxy components of a crude. Several national standards (in North America, ASTM D-87 and D-127) specify appropriate procedures for that purpose. Correlations between melting points and molecular weights can, however, be negated by molecular symmetry, and although branched-chain hydrocarbons usually melt at substantially lower temperatures than their straight-chain counterparts, reversals occur when they are highly symmetrical. Whereas *n*-pentane melts at −130°C, 2,2-dimethylpropane thus melts at −20°C; and whereas *n*-octane melts at −57°C, 2,2,3,3-tetramethylbutane does so at 104°C.

Boiling Ranges

In principle, detailed characterization of a crude oil or oil fraction requires determining a boiling *range*, which can be established by either (i) recording a distillation curve or, more simply, the temperatures at which 5% and 95% of the total distillate have accumulated in a receiver [4] or, subject to a calibration curve that relates retention times to boiling points (see Fig. 6.2.1), by (ii) recourse to temperature-programmed gas chromatography.

However, because average (volume or weight) boiling points, which can be calculated from distillation curves, commonly correlate better with many physical and chemical properties (Maxwell, 1950), so-called "mean average

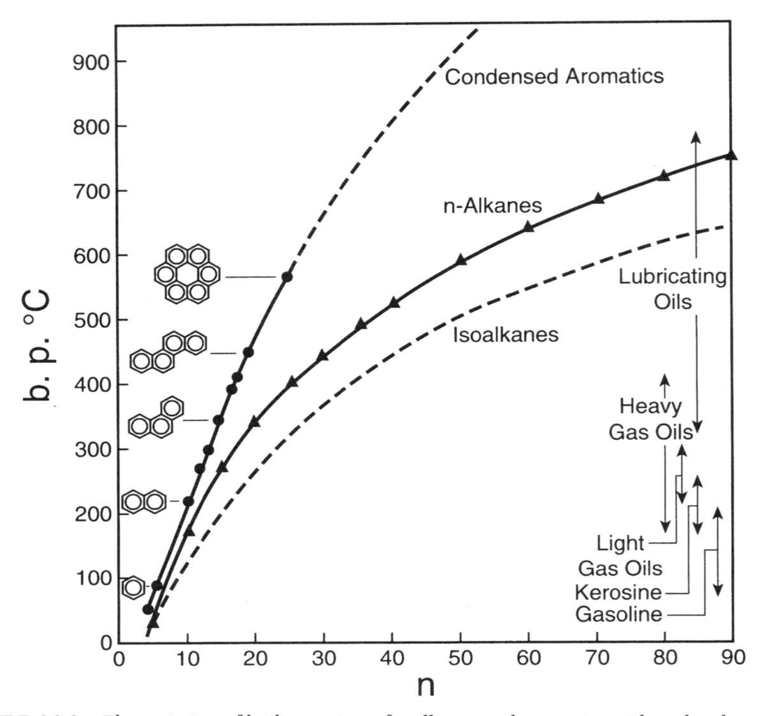

FIGURE 6.2.2 The variation of boiling points of *n*-alkanes and aromatics with molecular weight.

boiling points" (mbps)—by definition, the boiling points that best correlate with molecular weights—can be directly calculated (in °Rankine; [5]) from

$$\text{mbp}\ (^\circ\text{R}) = 0.5\left[\sum_{i=1}^{n} x_i T_{\text{bi}} + \left(\sum_{i=1}^{n} xv_i T_{\text{bi}}^{1/3}\right)^3\right],$$

where x_i is the mole fraction and xv_i the volume fraction of component i, and T_{bi} the normal boiling point of component i in °R (API, 1983).

Figure 6.2.2, which might be viewed against Figure 6.2.1, illustrates the dependence of boiling points on molecular weights among *n*-alkanes and condensed aromatics [6]; Table 6.2.2 lists the melting and boiling points (as well as the densities) of some hydrocarbons commonly found in crude oils; and the boiling ranges of major oil fractions are shown in Table 6.2.3, which also indicates the routine dispositions of these fractions.

Densities

Specific gravities of oils are usually expressed as API gravities (°API) and so given, vary *inversely* with viscosity, asphaltic matter content (which increases

TABLE 6.2.2 Melting Points, Boiling Points, and Densities of Hydrocarbons Commonly Present in Crude Oils

Component	m.p. (°C)	b.p. (°C)	Density (g/cm^3)
Paraffins			
Methane	−184	−161.5	
Ethane	−172	−88.3	
Propane	−189.9	−42.2	
n-Butane	−135	−0.6	
Isobutane	−145	−10.2	
n-Pentane	−131.5	36.2	0.626
n-Hexane	−94.3	69.0	0.659
Isooctane	−107.4	99.3	0.692
n-Decane	−30	174.0	0.730
Naphthenes			
Cyclopentane	−93.3	49.5	0.745
Me-cyclopentane	−142.4	71.8	0.754
Cyclohexane	6.5	81.4	0.779
Aromatics			
Benzene	5.5	80.1	0.885
Toluene	−95	110.6	0.967
o-Xylene	−29	144.4	0.880
Naphthalene	80.2	217.9	0.971

from 4–8% at 40 to ~50% at 10–15°API), and N-content (which rises from 0.08–0.20% to ~1% over the same interval). In European practice, specific gravities are often also cited in °Baumé, and Table 6.2.4 shows how this translates into °API. However, as the relationship between the two scales is not strictly linear [7], it is also helpful to note that

$$°\text{Baumé} = (140/\text{s.g.}_{60}) - 130$$
$$°\text{API} = (141.5/\text{s.g.}_{60}) - 131.5,$$

where s.g._{60} denotes the specific gravity at 15.5°C (60°F).

Dissolved C_1–C_5 hydrocarbons, exceptionally high resin contents, or removal of $<C_{16}$ hydrocarbons by biodegradation and/or water washing (see Chapter 3) can cause the gravities of light crudes to vary widely even when they are in close proximity to each other or otherwise appear to be very similar. But if not subject to these modifiers, specific gravities increase with average molecular weights, fall as temperatures rise (see Table 6.2.5), and consequently also fall as depths increase (see Table 6.2.6) and the oils are exposed to higher geothermal temperatures [8]. But there are important exceptions to these generalizations: in Kuwait's Burgan field, API gravities have been observed to *decrease* at ~1°/60 m (200 ft) depth, and similar reversals of the general rule have been reported from several fields of Russia's Apsheron peninsula and from the Hawkins pool of NE Texas.

TABLE 6.2.3 Boiling Ranges[a] and Disposition of Oil Fractions

	Boiling range	Disposition
Light straight-run gasoline	32–104°C	gasoline
Naphtha	82–204°C	reformed for gasoline
Kerosine	165–282°C	reformed for gasoline
Light gas oil (LGO)	215–337°C	diesel fuel
Heavy gas oil (HGO)	320–446°C	[b]
Vacuum gas oil (VGO)	398–565°C	[b]
Resid	>565°C	[c]

[a] Overlapping temperature ranges reflect different refinery practices and/or variations in product slate compositions.
[b] Processed as catalytic cracker feed.
[c] Used as asphalt or processed as coker or hydrocracker feed.

TABLE 6.2.4 Relationships between Specific Gravity, °API, and °Baumé

°Baume	s.g.[a]	°API
50.0	0.7778	50.4
45.0	0.8000	45.4
40.0	0.8235	40.3
35.0	0.8485	35.3
30.0	0.8750	30.2
25.0	0.9032	25.2
20.0	0.9333	20.1
15.0	0.9655	15.1
10.0	1.0000	10.0

[a] At 15.5°C.

TABLE 6.2.5 The Variation of Specific Gravity and API Gravity with Temperature[a,b]

15.5°C		37.8°C		93.3°C		148.9°C	
s.g.	°API	s.g.	°API	s.g.	°API	s.g.	°API
0.7	70.6	0.67	69.0				
0.8	45.4	0.78	49.9	0.74	59.8	0.69	73.6
0.9	25.7	0.88	29.3	0.85	35.0	0.82	41.0
1.0	10.0	0.98	12.9	0.96	15.6	0.92	22.3

[a] Bell (1945).
[b] Specific gravities measured at 15.5°C.

TABLE 6.2.6 Variations of Specific Gravity with Depth[a]

Depth (m)	s.g.	°API
Pennsylvanian NE Oklahoma crudes:		
150–600	0.88–0.85	30–35
600–1500	0.85–0.82	35–40
1500–1800	0.82–0.80	40–45
Tertiary Gulf Coast crudes:		
0–300	0.94	19
300–600	0.93	21
600–900	0.88	29
900–1200	0.86	33
1200–1500	0.85	34
1500–1800	0.81	44

[a] After Levorsen (1967).

Volume/Oil Mass Relationships

In-situ crude oils always contain some dissolved C_1–C_5 hydrocarbons, and the volume of a given mass of oil depends therefore on the formation gas–oil ratio and the reservoir pressure. As pressures increase, this volume increases because of dissolution of more gas until the saturation pressure or so-called "bubble point" is reached, and thereafter falls. For example,

1 m^3	original reservoir conditions (all liquid)
↓	
1.1 m^3	at saturation pressure (or "bubble point")
↓	
0.8 m^3 (+ gas)	en route to stock tank
↓	
0.625 m^3	(+ 178 m^3 gas) at stock tank at STP

Viscosities

The flow ability or fluidity (ϕ) of a liquid is generally cited as an *absolute* viscosity η (= $1/\phi$) or as a *kinematic* viscosity k (= η/d, where d = density in g/cm^3). In the cgs system, η and k are expressed in poise (P) or centipose (cP) and stokes (St) or centistokes (cSt) units, respectively, whereas in the SI system they are reported in Pa · s or mPa · s and m^2s^{-1} or mm^2s^{-1}. (In the latter case, 1 mm^2s^{-1} = 1 cSt.)

Viscosity measurements are conveniently made with capillary viscometers, of which several are specified in national standards (e.g., ASTM D-445), and η is then evaluated from

$$v = [\pi r^4 \Delta p]/8nL,$$

where v is the volume of liquid passing through the capillary in unit time, r the radius of the capillary tube, L its length, Δp the pressure drop along its length, and n a coefficient determined by calibration of the viscometer.

For some purposes, flow ability is on occasion also measured as the Saybolt universal viscosity—an empirical quantity defined as the time (in seconds) taken by 60 ml to flow through a calibrated orifice (ASTM D-88). This quantity can be related to the absolute viscosity by

$$1\ \text{mPa}\cdot\text{s} = [0.22 - (135/\text{SUs})][141.4/(^\circ\text{API} + 131.5)]$$

and to the kinematic viscosity by

$$k = a(\text{SUs}) + (b/\text{SUs})],$$

where a and b are constants, and SUs is the flow time (McCullough, 1955).

Absolute viscosities vary directly with average molecular weights of crude oils and oil fractions, and inversely with the oil temperature, API gravity, and volume (or amount) of dissolved gases. The temperature coefficient of η can be approximated by an expression of the form

$$\log \log(\eta + c) = A + B \log T,$$

where T is the temperature, A and B are constants, and c assumes a value of 0.6 for all oils with $k > 1.5$ cSt. The effect of pressure on η can be estimated from Kouzel's (1965) correlation

$$\log 10\ \eta/\eta_o = [(p - 14.696)/1000](0.0239 + 0.01638\ \eta_o^{0.278}),$$

which is claimed to be accurate to 35 MPa (5000 psi) and 250°C (475°F). In this expression η_o denotes the viscosity at the temperature of interest (T) and atmospheric pressure, and η its value at T and pressure p.

For lubricating oils, whose quality is often expressed by the temperature coefficient of the viscosity, it is sometimes more convenient to use a viscosity "index" instead of η *per se*. This index, the so-called "Dean–Davis scale" (see ASTM D-2270), assigns representative Gulf Coast and Pennsylvania crudes values of, respectively, 0 and 100 (which reflects the fact that naphthenic oils possess substantially higher temperature coefficients than corresponding paraffinic oils), and rates the oils by

$$\text{VI} = (a - b)/(a - c)$$

where a and c are the viscosities of the two reference oils, and b is the viscosity of the unknown. All three parameters are measured at 100°F (38°C) and

TABLE 6.2.7 The Variation of η with °API

°API	η (10^3 SUs)
8	50–500
10	20–50
14	2–5
16	1–1.5
18	0.5–0.6
20	0.3–0.35

expressed in Saybolt Universal seconds (SUs) or kinematic viscosity units (m^2s^{-1}) [9].

An illustrative example of the dependence of η on oil density, in this case expressed as API gravity and relating to some bitumens and resids, is shown in Table 6.2.7 [10].

Surface Tension

Although the surface tension σ of a substance is theoretically related to its structure and molecular weight, surface tensions change little from one oil component or fraction to another (Table 6.2.8) and are therefore in themselves of little interest.

Substantial changes do, however, accompany changes in temperature: for example, the surface tension of pentane drops from 16.0 dyn cm^{-1} at 20°C to ~8 at 90°C, and that of octane falls from 21.8 at 20°C to ~15 at 90°C. Significant reductions are also brought about by polar compounds such as fatty acids [11], and importance attaches therefore to the *interfacial* tension $\sigma_{a/b}$ between an oil and another immiscible fluid. This parameter can be evaluated under simulated reservoir conditions by a pendant-drop or a spinning-drop technique.

In the former method, a pressure sight glass or PVT cell is charged with a gas or brine at reservoir conditions, a drop of oil is suspended from a capillary tube that extends into the cell, and the oil/gas or oil/brine interfacial tension is derived from the geometry of the drop (Adamson, 1976; Girault *et al.*, 1984). In the spinning-drop technique, which can, however, only measure $\sigma_{a/b}$ between two *liquids*, the droplet is introduced into a rapidly spinning brine-filled tube and viewed with a strobe light. Here, too, $\sigma_{a/b}$ is evaluated from the droplet shape (Chatenay *et al.*, 1982).

TABLE 6.2.8 Surface Tension of Some Oil Components and Oil Fractions[a]

n-Alkanes	
Pentane	16.0
Hexane	18.4
Heptane	20.3
Octane	21.8
Cycloalkanes	
Cyclopentane	22.4
Cyclohexane	25.0
Aromatics	
Benzene	28.8
Toluene	28.5
Ethylbenzene	29.0
Butylbenzene	29.2
Fractions	
Gasoline	26
Kerosine	30
Lubricating oils	34

[a] In dyn cm^{-1} at 20°C.

Optical Properties

Color

In transmitted light, colors of crude oils vary from light yellow to reddish-brown and become progressively darker as specific gravities increase. Some very dark oils are opaque. Since color is usually ascribed to polynuclear aromatics or polyunsaturated alkanes with aromatic substituents in which π-electrons are conjugated, and the abundance of such molecules in crude oils varies inversely with the API gravity, oil color is deemed to be related to aromatics content—even though straight-run gasolines are believed to gain it from oxygenated compounds (fluvenes).

In any event, where color is of special interest, color intensities can be readily measured with a Saybolt colorimeter.

Fluorescence

Regardless of their origins, all oils exhibit a fluorescence or "bloom," which can be observed in ultraviolet light at λ = 254–365 nm, manifests itself in colors that range continuously from yellow through green to bluish, and is the more intense the higher the aromatics content.

TABLE 6.2.9 Optical Activity of Crude Oils

Geologic age	No. of oils	[α]
Tertiary	86	0.63
Cretaceous	18	0.28
Jurassic	20	0.20
Carboniferous	23	0.24
Devonian	21	0.18
Silurian	14	0.12

Because it is easily detected [12], fluorescence offers useful means for identifying oil shows in cores, cuttings and drilling muds, and as its intensity rapidly decreases when the oil is oxidized, it also differentiates between fresh and aged oils.

Optical Activity

Most oils rotate the plane of polarization of polarized light and are dextro- or levorotary. Such optical activity (α, expressed in °/mm) can be measured with a polaroscope at 20°C by using monochromatic Na–D light (λ = 589.3 nm), and is most pronounced in middle-range (250–300°C) oil fractions, in which it is associated with steranes such as cholestanes and/or triterpanes such as hopanes (Fenske *et al.*, 1942). Very little, if any, optical activity has been observed in $<$200°C and $>$300°C fractions.

In neutral lubricating oils, $[\alpha]$ increases, sometimes quite rapidly, with the average molecular weight: examples present themselves in blends from Gulf Coast and California crudes, in which it rises from ~0.5 to ≥3 as molecular weights rise from ~300 to 400. However, in other cases it remains virtually constant at 0.2–0.3 over the 330–480 range.

Amosov (1951) has shown optical activity to depend on, and vary inversely with, the geological age of a crude oil: Table 6.2.9, in which $[\alpha]$ denotes the extent of clockwise (dextro-) rotation, illustrates this.

Refraction

The refractive index of a substance is defined by $n = c_o/c_s$, where c_o and c_s are the velocities of light in a vacuum and when passing through a substance, and can be written as

$$n = \sin i/\sin r,$$

where i is the angle of incidence and r the angle of refraction, both determined with incident light normal to the surface of the substance.

TABLE 6.2.10 Relationships among °API, Density, and Refractive Index

°API	Density	n^a
6	1.029	1.566
22	0.918	1.509
44	0.802	1.448
58	0.742	1.417
72	0.691	1.390

[a] Measured at 20°C with Na–D light.

The refractive index of an oil is generally measured with a conventional Abbé refractometer and monochromatic Na–D light at 20°C (ASTM D-1218). Under these conditions n ranges from ~1.39 to 1.49, but depends on density or some covariant function thereof (see Table 6.2.10). Among n-alkanes, n thus increases with the molecular weight (see Table 6.2.11); among oil fractions of similar average molecular weight, it increases sequentially from paraffinic to naphthenic and aromatic species; and among naphthenes and aromatics, it tends to be significantly higher for polycyclics than for the corresponding monocyclics. Refractive indices—or the refractive *dispersion,* defined by

$$\Delta n = (n_1 - n_2)/d,$$

where Δn is the difference between refractive indices at two specified wave-

TABLE 6.2.11 Refractive Indices of Some Hydrocarbons Commonly Occurring in Crude Oils

C_5H_{12}	1.3579
C_6H_{14}	1.3749
C_7H_{16}	1.3876
C_8H_{18}	1.3975
C_9H_{20}	1.4054
$C_{10}H_{22}$	1.4119
Nonadecane	1.4409
Eicosane	1.4425
Cyclopentane	1.4064
Cyclohexane	1.4266
Cycloheptane	1.4449
Benzene	1.5011
Methylbenzene	1.4961
Ethylbenzene	1.4959
Propylbenzene	1.4920

TABLE 6.2.12 Thermal Conductivities of Some Rock Types[a]

Conglomerates	2.3–11.5
Dolomites	6–14
Limestones	5–7.5
Marls	2.3–7.3
Sandstones	3.5–13.2
Shales	3–7
Coal	0.3–1.2

[a] In cal cm^{-1} s^{-1} $°C^{-1}$.

lengths [13] and d is the density of the oil—can therefore provide useful information about oil compositions.

Thermal Properties

Thermal Conductivity

Regression analyses of data for the thermal conductivity of crude oils and oil fractions have led to

$$\kappa = (0.28/d)(1 - 0.00054) \cdot 10^{-3},$$

where κ expresses the conductivity in cgs units (cal cm^{-1} s^{-1} $°C^{-1}$), and d is the density. This equation holds for all oils and is applicable to paraffin waxes above their melting points. An equivalent expression (API, 1983),

$$\kappa = 0.07727 - 4.558 \times 10^{-5},$$

gives κ in Btu h^{-1} ft^{-1} $°F^{-1}$ and is considered adequate for most engineering tasks. It does, however, fail when critical temperatures are approached (and $\kappa \rightarrow$ infinity).

Some representative values reported by Kappelmeyer and Haenel (1974) for the thermal conductivities of rock types, which are important because they govern heat flow from the earth's interior and consequently affect geothermal gradients, are set out in Table 6.2.12.

Thermal Expansion

Thermal expansion coefficients (ζ) of crude oils and oil fractions vary with API gravity. Typical values, measured at 15.5°C and expressed as vol/vol per

TABLE 6.2.13 Variation of Thermal Expansion Coefficient ζ with API Gravity[a]

s.g. (15.5°C)	°API	Mean ζ
0.67	79	0.00144
0.67–0.72	78–65	0.00126
0.72–0.77	64–51	0.00108
0.78–0.85	50–35	0.00090
0.85–0.97	34–15	0.00072
>0.97	<14	0.00036

[a] Bell (1945).

°C, are listed in Table 6.2.13, preceded by the corresponding specific and API gravities.

"Quality" Parameters

Odor

Oils that contain significant concentrations of unsaturates, certain types of nitrogenous compounds, and/or sulfur-bearing moieties such as mercaptans tend to possess a pervasive H_2S-like odor [14]. In contrast, oils mainly composed of light hydrocarbons, containing high proportions of aromatics, or composed of a mix of paraffins and naphthenes possess a sweet gasoline-like odor.

Cloud and Pour Points

Both these "points" represent empirical measurements designed to assess the effects of low temperatures on an oil and to indicate the proportions of waxy matter in it.

As detailed in ASTM D-2500 and D-3117, the cloud point is determined by chilling a 30- to 40-ml oil sample in an ice bath, intermittently shaking it while it gradually regains room temperature, and noting the temperature at which a slight haze shows that paraffin waxes are beginning to solidify.

The pour point is determined by heating an oil sample in a tube to 46°C in order to dissolve all wax, slowly cooling it in a bath held at 11°C (20°F) below the estimated point, and observing the temperature at which the oil will not readily flow when the tube is held horizontally (ASTM D-97). As a rule, pour points are 2–5°C lower than the corresponding cloud points, and range from −56 to 32°C (−70 to +90°F).

Flash and Burning Points

Both parameters are designed to assess the volatility of an oil and thereby identify potential handling hazards posed by it. The flash point is the lowest temperature at which *vapors* from a heated oil will ignite with a short-duration flash when an open flame is passed across the surface of the sample. The burning point is the lowest temperature at which the *oil* ignites.

Aniline Points

Originally defined as the "consolute" temperature—i.e., the lowest temperature at which aniline is completely miscible with another liquid—the aniline point is now restated as the temperature at which *equal volumes* of the liquids are so miscible (ASTM D-611). For hydrocarbon liquids, aniline points increase slightly with the molecular weight, and among hydrocarbons with similar average molecular weights, they increase with paraffin contents. They are therefore occasionally used for roughly estimating the aromaticities of crude oils or oil fractions.

Carbon Residue

Two test procedures, the Conradson method (specified in ASTM D-129) and the Ramsbottom method (detailed in ASTM D-524), afford estimates of the amounts of residual carbonaceous matter (or "coke") formed by destructive distillation of oils that are substantially nonvolatile at atmospheric pressure [15]. Both procedures involve heating the test sample in an inert gas for ~20 minutes until all volatile material has been discharged and the crucible bottom has become red hot. The residue is then cooled, weighed to obtain the coke yield, and finally incinerated at 750 ± 25°C, cooled, and reweighed to determine the ash content, if any (Anon., 1979).

3. THE HEAVY HYDROCARBONS

BITUMINOUS SUBSTANCES

As already noted, data on physical properties of bitumens [16] are frequently suspect. More often than not they relate to single samples, are unrepresentative, and/or are flawed by improper subsampling [17] (which partly accounts for variability among replicated analyses of bulk specimens; Hepler and Hsi, 1989). The lack of a consensual boundary between heavy oils and bitumens permits improper reportage of "bitumen" compositions, and failure, or inability, to

TABLE 6.3.1 Densities of Some Bitumens and Related Heavy Oils[a]

Country	Field	Density (g/cm^3)
Australia	Lakes Entrance	0.959
Brunei	Seria	0.85–0.95
China	Sheng-Li	0.91–0.93
India	Badarpur	0.91–0.97
	Digboi	0.83–0.98
Indonesia	Duri	0.935
	Kulin	0.988
	Sanga Sanga	0.85–0.95
Iran	Cyrus	0.944
Japan	Ishinazaka	0.938
Malaysia	Miri	0.945
Pakistan	Yoya Mair	0.959

[a] Sastri and Khan (1982).

record or cite data with appropriate regard for the extent of molecular association creates uncertainties about densities, viscosities, average molecular weights, etc. As matters stand, information on bitumens and related heavy oils—mostly relating to samples drawn from Alberta's heavy oil and oil sand deposits (see Chapter 3)—allows therefore no more than tentative characterization of bituminous materials. But even that suffices to show that they continue the property trends observed in the progression from light to medium (and heavier) oils.

Densities

The most detailed data relate to Athabasca bitumens, for which densities measured at 15.5°C (60°F) vary between 0.99 and 1.03 g/cm^3. Since these values stand in close accord with densities for bitumens and related heavy oils reported from other jurisdictions (see Table 6.3.1), this range probably reflects true compositional variations.

However, a variation of densities with temperature at atmospheric pressure, which has been generalized from measurements on four Athabasca bitumens, formulated as

$$dT = d_o - 0.62\ T,$$

where d_o = density at 0°C, and claimed to be valid between 0 and 150°C (Bulkowski and Prill, 1978), is unlikely to be other than illustrative [18]. A better indication of responses to changes in temperature and pressure is pro-

TABLE 6.3.2 Illustrative Variation of Bitumen Density with Pressure and Temperature[a]

			Density (g/cm³)					
Bitumen	T (°C)	At:	0.098 MPa	19.6 MPa	39.2 MPa	58.8 MPa	78.4 MPa	98.0 MPa
Borneo	30		1.031	1.040	1048	1.055	1.061	
	45		1.022	1.031	1.039	1.046	1.054	1.060
	60		1.013	1.022	1.030	1.039	1.046	1.053
California	25		1.014	1.023	1.031	1.038	1.045	1.051
	45		1.002	1.011	1.020	1.028	1.035	1.041
	65		0.990	1.000	1.009	1.017	1.025	1.032
Mexico	25		1.042					
	45		1.029	1.039	1.047	1.056	1.063	1.069
	65		1.017	1.027	1.036	1.044	1.052	1.059
Venezuela	25		1.024	1.032	1.040	1.048	1.054	1.061
	45		1.012	1.020	1.029	1.037	1.044	1.051
	65		1.000	1.009	1.018	1.027	1.034	1.041

[a] Pfeiffer (1950).

vided by the data in Table 6.3.2, which relate to "residual" bitumens from which volatile (crude oil) components had been removed by prior distillation.

Viscosities

Like the viscosities of conventional crude oils, the viscosities of bitumens vary directly with density and average molecular weight, and inversely with temperature.

Measurements on a series of Athabasca bitumens whose densities ranged from ~1.000 to 1.029 g/cm³ showed η to increase from ~20 Pa · s to 3000 Pa · s (Ward and Clark, 1950), and similar variations were later reported by others. But the viscosities of *altered* bitumens [19], whose asphaltene concentrations ranged from ~11 to 25% and whose densities might therefore be expected to differ proportionately, could not be meaningfully correlated with asphaltene contents (Waxman *et al.*, 1980).

The characteristic inverse relationship between viscosity and temperature is exemplified by dewatered Peace River (Alberta) bitumens, for which η falls from 80–350 $\times$ 10³ mPa · s at 20°C to 200–500 mPa · s at 100°C and <50 mPa · s at 150°C. However, the presence of dispersed water in a bitumen can promote non-Newtonian flow behavior, and where it does so, it will substantially increase η. For water-bearing Cold Lake (Alberta) bitumens at 25°C/1 atm, values to almost 2 $\times$ 10⁷ mPa · s have thus been reported (Dealy, 1979).

TABLE 6.3.3 Apparent Molecular Weights of Four Alberta Residua[a]

Residua no.[b]	Molecular weight 1	2	3	4
424–525°C fraction				
VPO, average	460	490	420	470
gel permeation				
M_n	290	330	340	290
M_w	370	420	430	380
M_z	450	500	510	450
>525°C fraction				
VPO, average	1600	1325	1295	1250
gel permeation				
M_n	960	950	960	820
M_w	3220	3080	3140	2990
M_z	—not reported—			

[a] Gray (1994).
[b] Residua from 1, Athabasca; 2, Cold Lake; 3, Lloydminster; 4, Peace R.

Molecular Weights

Uncertainties about molecular structure and crystallographic ordering in bituminous substances and related asphaltics (see Chapter 7) inevitably cast doubts on average molecular weights reported for these materials, and there is, indeed, often little consensus about values appropriately assigned a given sample. Table 6.3.3 illustrates this by juxtaposing some molecular weights recorded by vapor-pressure osmometry with data from gel permeation that yield molecular weight *distributions* and therefore allow calculation of

1. A number-average molecular weight from

$$M_n = \sum w_i \Big/ \sum (w_i/M_i)$$

2. A weight-average molecular weight from

$$M_w = \sum (w_i M_i) \Big/ \sum w_i$$

3. A z-average molecular weight from

$$M_z = \sum (w_i M_i^2) \Big/ \sum (w_i M_i)$$

where M_i is the molecular weight of the weight fraction w_i.

Of these parameters, M_n is grossly affected by molecular aggregation, M_z by heavy moieties, and all three by the molecular volume and polarity of the test substance [20].

The data in Table 6.3.3 do not seriously conflict with average molecular weights between 540 and 800 for Athabasca bitumens (Berkowitz and Speight, 1975), or with a value of ~1600 proposed for completely *disaggregated* asphaltene molecules (Moschopedis and Speight, 1975)—but contrast sharply with molecular weights of 2–10 × 10^3 reported from cryoscopy, VPO, and gel permeation (Moschopedis *et al.*, 1976; Yen, 1979; Speight *et al.*, 1985) and with values of 20 × 10^3 or higher inferred from neutron scattering (Ravey *et al.*, 1988) [21]. Correlations of other properties with average molecular weights are therefore at least questionable in their *details*.

Coals

Transitions from liquid and semisolid to solid substances imply changes in physical properties that become the more pronounced the more overlying strata compact the solid during catagenesis.

It might reasonably be supposed that this is true of all kerogens in early stages of catagenesis, and even more so of refractory kerogens and the carbonaceous kerogen residues left after maximum delivery of oils, and if so, kerogens and kerogen-like substances would represent the bridge between bituminous substances and coal. But little other than its substantial insolubility is currently known about the physical properties of kerogen, and coals appear therefore to be sharply differentiated from oils and asphaltics without necessarily (or actually?) being so.

Porosity

A dominant property of coal is an intricately structured pore system, which endows it with the characteristics of a molecular sieve and creates a large *internal surface* that governs its responses to almost all facets of production, preparation, and processing.

Expressed as gross void volumes, porosities range from >25% among coals with <82% C to <2% among bituminous coals with 89–91% C and then increase again among anthracites (Fig. 6.3.1; [22]). However, efforts to measure the surface areas (S) associated with these properties have yielded inconsistent results (see Berkowitz, 1985, 1994) and permit no more than a broad statement that S falls from 100–200 m^2g^{-1} or more among lignites and subbituminous coals to 20–50 m^2g^{-1} among mvb coals before increasing to 100 m^2g^{-1} or more among anthracites.

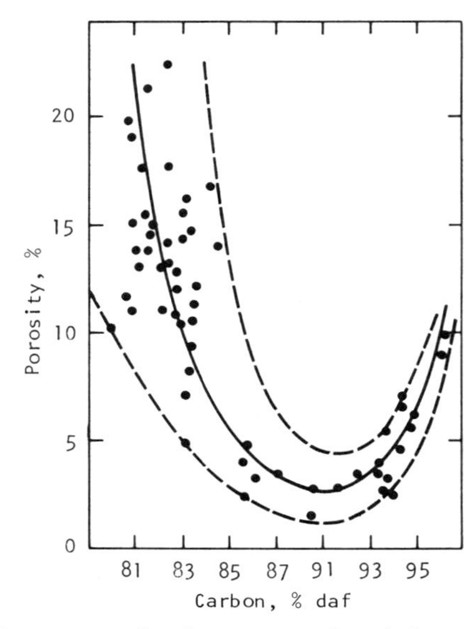

FIGURE 6.3.1 The variation of coal porosities with rank (King and Wilkins, 1944).

Uncertainties about *S* arise from inherent difficulties in measuring this parameter [23] and, more particularly, from a preponderance of micropores [24]: pore size distributions indicate that >90% of the total internal surface, or 50–80% of the total void space, accrue from a capillary system in which <40-nm pores are linked by 5- to 8-nm passages.

Density

The extent and structure of the pore system requires differentiation between

in-place densities, which represent weight-to-volume ratios of undisturbed coal in its seams;
bulk densities, which reflect weight-to-volume ratios of broken coal and are consequently functions of particle size and packing density;
apparent specific densities, which are determined by liquid displacement, but influenced by the molar volume of the liquid;
absolute densities, the putative true densities that have been mostly measured by displacement of helium [25].

Of these parameters, the first is *not* corrected for moisture and mineral matter, and corresponds to the density of crude oil in its pristine reservoir. The second is likewise uncorrected. However, the apparent and absolute densities, which

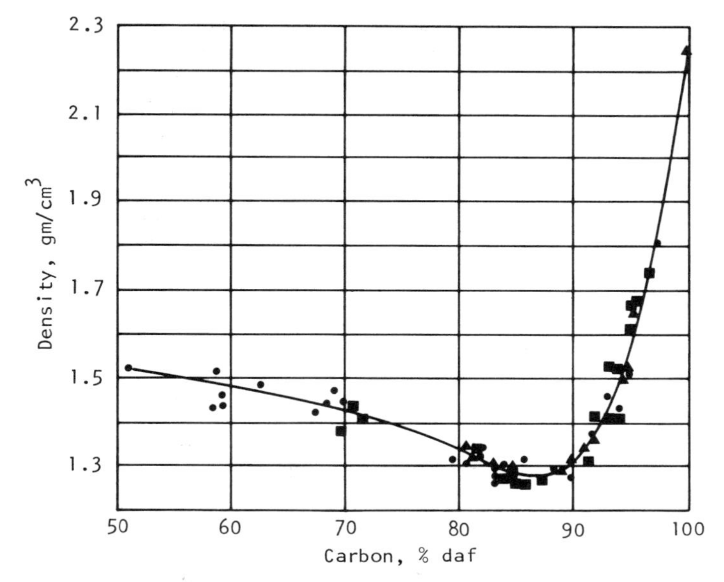

FIGURE 6.3.2 The variation of "true" coal densities with rank. Measured in helium [▲, ■] (Franklin, 1949; van Krevelen, 1961) and water [●] (Dulhunty and Penrose, 1951).

purport to be the densities of the organic material *per se* and correspond to the density of a crude oil at a stock tank [26], must be corrected for H_2O and mineral matter by

$$d_{corr} = [d_a d_{exp}(100 - A)]/[100\ d_a - d_{exp}A],$$

where A and d_a denote the amount (%) and density of ash obtained by combustion of the coal [27], and d_{exp} is the observed coal density. This latter measurement requires some care: useful data can only accrue from displacement of water, benzene, *n*-hexane, etc., if substantially dust-free $<$3-mm coal is used, and at least 12 h are allowed for equilibration.

Figure 6.3.2 shows how the absolute density varies with coal rank. But this is, in effect, a mean that corresponds to the average density of an oil; as with the class components of a bitumen (see Chapter 5), each component maceral of coal is characterized by its own rank-dependent density. Table 6.3.4 illustrates this with densities of some vitrinites, exinites, and inertinites, all measured by displacement of methanol.

Mechanical Properties

Few coals are sufficiently homogeneous and cohesive to allow them to be fashioned into the substantially homogeneous test pieces required for conven-

TABLE 6.3.4 Apparent Densities of Macerals[a]

Maceral	Carbon (% daf)	Density (g/cm^3)
Vitrinite	83.5	1.345
	85.7	1.334
	88.4	1.317
	88.8	1.368
Exinite	85.5	1.201
	87.4	1.213
	89.1	1.288
	89.3	1.347
Inertinite	86.8	1.463
	88.0	1.415
	89.6	1.414
	89.8	1.413

[a] Kröger and Badenecker (1957).

tional measurements of strength and/or hardness, and these parameters, formally analogous to the viscosity of a fluid and of paramount practical importance, are therefore generally assessed by simulations.

Problems in determining the *in situ* strength of coal—a parameter that bears on mine design and operation [28]—are underscored by the fact that random internal cracks cause measured compressive strengths to vary by as much as ±30%. But since the probability of survival of a cube with side n tends to vary inversely with n (Evans and Pomeroy, 1958), the size dependence of strength has been statistically described by a generalized expression of the form

$$C(n) = e^{n \ln C}$$

(Berenbaum, 1961, 1962), where C and $C(n)$ are probabilities of survival of cubes with unit side and side n, or by

$$s^* = k/\sqrt{n}$$

(Brown and Hiorns, 1963), where s^* is the compressive strength of a cube with side n, and k is a coal-characteristic constant that is very nearly independent of the size of the test specimen (Hustrulid, 1976) and is now often quoted in lieu of a specific value for s^* [30].

Hardness, which is also deemed to reflect abrasiveness, has been calculated by means of

$$H = L/A = (2L/d^2) \sin(\theta/2)$$

from permanent indentations produced in a coal surface by a Vickers or

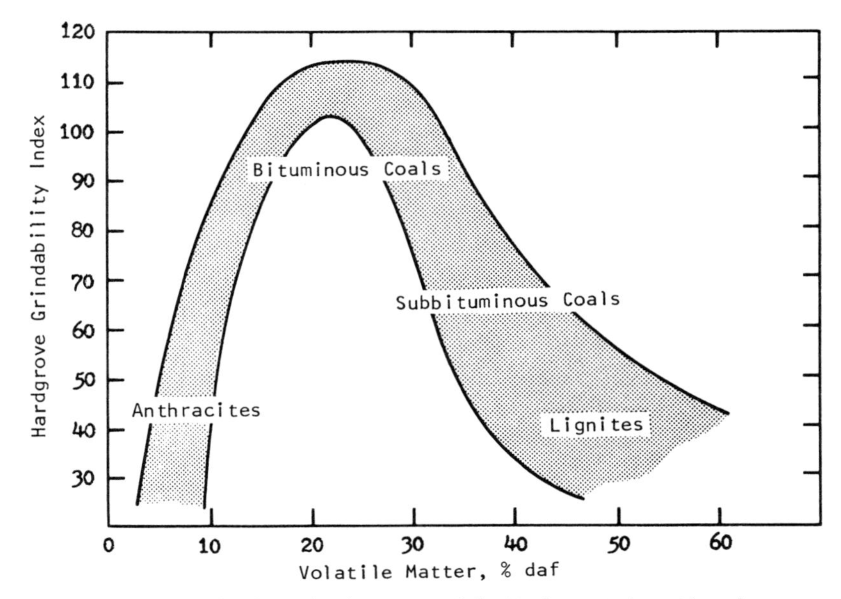

FIGURE 6.3.3 Generalized variation of the Hardgrove index with rank.

Knoop scleroscope. In this expression, L is the applied load, A the area of the impression (mm^2), d the diameter of the impression (mm^2) and θ the angle between opposite faces of a pyramid (136°).

However, such measurements command little practical interest, and more has been gained from simulations—in particular from use of the ASTM drop shatter test (D-440) and/or the ASTM tumbler test (D-441) [31]. Both furnish a measure of hardness or strength by defining x and y as average screen sizes of the sample before and after the test, and expressing the results as *size stability* ($s = 100\ y/x$) and *friability* ($f = 100 - s$).

Where interests center on energy expenditures incurred by comminution, recourse is had to the Hardgrove test (ASTM D-409), which measures the amount of comminuted material produced by milling under standard conditions [32]. Since the "grindability index" (GI) is defined as

$$GI = 13 + 6.93\ w,$$

where w is the weight of -200 mesh ($<76\ \mu m$) material produced during the test, energy expenditures are inverse functions of GI and can be directly correlated with rank dependence (Fig. 6.3.3).

Related elastic constants have been less closely studied, but available data show some distinctive trends. Values for the Young's (dynamic) modulus E

TABLE 6.3.5 Variation of Young's Modulus with Rank[a]

% Carbon (daf)	$E \times 10^{10}$ dyn cm^{-2} [b]	
	a[c]	b[c]
81.5	4.50	
85.0	5.27	
89.0	4.53	
91.2	4.65	
92.5	5.19	
93.4	5.69	4.89
95.0	8.69	4.82
96.0	14.33	6.61

[a] Schuyer *et al.* (1954).
[b] In the SI system, 1 dyn = 10 mN.
[c] Parallel and (*b*) perpendicular to the bedding plane; single values indicate absence of anisotropy.

have been estimated from compression and bending of rods (Inouye 1951; Morgans and Terry, 1958) and have also, perhaps more reliably, been calculated from sound velocity measurements via

$$v = (E/d)^{0.5},$$

where d is the density of the test specimen (Schuyer *et al.*, 1954). These data show E to be virtually insensitive to rank and specimen orientation up to 92–93% carbon, but to be markedly anisotropic among anthracites (see Table 6.3.5).

The same pattern is exhibited by the shear modulus G ($\sim 0.4E$) and by Poisson's ratio π, which was estimated to be fairly constant at 0.35 up to 92–93% C (van Krevelen, 1981), but was actually found to vary between 0.37 and 0.44 (Morgans and Terry, 1958).

The compressibility k of "whole" coal has been reported as $\sim 2 \times 10^{-11}$ cm^2/dyn up to 92–93% C, but varies significantly among different petrographic components (see Table 6.3.6). A more recent study, which found marked hysteresis between compression and decompression (Suuberg *et al.*, 1995), may, however, require some revision of these data.

Thermal and Electrical Properties

Partly to meet "needs to know" for coal processing, and partly as means for elucidating molecular structure, considerable attention has been directed to thermal and electrical properties of coal.

TABLE 6.3.6 Compressibility of Coal and Coal Macerals[a]

Carbon (% daf)	Compressibility κ (10^{-11} cm²/dyne)			
	Whole coal	Vitrinite	Exinite	Micrinite
81.5	2.07			
83.5		2.62	4.58	1.71
85.0	1.75			
85.7		2.36	3.83	1.94
88.4		2.09	2.92	1.83
88.8		1.94	1.83	1.66
89.0	2.07			
91.2	1.96			
92.5	1.84			
95.0	1.27			
96.0	1.02			

[a] Data for "whole" coal from Schuyer *et al.* (1954); data for coal macerals from Kröger and Ruland (1958).

Specific Heats

Normally measured at constant pressure and then designated by C_p, specific heats are conveniently determined for dry coal [33] by the method of mixtures (ASTM C-351) or with a Bunsen ice calorimeter and, measured at 25°C, have been reported to decrease linearly from ~1.79 J g^{-1} at 70% C to 1.49 J g^{-1} at 90% before falling more rapidly to 1.19 J g^{-1} at 95% C. (For graphite at 25°C, C_p = 0.691 J g^{-1}.) An acceptable approximation is given by

$$C_p = 0.242(1 + 0.008\% \text{ vm}),$$

where vm is the volatile matter content. However, better agreement has been claimed from evaluating C_p from an expression of the form

$$C_p = fR/2M,$$

where f is the number of vibrational degree of freedom for each atom, R the gas constant, and M the molecular weight. Assuming each C and H atom on average to contribute with only one degree of freedom [34], this yields values within ±5% of experimentally observed heats (van Krevelen, 1961).

Of interest in this connection, because they exemplify the chemical linkages between heavy hydrocarbons, are some measurements of specific heats by a heat flux technique at 300 K (Lindberg *et al.*, 1985): for bitumen from Utah's NW Asphalt Ridge (see Chapter 3), this returned a value of 1.85 J g^{-1}, in excellent agreement with 1.79 J g^{-1} for a more highly aromatized lignite with ~70% C (and ~60% C_{ar}). But the significance of this accord is tempered by other work that evaluated C_p of powdered coal at 25°C by differential scanning calorimetry (Vargha-Butler *et al.*, 1982), and led to substantially lower values—0.98 ± 0.02, 1.03 ± 0.03, and 1.04 ± 0.03 for three coals in the 82–88% C range—as well as to the suggestion that observed specific heats may depend on particle size and sample preparation [35].

Thermal Conductivities

Thermal conductivities (κ), which can be measured by either of the two methods specified in ASTM C-177 and C-518, have been reported to range from 3.3×10^{-4} to 8.6×10^{-4} cal s^{-1} cm^{-1} °C^{-1}. Systematic variation with coal rank is, however, obscured by variously composed and disseminated mineral matter, and only fair correlation with uncorrected densities could be noted (Fritz and Diemke, 1941). The porosity of coal also makes κ depend intimately on the nature of the gas or vapor that suffuses the coal.

Thermal Diffusivity

The thermal diffusivity α is numerically equal to the proportionality constant in Fourier's second law equation of unsteady-state conduction,

$$dT/dt = \alpha \nabla^2 T,$$

and related to C_p, κ, and density d by

$$\alpha = \kappa / C_p d,$$

from which it is conveniently evaluated [36].

Thermal Expansion

Isolated measurements of the linear thermal expansion coefficient have been reported by Bangham and Franklin (1946). For two bituminous coal artifacts with 82.4 and 89.7% C, this increased from 32×10^{-6}/°C at 30°C to 58×10^{-6}/°C at 330°C, and for an anthracite monolith with 94.2% C it changed little, but depended on orientation: over the 30–330°C interval, it rose from $\sim 27 \times 10^{-6}$/°C to 29×10^{-6}/°C when measured perpendicularly to the bedding plane, and from 15×10^{-6}/°C to 18×10^{-6}/°C when measured parallel to it.

Virtually identical results, but showing pronounced anisotropy in coals with >87% C, were subsequently recorded in a study of coals with 85–96% C between 20 and 40°C (Schuyer and van Krevelen, 1955; Joy, 1955).

Electrical Conductivity

Excepting anthracites—which are semiconductors whose conductivities can be measured with a Wheatstone bridge, an ammeter–voltmeter system, or,

TABLE 6.3.7 Resistivity of Anthracites[a]

Carbon (% daf)	ρ^b (Ω/cm)	Δe (eV)
93.7	4.01×10^7	0.5
94.2	(a) 6.09×10^4	0.34
	(b) 4.97×10^4	
95.0	(a) 1.71×10^3	0.27
	(b) 0.70×10^3	
96.0	(a) 6.43	0.17
	(b) 3.73	

[a] Schuyer and van Krevelen (1955).
[b] (a) perpendicular, (b) parallel to bedding plane.

when resistivities ($\rho = 1/\kappa$) exceed 10^9 Ω, with a fluxmeter—coals are or behave as electrical insulators. However, since κ is extremely sensitive to ambient conditions and greatly affected by moisture, it is generally preferable to measure the electron excitation energy Δe from the temperature variation of ρ:

$$1/\kappa = \rho = \rho_o \exp(\Delta e/2kT),$$

where ρ_o is the resistivity at $1/T = 0$, k is the Boltzmann constant (1.3805×10^{-17} μJ/K) and T is the temperature of measurement. Δe can then be obtained from a plot of ρ vs $1/T$.

Typical values for ρ and Δe of monolithic specimens (see Table 6.3.7) show that ρ becomes increasingly dependent on orientation when C > 93–94%, but that Δe displays no such anisotropy. Conduction is therefore believed to be associated with charge transfer across aromatic carbon platelets in the coal (see Chapter 7).

NOTES

[1] Coal is something of an exception: many of its physical properties were studied in the expectation that they would provide information about its chemical structure. That expectation has remained largely unfulfilled. But the data have contributed much to a data base that allows a wide range of physical properties to be broadly inferred from knowledge of one or two *analytical* parameters.

[2] "Oil equivalents" are generally only used for estimating the total energy reserves of a region or a society's total demand for energy.

[3] By definition, such properties do not depend on the *quantity* of material.

[4] See, e.g., ASTM D-86, D-216, D-285, D-447, and D-2892, which specify atmospheric-pressure distillation methods, and D-1160, which details distillation under reduced pressure.

[5] The Rankin scale is an absolute temperature scale in which °R = O at absolute zero, and the ice point of water (at 1 atm) lies at 459.67°R. The degree increments are the same as those of the Fahrenheit scale.

[6] This variation arises in part from increasing electron correlation and consequent weak electrostatic (London-force) interactions between larger molecules.

[7] The API scale is arbitrary, but simplifies the construction of hydrometers because it allows linear calibration of the stems.

[8] The dependence of API gravities on reservoir depth lends importance to estimates of downhole temperatures in a formation, and these can be calculated with fair accuracy from

$$T_f = T_o + [g_{gr} d_f/100],$$

where T_o is the surface temperature (°F), T_f the formation temperature (°F), d_f the formation depth (ft), and g_{gr} the geothermal gradient (°F/100 ft).

[9] Lubricating oils are in fact *classified* in terms of viscosity. A case in point is the American Society of Automotive Engineers' (ASAE) scheme, which defines oils in terms of Saybolt universal seconds.

[10] A more complete tabulation of viscosity data for heavy oils and bitumens is contained in *Thermodynamic and Transport Properties of Bitumens and Heavy Oils*, AOSTRA Edmonton, AB (July 5, 1984), but this report acknowledges that sources of data, origin and/or quality of samples, and the reliability of measurements are often unknown.

[11] In most cases, optimum concentrations do not exceed amounts corresponding to the monolayer capacities of the exposed surfaces. Further additions have little effect.

[12] At wavelengths between 254 and 365 nm, the human eye can detect one part of oil in 10^5 parts of CCl_4.

[13] Δn is usually obtained from the difference between n at $\lambda = 6.563$ and 4.861 mm^{-7} (the C and F lines in the H spectrum).

[14] Domestic gas supplies possess a similar odor that is due to traces of H_2S which are added as a safety measure to help detect leakage.

[15] The Conradson method is generally preferred for bitumens.

[16] The voluminous data for variously processed tars and asphalts are not relevant in this context.

[17] Such errors are particularly acute when sampling oil sands, which are usually characterized by very heterogeneous distribution of bitumen, water, and solids. Wallace and Kratochvil (1984) have stressed that for bulk samples in the order of 325 kg, a minimum of 20 120-g subsamples need to be prepared and analyzed if sampling errors are to be smaller than 3× the analytical standard deviation.

[18] Analogous, equally limited formulae for some Venezuelan heavy oils thus use

$$d = 1.0227 - 0.000635\,T$$
$$d = 0.9937 - 0.000468\,T$$
$$d = 0.9967 - 0.00072\,T, \text{etc.}$$

[19] These were samples of *produced* Peace River (Alberta) bitumen collected during the backflow stage of a steam simulation pilot test. Separate samples were taken downhole, from production rods and tubing, and from a field separator. The study sought to elucidate the impact of temperature on asphaltenes in the bitumen.

[20] Since the analytical standards most often used for calibrating a GPC column are polystyrenes whose molecular volume/mass relationship could differ substantially from that characterizing the test substance, molecular weight data from GPC can be as doubtful as molecular weights obtained by other means. Reasonably reliable results depend on selecting calibration compounds that, as far as possible, reflect what is known or can be legitimately inferred about the test sample.

[21] Molecular weights inferred from neutron scattering are interpreted in terms of flat sheets with 6- to 20-nm diameter and an average 0.6- to 0.8-nm thickness, and thus point, much like the solvent dependency of average aggregate sizes, to extensive polydispersity (see Chapter 7).

[22] The form of this relationship reflects progressive compaction during maturation to low volatile bituminous (lVB) rank, and thereafter structural changes that allow crystallographic reordering and gradual development of pseudo-graphitic "crystallites." Significantly, density/rank relationships follow a similar path, and coals become increasingly anisotropic as carbon contents move beyond ~87–88%.

[23] Measurements of surface areas of finely pored solids by gas or vapor sorption techniques commonly have error limits of ±30–40%, and similar, if not larger, errors attend measurements by small-angle X-ray scattering.

[24] A definition of pore sizes by the International Union of Pure and Applied Chemistry (IUPAC) shows the diameters of *macropores* as >500 nm, of *mesopores* as 20–500 nm, of *micropores* as 8–20 nm, and of *submicropores* as <8 nm. However, in publications predating 1972, all <20-nm pores are referred to as micropores.

[25] The use of helium was prompted by its small effective diameter (0.18 nm) and by the assumption that it is not appreciably sorbed. On the further supposition, later shown to be questionable, that coals contain few blind pores, it was therefore expected to almost completely suffuse a sample. In any event, as shown in Figure 6.3.2, *water* to which a trace of wetting agent has been added will yield densities that are virtually identical with those obtained with He.

[26] That is, densities of oils freed of formation waters, substantially stripped of dissolved C_1–C_5 hydrocarbon gases, and tanked at STP.

[27] This expression requires separate determination of ash densities and recognizes that densities of mineral matter and of the ash derived from it differ little. Where d_a cannot be determined, it may be assumed to lie between 2.7 and 3.0 g/cm^3.

[28] Among other matters, such data make it possible to estimate the bearing strengths of coal pillars in underground room-and-pillar mining.

[29] This is consistent with the formal Rosin–Rammler size distribution,

$$100 - U(i) = 100 \exp(-bin),$$

where $U(i)$ is the percentage of undersize material, and b, i, and n are constants.

[30] Reporting a value for k avoids the ambiguities that commonly attach to specific strength data. The SI units for k are Pamr, with r normally lying between 0 and 0.5.

[31] A useful alternative to the ASTM tumbler test presents itself in the Micum drum test (ISO 1959, 1974).

[32] The Hardgrove test employs a mill in which a 50-g sample of 16–30 mesh (≡0.6–1.18 mm) coal is crushed in the course of 60 revolutions of eight steel balls, each loaded by $64\pm\frac{1}{2}$ lbs (29 ± 0.22 kg). The procedure is loosely based on Rittinger's law, which posits that work done in comminution is proportional to new surface created. But this is questionable if the solid contains internal flaws—and invalidated when comminution produces small particles that, like coal, deform plastically and reaggregate in tightly cohering clusters (Boddy, 1943; Bangham and Berkowitz, 1945, 1947). Because lignites and subbituminous coals are much softer than more mature coals, the variation of Hardgrove indices with rank can, in fact, only be understood if it is assumed that comminution involves two mutually opposed processes, viz., (i) progressive size diminution, and (ii) concurrent pronounced plastic deformation and reaggregation of small particles. In such circumstances, even slightly prolonged grinding could prove counterproductive.

[33] This is, in fact, imperative for comparison of different coals. Heat capacities of a *moist* coal can then be calculated by assuming that the heat capacities of dry coal and water are additive.

[34] This assumption reflects the fact that the specific heat of graphite corresponds to the equipartition value of one vibrational degree of freedom. But some assumptions are also necessary with respect to M.

[35] These observations do, however, still need to be validated. So, for that matter, does the calorimetric technique itself.

[36] A procedure for directly determining α is detailed in ASTM C-714-73, but is difficult: It requires exposing a coal disc to a short high-intensity thermal pulse from a flash lamp and recording the temperature on the other side of the disc. Assuming that there is no heat loss from the edge of the disc, Fourier's second law becomes

$$\delta T/\delta t = \alpha \delta^2 T/\delta x^2,$$

and this can be integrated to give the half-time needed for the temperature at the rear face to rise to half its maximum value; α is then given by

$$\alpha = 0.139\ L^2/t^{1/2},$$

where L is the thickness of the disc.

REFERENCES

Adamson, A. W. *The Physical Chemistry of Surfaces,* 1976. New York: Wiley & Sons.

Amosov, G. A. *Geokhimicheskii Sbornik, Trudy VNIGRI, New Series #5,* 2–3 (1951); *Engl. Transl., Israel Progr. for Sci. Transl.* (1965).

Anon. *Syncrude Analytical Methods for Oil Sand and Bitumen Processing,* 1979. Edmonton, AB: AOSTRA.

American Petroleum Institute (API). *Technical Data Book—Petroleum Refining,* 4th ed., 1983. Washington, D.C.: API.

Bangham, D. H., and N. Berkowitz. *Bit. Coal Res.* **2**, 139 (1945); *Research (London)* **1**, 86 (1947).

Bangham, D. H., and R. E. Franklin. *Trans. Faraday Soc. B 42,* 289 (1946).

Bell, H. S. *American Petroleum Refining,* 3rd ed., 1945. New York: Van Nostrand.

Berenbaum, R. *J. Inst. Fuel (London)* **34**, 367 (1961); **35**, 346 (1962).

Berkowitz, N. *The Chemistry of Coal,* 1985. Amsterdam: Elsevier.

Berkowitz, N. *Introduction to Coal Technology,* 2nd ed., 1994. New York: Academic Press.

Berkowitz, N., and J. G. Speight. *Fuel* **54**, 138 (1975).

Boddy, R. G. H. B. *Nature (London)* **151**, 279 (1943).

Brown, R. L., and F. J. Hiorns. In *Chemistry of Coal Utilization* (H. H. Lowry, ed.), Suppl. Vol., p. 135, 1963. New York: Wiley.

Bulkowski, P., and G. Prill. *Internal Rept., Alberta Research Council,* Feb. 1978.

Chatenay, D., D. Langevin, and J. Meunier. *J. Dispersion Sci. Technol.* **3**, 245 (1982).

Dealy, J. M. *Can. J. Chem. Eng.* **57**, 677 (1979).

Dulhunty, J. A., and R. E. Penrose. *Fuel* **30**, 109 (1951).

Fenske, M. R., F. L. Carnahan, J. N. Breston, A. H. Caser, and A. R. Rescorla. *Ind. Eng. Chem.* **34**, 638 (1942).

Franklin, R. E. *Trans. Faraday Soc.* **45**, 274 (1949).

Fritz, F. S., and H. Diemke. *Chem. Abstr.* **35**, 4939 (1941).

Girault, H. H. J., D. J. Schiffrin, and D. D. V. Smith. *J. Coll. Interface Sci.* **101**, 257 (1984).

Gray, M. R. *Upgrading Petroleum Residues and Heavy Oils,* 1994, p. 34. New York: Marcel Dekker.

Hepler, L. G., and Chu Hsi (eds.). AOSTRA Technical Handbook 1 on Oil Sands, Bitumens and Heavy Oils, AOSTRA Tech. Publ. Ser. #6, 1989. Edmonton, AB: AOSTRA.
Hustrulid, W. A. *Rock Mech.* **8**(2), 115 (1976).
Inouye, K. *J. Coll. Sci.* **6**, 190 (1951).
International Standards Organization (ISO). *Tech. Comm. #27* (1959); *TC 27/SC 3, Document #13* (1974).
Joy, A. S. *Ann. Repts., Fuel Research Stn., Greenwich, Britain* (1955).
Kappelmeyer, O., and R. Haenel. *Geothermics,* 1974. Berlin/Stuttgart: Borntraeger.
King, J. G., and E. T. Wilkins. In "Proc. Conf. Ultrafine Struct. Coals and Cokes," p. 46. BCURA, London (1944).
Kröger, C., and J. Badenecker. *Brennst. Chem.* **38**, 82 (1957).
Kröger, C., and H. Ruland. *Brennst. Chem.* **39**, 1 (1958).
Kouzel, B. *Hydrocarbon Process.* **44**(3), 120 (1965).
Levorsen, A. I. *Geology of Petroleum,* 2nd ed., 1967. San Francisco: W.H. Freeman & Co.
Lindberg, W. R., R. R. Thomas and R. J. Christensen. *Fuel* **64**, 80 (1985).
Maxwell, J. B. *Data Book on Hydrocarbons,* 1950. Princeton, NJ: Van Nostrand.
McCullough, J. J. *Oil Gas J.* **54**(12), 201 (1955).
Morgans, W. T. A., and N. B. Terry. *Fuel* **37**, 201 (1958).
Moschopedis, S. E., and J. G. Speight. *Fuel* **54**, 210 (1975).
Moschopedis, S. E., J. F. Fryer, and J. G. Speight. *Fuel* **55**, 228 (1976).
Pederson, K. S., Aa. Fredenslund, and P. Thomassen. *Properties of Oils and Natural Gas,* 1989. Houston: Gulf Publ. Co.
Pfeiffer, J.Ph. (ed.). *The Properties of Asphaltic Bitumen,* 1950. New York: Elsevier.
Ravey, J. C., G. Ducouret, and D. Espinat. *Fuel* **67**, 1560 (1988).
Sastri, V. V., and M. H. Khan. *Proc. 2nd Int. Conf. Hvy. Crude and Tar Sands, Caracas, Venezuela, Feb. 1982.*
Schuyer, J., H. Dickstra, and D. W. van Krevelen. *Fuel* **33**, 409 (1954).
Schuyer, J., and D. W. van Krevelen. *Fuel* **34**, 345 (1955).
Speight, J. G., D. L. Wernick, K. A. Gould, R. E. Overfield, B. M. L. Rao, and D. W. Savage. *Rev. l'Inst. Français Petrole* **40**, 52 (1985).
Suuberg, E. M., S. C. Deevi, and Y. Yun. *Fuel* **74**, 1522 (1995).
Ward, T. E., and K. A. Clark. *Research Council of Alberta Rept. #57* (1950).
Wallace, D., and B. Kratochvil. *AOSTRA J. Res.* **1**, 31 (1984).
Waxman, M. H., C. T. Deeds, and P. J. Closmann. *Paper #9510, 55th Ann. Tech. Mtg. SPE, Dallas, TX,* Sept. 1980.
van Krevelen, D. W. *Coal,* 1st ed., 1961. Amsterdam: Elsevier; 2nd ed., 1981. Amsterdam: Elsevier.
Vargha-Butler, E. I., M. R. Soulard, H. A. Hamza, and A. W. Neumann. *Fuel* **61**, 437 (1982).
Yen, T. F. *ACS Div. Petrol. Chem., Prepr.* **24**(4), 901 (1979).

CHAPTER 7

The Molecular Structure of Heavy Hydrocarbons

Ongoing refinement of classic chemical methods for elucidating molecular structures and, since the 1950s, rapid development of new spectroscopic techniques for such studies have enabled important progress in defining molecular configurations of asphaltics, kerogens, and coals. What can now be said about the structure of these substances is still far from definitive. Much is, in fact, little better than speculative, and as few of the relevant data bases are entirely unequivocal, some of the graphic constructs put forward to illustrate hypothesized configurations reflect personal preferences. However, drawing on information from diverse lines of inquiry has permitted formulation of structure "elements" and statistically favored "average configurations"; and that, given the compositional and structural heterogeneity of the heavy fossil hydrocarbons, is perhaps all that can reasonably be expected.

How plausible current formulations are can, however, only be judged in light of the data from which they were assembled. It may therefore be useful to summarize briefly the essential features of the experimental methods by which these data were generated—and, equally important, to note their limitations.

1. SOURCES OF INFORMATION

Fractionation

Investigations of complex mixtures generally require some form of fractionation of the material under study to resolve it into more tractable or better-defined components, and depending on what is to be resolved, simple distillation, solvent extraction, chromatography and/or sorption can be used for that purpose. But although such fractionation is *prima facie* trivial, downstream data relating to the separated components can be seriously prejudiced by inappropriately conducted fractionation. Separation of asphaltenes from bitu-

TABLE 7.1.1 Carbon/Hydrogen Contents of Some Coals[a]

Carbon (% daf)	H/C	Atoms/ 100 atoms C	H
80.8	0.767	52.8	40.5
81.5	0.788	52.4	41.3
82.9	0.784	53.3	41.8
83.1	0.802	52.6	42.2
85.5	0.725	56.0	40.6
86.5	0.772	54.4	42.0
87.1	0.765	54.4	41.6
89.3	0.721	56.7	40.9
89.8	0.613	60.5	37.1

[a] Abridged from Table 5.2.16 (Chapter 5); in atom weight percent.

mens is thus greatly affected by the nature and volume of the solvent used to precipitate asphaltenes from extract solutions, and solvents used for extraction of nonvolatile hydrocarbons such as coal often peptize them, and consequently deliver colloidally dispersed instead of molecularly dissolved material into the extractor. In either case, procedural shortcomings critically affect subsequent measurements of molecular weight and other properties and can seriously jeopardize inferences about the structure of the extract [1].

Analysis

Elemental Composition

Classic elemental analyses offer useful preliminary information when expressed as atom-weight percentages. Table 7.1.1 illustrates this with some C/H data that, by comparison with specific CH compounds, disallow simple alkanes, catacondensed aliphatics, and pericondensed aromatics as possible components, and suggest that *statistically preferred* skeletal configurations are based on

(i) dibenzyl, 1,2,3,4-tetrahydronaphthalene, or di-, tri-, and tetraphenylmethanes,
(ii) catacondensed three- or four-ring aromatics with short side chains,
(iii) pericondensed four-ring or larger aromatic systems with relatively long aliphatic chains, and/or
(iv) pericondensed aliphatic systems like, or related to, polyamantanes

Such tentative inferences can be refined by eliminating all seemingly forbidden configurations with, e.g., $1.0 < H/C < 0.5$, and comparing the heats of combustion of theoretically allowed structures, expressed as ΔQ per C atom, with that of the substance under study (Berkowitz, 1985).

Chromatographic Analysis

The constituents of liquid hydrocarbons can be identified by relatively simple chromatographic separation. Hydrocarbon mixtures $< C_{10}$ can thus be conveniently analyzed by using capillary columns (or packed and capillary columns) and recording outputs with a thermal or flame ionization detector (Osjord and Malthe-Sorenssen, 1983; Osjord *et al.*, 1985), and for liquids with up to C_{20} moieties, high-pressure liquid chromatography (HPLC) or a version of automated "mini-distillation" based on ASTM D-2892 (Pederson *et al.*, 1989) can be employed. Identification of chromatographically separated components does, however, require careful calibration of the system with known standards that reflect what is *suspected* to exist in the test liquid.

Functional Group Analysis

The disposition of heteroatoms in molecular configurations can often be determined from concentrations of O-bearing and other functional groups (Ignasiuk *et al.*, 1975). Examples of such determinations are:

1. [COOH] by ion exchange against $(CH_3COO)_2Ca$ and titration of free CH_3COOH formed by

$$2RCOOH + (CH_3COOH)_2Ca \rightarrow (RCOO)_2Ca + 2CH_3COOH$$

or by catalyzed thermal decarboxylation and gravimetric measurement of CO_2.

2. [OH] by acetylation with acetic anhydride, followed by hydrolysis with $Ba(OH)_2$, acidification with phosphoric acid, heating, and titration of CH_3COOH. Reaction with diazomethane, saponification to remove —COOH, and measurement of $[OCH_3]$ by a modified Zeisl procedure (Blom, 1960) will furnish *phenolic* [OH], and simple subtraction from total [OH] yields *alcoholic* [OH].

3. [=CO] by oxime formation (Blom *et al.*, 1957) or interaction with phenylhydrazine hydrochloride (Ihnatowicz, 1952); *reducible* [=CO] by reduction with titanous hydrochloride, and *nonreducible* [=CO] by reaction with dinitrophenyl hydrazine (Kröger *et al.*, 1965).

4. $[RCONH_2]$ by refluxing with NaOH and measuring NH_3 generated by

$$RCONH_2 \rightarrow RCOONa + NH_3.$$

5. [R—CHO] or [R—CO—R′] by refluxing with sodium borohydride ($NaBH_4$) in dilute aqueous NaOH and measuring infrared absorption by the reaction product near 1700 cm^{-1}.

However, of these determinations, only [O_{COOH}] and [O_{OH}] are considered to be accurate; the others merely provide acceptable estimates.

Molecular Weights

Reliable data allow corroboration of molecular configurations inferred from other measurements, plausible speciation of oils, and conversion of weight fractions (from distillation or chromatographic measurements) into mole fractions. But there is often disagreement about the average molecular weights that should be assigned to components of heavy hydrocarbons such as bitumens and residua.

Conventional crude oils, oil fractions, and residua from conventional crudes are characterized by molecular weights between 100 and 500—a range that encompasses C–C_{35} hydrocarbons and can be accurately covered by measuring freezing point depressions; extensions to C_{35}+ are possible by making use of low-pressure distillation at ~2 mm Hg. However, for bitumens and other asphaltics composed of very large and complex moieties, it is generally considered preferable, as well as more convenient, to use vapor-pressure osmometry (VPO) or gel permeation chromatography (GPC).

VPO (see ASTM D-2053; [2]) compares the vapor pressure of a solution of the test sample with the vapor pressure of the pure solvent, and if that comparison is made with a thermistor, the vapor pressure reductions (Δp) caused by the solute are expressed as temperature differences. A molecular weight determined in this manner is a molar or number average (M_n), which depends on the polarity of the solvent and is affected by aggregation of solute molecules [3].

For GPC measurements, a dilute solution of the test substance is passed through a chromatographic column, and elution times are compared with those of analytical standards whose molecular weights are accurately known. From the molecular weight distributions one can then, as already noted in Chapter 6, calculate:

1. the number-average molecular weight $M_n = \Sigma w_i / \Sigma(w_i/M_i)$;
2. the weight-average molecular weight $M_w = \Sigma(w_i M_i)/\Sigma w_i$; and
3. the z-average molecular weight $M_z = \Sigma(w_i M_i^2)/\Sigma(w_i/M_i)$

where M_i is the molecular weight of weight fraction w_i. But here, too, M_n can be grossly affected by molecular aggregation, M_z by heavy moieties, and all three parameters by the molecular volume and polarity of the test substance [4].

That molecular weights are now commonly determined by VPO or GPC, which are sometimes thought to be superior to measurements of freezing- or boiling-point changes, does not therefore necessarily guarantee greater accuracy or reliability.

Pyrolysis

Coupled with mass spectroscopy in techniques in which volatilized matter released from a material is directly transferred to a mass spectrometer (Durfee and Voorhees, 1985; Baldwin *et al.*, 1987), pyrolysis can yield much information about aromatic units in molecular core configurations and incidentally also validate inferences from oxidative degradation (see later discussion). Aromatics definitively identified in the pyrolysis products of aromatic pitches and coals have thus also been detected in product slates from $Na_2Cr_2O_7$ oxidation of these substances.

The usefulness of mass spectrometry in this context has been significantly improved by employing a continuous ion source that enhances sensitivity, and by incorporating a "blanking generator" that allows a particular mass region to be selected for detection. With careful temperature programming, such modifications have made it possible to generate very detailed PY/MS spectra of heavy hydrocarbons. But since generation of volatiles at high temperatures frequently raises questions about their chemical origins, PY/MS will generally only furnish unequivocal information when it is applied to *inherently volatile* fractions of heavy hydrocarbons.

Oxidative Degradation

Important information about structural features of heavy hydrocarbons has accrued from their controlled degradation with liquid oxidants [5]—in particular, HNO_3, $KMnO_4$, H_2O_2, NaOCl, $Na_2Cr_2O_7$, RuO_4, and HCO_3H [6]. However, results from oxidative degradation of coal (see Berkowitz, 1985) make it evident that product slates are critically influenced by the reaction conditions and do not necessarily bear a reasonably close resemblance to their progenitors. Thus, if an aromatic ring does not carry an electron-donor function such as —OH or $—NH_2$, most oxidants will degrade alkyl benzenes, alkyl pyridines, and alkyl thiophenes to the corresponding carboxy acids. But if it *does* carry such a function, the ring, too, will be quickly broken down, with pericondensed aromatics furnishing benzenoid acids ranging from benzoic and benzene dicarboxylic acids to tricarboxy- and hexacarboxy acids [7]. Olefins will similarly be degraded to glycols, which then split into carboxy acids, as in

$$\mathrm{R{-}CH{=}CH{-}R'} \rightarrow \mathrm{R{-}\underset{\displaystyle HO}{\underset{|}{C}H}{-}\underset{\displaystyle OH}{\underset{|}{C}H}{-}R'} \rightarrow \mathrm{R{-}COOH + HOOC{-}R'}.$$

Structures inferred from oxidation products consequently are not definitive unless they are independently validated from other sources.

SPECTROSCOPY

Independent and often *definitive* structural information derives from various nondestructive spectroscopic measurements.

Nuclear Magnetic Resonance Spectroscopy

Arguably the most powerful tool now at hand for structural studies of complex hydrocarbons offers itself in high-resolution ^{1}H and ^{13}C nuclear magnetic resonance (NMR) spectroscopy [8], which can furnish detailed fine-structure spectra by detecting two quite separate effects of interaction between a strong external magnetic field (B_o) and a nucleus.

One is a "chemical shift" that comes about becomes B_o causes electrons to circulate around the nucleus in a plane perpendicular to itself, and thereby creates a small magnetic field that, depending on whether it opposes or augments B_o, shields or "deshields" the nucleus and shifts resonance to a frequency ν_o defined by

$$2\pi\nu_o = -\gamma B_o(1 - s) \text{ or } -\gamma B_o(1 + s).$$

The screening constant s depends on the particular chemical environment of the nucleus, and consequently allows separating ^{1}H in, e.g., $-CH_3$ and $-CH_2-$.

The other effect of interaction between B_o and a nucleus is a nuclear spin–spin decoupling due to indirect interaction of near-neighbor nuclei via the spins of their bonding electrons. This decoupling is insensitive to the magnitude of B_o, but depends on the number and types of nuclei and on the bonds between them, and finds expression in fine spectral lines from which the nature of linkages between different components of a molecule can be inferred.

Since the late 1950s, when the first ^{1}H NMR measurements on coal were recorded (Bell *et al.*, 1958; Friedel, 1959), the quality of NMR spectra has been vastly improved through the use of "magic-angle" spinning (Andrew, 1971), cross-polarization (Pines *et al.*, 1972, 1973), several versions of multi-

TABLE 7.1.2 Assignments to ^{13}C Signals in High-Resolution NMR Spectra[a]

^{13}C shift range (ppm from TMS)	Assigned to C in
165–158	aromatic ether C—O
158–148	phenolic C—O
148–129	aromatic C—C
129–118	aromatic C—H
124	C_4 of phenanthrene
121	C_7 of fluoranthene
120	C_1 of fluorene
118–108	aromatic C—H *ortho* to ether—O and —OH
57–37	—CH_2— bridge structure; bridge-head C of naphthenes and branched alkyls
37–30	—CH α to an aromatic ring[b]
30–23	some C β to an aromatic ring
23–19	—CH_3 α to an aromatic ring
19–17	—CH_3 of ethyl group
17–13	—CH_3 γ or further from an aromatic ring

[a] Bartle *et al.* (1975).
[b] Superimposed on band peaking at 29.7 ppm, assigned to CH_2 in long chains.

plepulse spectroscopy (Vaughan *et al.*, 1976; Pembleton *et al.*, 1977; Gerstein *et al.*, 1977) and Fourier transform spectroscopy (Shaw, 1976; [9]). Spectra are displayed as variations of reasonance (≡ induced voltage) with a chemical shift δ that refers to an arbitrary standard that is now almost always tetramethylsilane, $(CH_3)_4Si$. Structural detail (Tables 7.1.2 and 7.1.3) is, however, only

TABLE 7.1.3 Assignments to ^{1}H Signals in High-Resolution NMR Spectra[a]

^{1}H shift range (ppm from TMS)[b]	Assignment[c]	Further subdivision at 220 MHz
5.5–9.0	aromatic H phenolic H	sterically hindered aromatic H: $\delta > 8.1$
4.7–5.5	olefinic H	
3.3–4.5	—CH_2— α to two rings	
2.0–3.3	C—H α to a ring	—CH_2— of acenaphthenes and indenes: δ = 3.0–3.3
1.6–2.0	naphthenic CH_2; methine[d]	
1.0–1.6	—CH_3 β to a ring; —CH_2— β or further from a ring	
0.5–1.0	—CH_3 γ or further from a ring	

[a] Bartle and Jones (1979).
[b] Tetramethylsilane $(CH_3)_4Si$.
[c] In 60- and 100-MHz spectra.
[d] At other than α to a ring.

obtainable from the ^{1}H and ^{13}C NMR spectra of liquids or of solids dissolved in a suitable solvent. ^{1}H NMR spectra of semisolid and solid hydrocarbons will, as a rule, merely show a single broad band that cannot be resolved without recourse to spectral reconstruction (Brown and Ladner, 1960), and ^{13}C NMR spectra of such materials cannot furnish much more than the ratio of aliphatic to aromatic carbon (C_{al}/C_{ar}) unless recorded with a time delay in the radiofrequency pulse sequence that discriminates against broad signals (Zilm *et al.*, 1979). Even so, there is some evidence that spectral intensities are not always entirely quantitative [9].

Infrared Spectroscopy

Discrete absorptions of infrared (IR) radiation by hydrocarbons result from excitation of specific rotational–vibrational bond deformations, and IR spectra covering wavelengths between 750 and 3300 cm^{-1} thus offer important means of identifying individual bond configurations. Diagnostic bands accruing from absorption by aromatic and nonaromatic CH moieties (Table 7.1.4) have, in fact, yielded more direct information about chemical structures in heavy hydrocarbons than any other investigative method except NMR spectroscopy. They have also provided information about common mineral species associated with them (see Table 5.2.17).

Difficulties arising from the relatively poor optical transmissivity of heavy hydrocarbons can be sidestepped by recording spectra on alkali halide (usually KBr) discs in which the substance under study has been incorporated. Concentrations can be as low as 2–3 mg gm^{-1} KBr.

Major improvements of spectral quality have accrued from introduction of Fourier-transform IR spectroscopy [11], which allows rapid signal averaging, computer-performed spectral subtraction, and least-squares fitting. Together, these refinements offer superior signal-to-noise ratio, greater sensitivity, and improved spectral resolution, and thereby substantially eliminate earlier problems from unacceptably high IR scattering and relatively high diffuse background absorption, which were encountered whenever the material under study was characterized by high carbon content [12].

Quantitative evaluation of IR spectra thus offers information that can be correlated with information from functional-group analyses and NMR spectroscopy or independently used to, e.g., determine H_{al}, derive C_{al} from the stoichiometric $H_{al}/C_{al} = 1.8$, and in this manner estimate the aromaticity of heavy hydrocarbons by difference (Solomon, 1979).

X-Ray Diffraction

Coherent scattering of X-rays by matter—i.e., reflection of incident X-radiation on the same wavelength λ—causes interference patterns that mirror the geome-

TABLE 7.1.4 Some Assignments to Absorption Bands in IR Spectra

Band wavelength λ		
cm^{-1}	μm	Assignments
3300	3.0	—OH and —NH stretching
3030	3.3	aromatic C—H stretching
2940	3.4	aliphatic C—H stretching
2925	3.42	$—CH_3$ and $—CH_2$ stretching
2860	3.5	aliphatic C—H stretching
1700	5.9	C=O stretching
1600[a]	6.25	aromatic C=C stretching; H-bonded C=O
1500	6.65	aromatic C=C stretching
1450	6.9	aromatic C=C stretching; $—CH_3$ asymm. deform.; $—CH_2$ scissor deformation
1380	7.25	$—CH_3$ symm. deformation; cyclic $—CH_2—$
1300–1000	7.7–10.0	$C_{ar}—O—C_{ar}$, $C_{al}—O—C_{al}$, and $C_{ar}—O—C_{al}$ stretching
900–700	11.1–14.3	aromatic bands
Tentative assignments in the 900–700 cm^{-1} region:		
800	11.6	[arom. HCC rocking in single
820	12.2	[
750	13.3	[condensed rings
873	11.5	[subst. benzene ring with isolated or two
816	12.3	[neighboring H atoms;
751	13.3	o-substituted benzene ring
893	11.2	angular condensed ring system
758	13.2	mono-subst. benzene ring; condensed system

[a] See Note [13].

try of the scattering matter. From well-ordered crystalline or microcrystalline materials, in which ordered regions are very large compared to λ, these patterns consist of diffraction lines or narrow bands that obey Bragg's law,

$$2d \sin \theta = n\lambda,$$

where d is the interplanar spacing, θ the Bragg angle, and n the "order" of the diffraction. In such cases it is possible not only to define the structure of the crystalline region, but also to infer the size of that region, because diffraction bands broaden as crystallite dimensions decrease, and the half-peak widths of the bands are inversely proportional to these dimensions.

But substantially amorphous materials produce continuous scatter curves that, instead of well-defined discrete bands, show grossly broadened maxima, and these are difficult to interpret—in part because θ, which determines the shapes of the scatter curves, also determines whether the spectrum reflects atomic arrangements or molecular structure (which it does if θ is relatively large), or organization on a macromolecular or micellar scale (which is the case when θ approaches zero). Efforts to assess the crystallographic organization and structure of heavy hydrocarbons from X-ray diffraction spectra have therefore proven rather less successful than other spectroscopic studies.

Since the scatter curves of such hydrocarbons—which are usually recorded with CuK_α (λ = 1.540 Å) or CoK_α radiation (λ = 1.785 Å) and then plotted as beam intensities vs $2 \sin \theta/\lambda$ (expressed in $Å^{-1}$)—resemble the diffraction patterns of relatively disordered microcrystalline graphites, they are routinely indexed as for graphite; and assuming the scattering substance to be fairly highly aromatized amorphous matter, the spectral data are then interpreted in terms of ordered regions or "crystallites" defined by an average diameter L, an average stack height n, which is equivalent to the number of layers per crystallite, and an interlayer spacing d.

However, as early estimates of n from half-peak widths (β) of the $\langle 002 \rangle$ or $\langle 004 \rangle$ bands by means of the Laue equation

$$\beta = C\lambda\sqrt{n} \cos \theta,$$

where C is a constant (~1.0), or estimates of L from the half-peak widths of the $\langle 100 \rangle$ or $\langle 110 \rangle$ bands by the same expression (Blayden *et al.*, 1944) proved little better than qualitative, subsequent work has centered on more refined procedures, the most useful of which turned out to be:

1. Fourier transforms giving radial distribution functions that show time- and space-averaged numbers of atoms at different distances from any arbitrarily chosen atom (Franklin, 1950; Nelson, 1954; Peiser *et al.*, 1955), and
2. cosine transforms (equivalent to one-dimensional Patterson projections), which indicate average numbers of aromatic layers at various perpendicular distances from any one layer, and therefore allow estimates of packing distributions [14] (see Hirsh, 1954, 1960; Diamond, 1957, 1958).

Some diffraction spectra have also been matched to theoretical scatter curves computed from the Debye formula for X-ray scattering by individual molecules (Ergun, 1958; Ergun and Tiensuu, 1959).

Even so, however, current information on crystallographic organization in heavy hydrocarbons must be viewed cautiously. This is particularly true of size distribution histograms, in which occasional negative terms appear to be artifacts that accrue from the mathematical treatment and from *amorphous* carbon—possibly associated with C atoms in alkyl groups, COOH, OCH_3, etc.—that is, at least partly, an arithmetic device for bringing the histogram total to 100%.

Statistical Constitution Analysis

In principle, the more important features of an average structural unit of a complex macromolecule can be derived by a statistical constitution analysis. This provides plausible estimates of aromatic carbon (C_{ar}/C), H bonded to aromatic carbon (H_{ar}/H), and ring condensation, and subject to some reasonable assumptions, it can also afford a measure of the mean size of a hypothetical aromatic cluster from which the number of rings per structural unit can be derived.

Such analyses have been employed in early studies of heavy oils (Vluchter *et al.*, 1935) and were later amplified for similar studies of coals (van Krevelen, 1954, 1961). But they have now been effectively superseded by more direct and reliable instrumental techniques (see earlier discussion) and command little more than historical interest [15].

2. MOLECULAR STRUCTURE IN HEAVY HYDROCARBONS

Like natural gas and conventional petroleums, heavy hydrocarbons are mixtures, and meaningful studies of molecular structure must therefore be centered on the *dominant* components. This has so far been synonymous with structural studies on asphaltenes and vitrinite-rich coals. Despite a considerable volume of relevant analytical data, very little is known about molecular configurations in resins, kerogens, and preasphaltenes. The structural models put forward for asphaltenes and coal can, however, serve as anchors for also deducing basic aspects of other substances—and what can be inferred in this manner is a remarkably close structural relationship between heavy hydrocarbons, regardless of the minutiae of their origins.

The Structure of Asphaltenes

Do Asphaltenes Possess "a Structure"?

Although frequently referred to as if forming a clearly delimited class of compounds, asphaltenes are poorly defined assemblies of diverse molecular species that, by their composition and chromatographic behavior, often even negate the solubility criteria (see Chapter 5) which define them:

1. Yields of asphaltenes precipitated from benzene solution by addition of *n*-hexane depend on, and vary with, the volume of the precipitant (Speight, 1991)

2. Asphaltenes from Utah bitumens appear to be characterized by significantly higher (number-average) molecular weights, but appreciably lower aromaticities and aromatic polycondensation than their Athabasca counterparts (Kotlyar *et al.*, 1988)
3. Sequential elution of asphaltenes from Athabasca bitumens on silica gel with 15% benzene/n-C_5H_{12}, 50% benzene/n-C_5H_{12}, and tetrahydrofuran delivered three polar fractions, of which one could be separated further into five subfractions by sorption on Al_2O_3 and successive elution with n-C_5H_{12}, n-C_5H_{12}/15% C_6H_6, 100% C_6H_6, C_6H_6/Et_2O/MeOH, and 100% CH_2Cl_2 (Selucky *et al.*, 1978; [16])
4. Ion-exchange and adsorption chromatography resolved a pentane-soluble asphaltene fraction [17] into acids (carboxy acids, phenols), bases (amides, carbazoles, pyridine benzologs), saturated hydrocarbons, and aromatics (McKay *et al.*, 1978)

Molecular weights of asphaltenes are similarly beset by uncertainties. Cryoscopy, vapor-pressure osmometry, and gel permeation chromatography have furnished values between 200 and 10^4 (Moschopedis *et al.*, 1976; Speight *et al.*, 1985); and neutron scattering, which varied appreciably from one asphaltene specimen to another, indicated molecular weights in excess of 2×10^5 (Ravey *et al.*, 1988).

However, unlike VPO measurements with nonpolar or weakly polar solvents such as benzene or toluene, which yielded values between 4500 and 5000, strongly polar solvents (pyridine, nitrobenzene) consistently returned molecular weights near ~1800 (Moschopedis *et al.*, 1976), and it has therefore been argued that these values were "correct" for disaggregated asphaltenes.

That similar low asphaltene molecular weights have been independently reported by others (Winniford, 1960; Bunger, 1977) lends some support to this argument. However, relationships between molecular weights and structure (see later discussion) also support the contention (Yen, 1982) that asphaltic molecules cannot be characterized by a single "true" molecular weight: in any one case, the observed value could be the weight of (i) a unit sheet (the lowest and, from some points of view, the "best" value), (ii) a stacked sheet weight, or (iii) an associated stacked sheet (or cluster); and this harmonizes with observations that vacuum residua are colloidal dispersions of heptane-soluble asphaltenes in a matrix of other molecules (Storm *et al.*, 1991) [18].

In these circumstances, an asphaltene "average molecular structure," properly understood, cannot be more than, at best, a semiquantitative collage of arbitrarily associated "structure elements" that represent, or are more or less directly derived from, individual molecular species existing in a particular asphaltene melange.

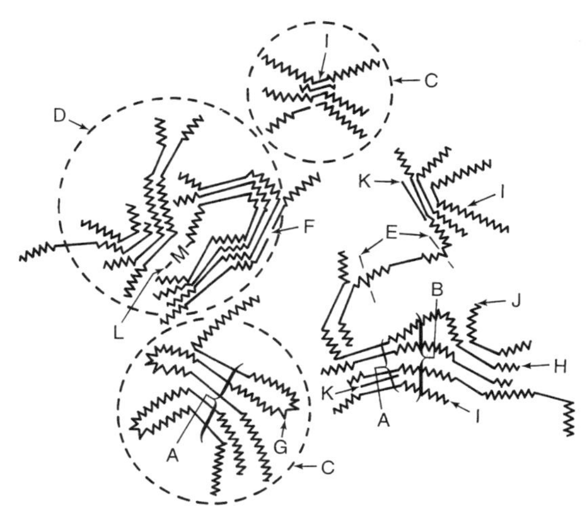

FIGURE 7.2.1 Conceptual macromolecular structure of asphaltenes (Dickie and Yen 1967). A, crystallite; B, chain bundle; C, particle; D, micelle; E, weak link; F, gap and hole; G, intracluster; H, intercluster; I, resin; J, single layer; K, porphyrin; L, metal (M).

Asphaltene Macrostructure

From X-ray diffraction patterns of solid asphaltenes (Dickie and Yen, 1967; Yen, 1979) it has been inferred that crystallographic organization can be represented by an asphaltene "macromolecule" (Fig. 7.2.1) in which clusters of partly ordered aromatic matter, dimensioned as in Fig. 7.2.2 and carrying aliphatic chains of varying lengths, are associated in micelles or particles.

A more recent study of dispersions of asphaltenes in liquids by small-angle neutron scattering—like small-angle X-ray scattering (SAXS), a classic method for characterizing colloid systems—furnished diffraction patterns that gener-

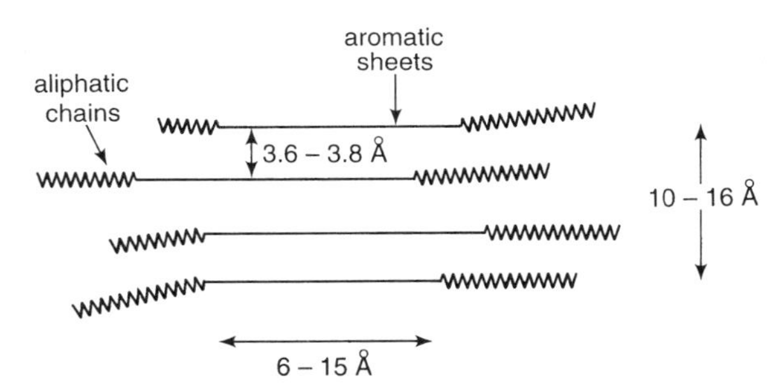

FIGURE 7.2.2 Dimensions of an asphaltene "crystallite" (Dickie and Yen, 1967).

ally confirm the earlier conclusions, but also provide some evidence for colloidal polydispersion in a marked solvent-dependence of average cluster, micelle, and particle sizes (Ravey *et al.*, 1988).

Aspaltene Molecular Structure

Information from which average molecular structure models have been assembled was, as in other similar studies, generated by solvent fractionation and chromatographic separation, and by chemical and/or spectroscopic analyses of products from thermolysis and/or degradation by liquid oxidants. (Particular attention was frequently given to Ru(VIII)-catalyzed degradation with Na metaperiodate in acetonitrile/CCl_4/H_2)O, a reagent that can selectively oxidize aromatic carbon to CO_2 without significantly affecting aliphatic and naphthenic structures.)

Mostly relating to asphaltenes isolated from Alberta's Athabasca and Peace R. oil sands bitumens, this information showed that

1. aromatic entities are linked by CH_2 bridges which range in length from C_2 to C_{22}, but are mainly represented by C_3 and C_4;
2. benzene di-, tri-, and tetracarboxylic acids—presumably formed from naphthalene, anthracene, and phenanthrene nuclei—generally represent less than 1–2% of the asphaltene specimen (and thereby indicating the dominant aromatic moieties to be substituted benzenes); and
3. Naphthenics are alkyl substituted.

However, prominent sulfone absorptions in IR spectra also indicated aliphatic sulfide linkages, and acetone-soluble fractions of asphaltenes, which were precipitated with *n*-pentane and represented up to 22% of the total, were found to contain alkyl carbazoles, variously substituted benzocarbazoles, dibenzocarbazoles, and tetrahydrodibenzocarbazoles that accounted for ~0.4% of the asphaltene samples. (Some of these entities were also detected in asphaltenes extracted from Utah tar sands; see Holmes, 1986). The specific structure elements, identified in oils formed by mild pyrolysis of Athabasca asphaltenes (Strausz *et al.*, 1992), are listed in Table 7.2.1.

Proceedings from this database [19], Strausz *et al.* (1992) have developed a structural model (see Fig. 7.2.3) in which entities marked A, B, and C represent hypothetical aromatic clusters, and information from detailed chemical studies is used to identify other components—notably pentacyclic naphthenic rings, condensation products of aromatic structure elements, biphenyls, and linkages that, when oxidatively ruptured, yield benzene di-, tri-, and tetracarboxylic acids. Expressed in weight percentages, the elemental composi-

TABLE 7.2.1 "Structure Elements" in Pyrolysis Products of Asphaltenes[a]

n-C_x substituted thiophenes $x = 9$–28 (10–15)
CH_3 and n-C_x substituted benzo- and dibenzothiophenes $x = 10$–25 (10, 11)
n-C_x substituted fluorenes $x \rightarrow >25$ (14)
n-C_x substituted cyclic sulfides $x = 10$–29 (3–17)
C_{12}–C_{34} alkanes (15, 16)
n-C_x substituted alkyl benzenes (12–28)
Alkylated naphthalenes and biphenyls
Alkylated anthracenes and phenanthrenes

[a] Numbers in parentheses identify the most abundant entities.

tion of this model is [C] 81, [H] 8.0, [N] 1.4, [S] 7.3, [O] 1, [V] 0.8; its empirical formula is

$$C_{420}H_{496}N_6S_{14}O_4V;$$

its H/C ratio is 1.18; and its molecular weight, measured by VPO in benzene, is 6190. Except for the molecular weight, which might suggest some degree of molecular aggregation, these values lie close to values for acetone-extracted asphaltenes, and the molecular weight could, if one chose to do so, be brought into accord with proposed lower values by breaking the model down into smaller fragments [20].

However, two other constructs—also based on chemical and spectroscopic studies, but put forward as structure elements from which asphaltene molecules might develop by polymerization—posit different configurations and exemplify diversity between asphaltenes. Figure 7.2.4a (Witherspoon and Winniford, 1967) shows the hypothesized monomeric unit of asphaltenes from a Venezuelan crude, and Figure 7.2.4b (Witherspoon and Winniford, 1968) represents its counterpart for asphaltenes from a California crude oil.

An earlier, and perhaps more tentative, formulation of "average structure" of Athabasca asphaltenes (Strausz, 1989), reproduced in Fig. 7.2.5, is *a priori* improbable because coking of the bitumen (Chapter 9) yields substantially more residual coke than this model would allow [21], and an inference (Speight, 1975, 1991) that aromatic units in molecules of asphaltenes from Athabasca bitumens are structured as in Figure 7.2.6 also lacks credibility—if

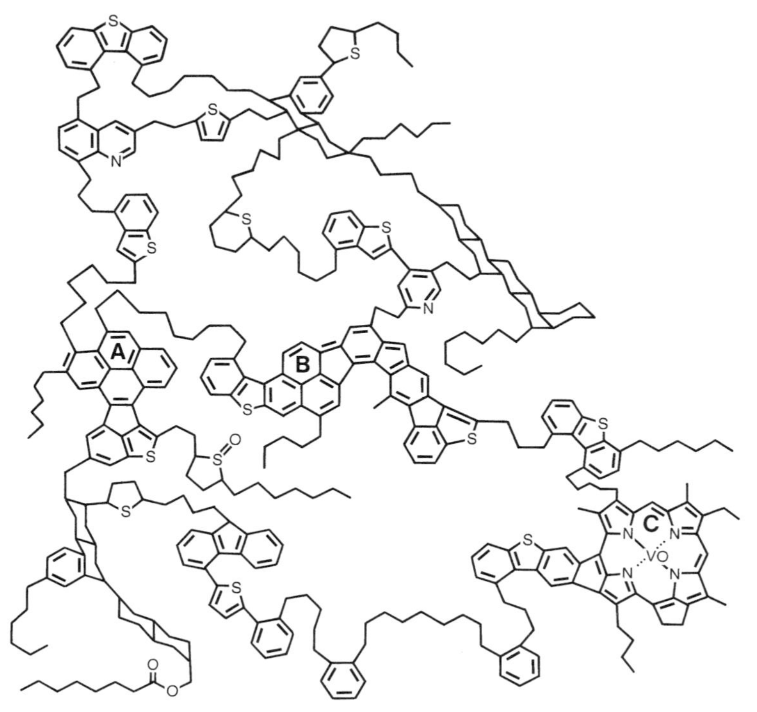

FIGURE 7.2.3 Hypothesized average molecular structure of Athabasca asphaltenes (Strausz *et al.* 1992). A and B represent larger atomic clusters, and C is a porphyrin moiety.

only because it is entirely based on an evaluation of *x*-ray diffraction patterns by Diamond's (1957, 1958) seriously flawed method for estimating size distributions.

THE LIGHTER COMPONENTS OF BITUMEN: STRUCTURAL FEATURES OF MALTENES

Bitumen components soluble in *n*-pentane, and by that criterion defined as *maltenes*, can be resolved by column chromatography into oils composed of saturated and aromatic hydrocarbons, and resins (Chapter 5). Very little is known about the molecular structures of these components [22]. However, Kotlyar *et al.* (1991) have reported a structural analysis of maltenes from Athabasca bitmuen by ^{13}C NMR techniques that discriminated between different carbon types and differentiated between two fractions with molecular

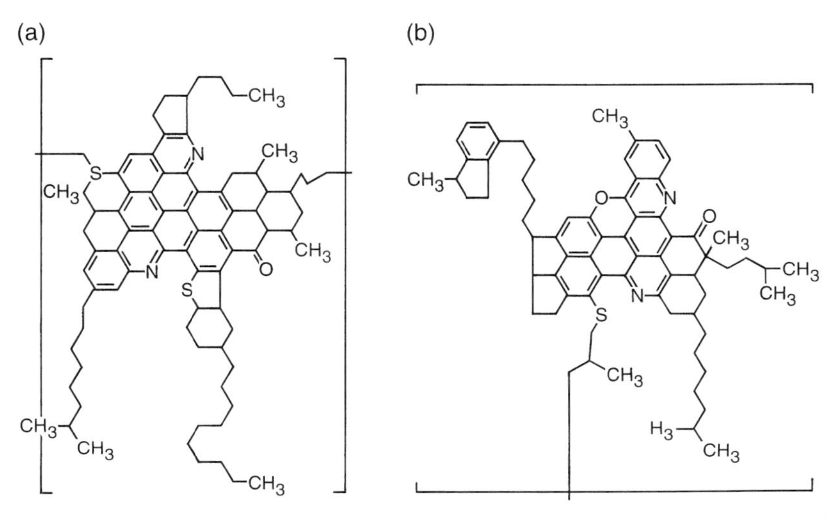

FIGURE 7.2.4 Hypothesized average molecular structures of asphaltenes in (a) Venezuelan crude oil, (b) Californian crude oil (Witherspoon and Winniford, 1967, 1968).

weights of <800 (≡ oils) and >800 (≡ resins). Of the lighter fraction (MW ~315), over 35% was accounted for by *tertiary* and *quaternary* aromatic carbon, and almost 30% by cyclic methylene and methine, but in the heavier fractions, with molecular weights near 800, some losses of C_{ar} to naphthenic moieties were indicated.

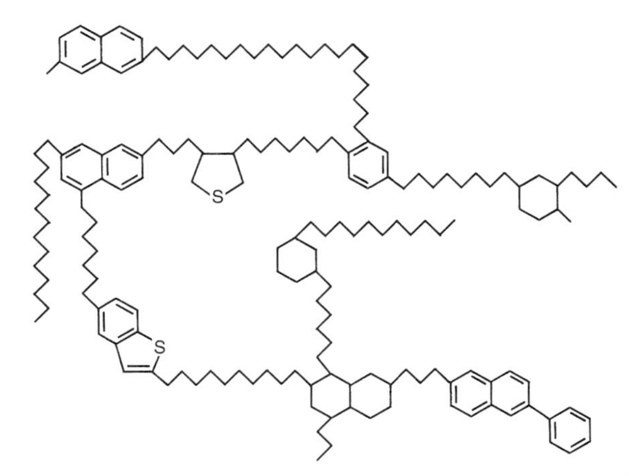

FIGURE 7.2.5 Another hypothesized average structural element proposed for Athabasca asphaltenes (Strausz, 1989).

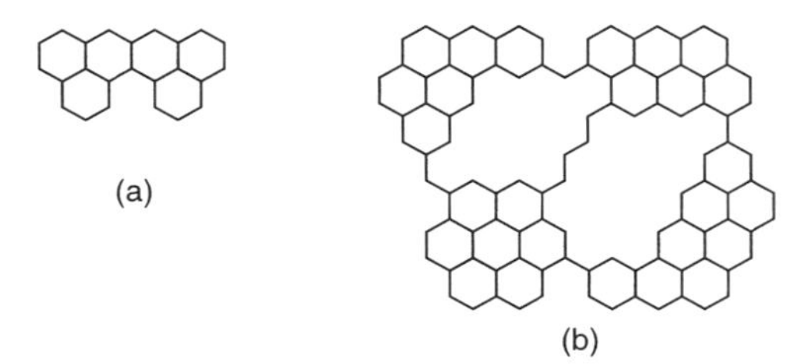

FIGURE 7.2.6 Hypothesized average *aromatic* structural element in Athabasca asphaltenes (Speight, 1975, 1991). (a) Decane-soluble asphaltenes; (b) decane-insoluble asphaltenes.

Structure in Kerogens

In principle, plausible basic features of an average molecular structure of kerogens can be inferred from C_{al}/C_{ar}, H/C, and the chemically identified structure elements in asphaltenes and coal—i.e., by viewing kerogens as hydrocarbon assemblages that, chemically as well as by their physical properties, represent fossil hydrocarbons *between* asphaltenes and coals.

Proceeding from the elemental compositions and atomic H/C ratios of kerogens, this would lead to macrostructures that are qualitatively quite similar to those of asphaltenes—but to "average molecular" structures characterized by fewer and/or shorter alkyl chains, to a higher C_{ar}/C ratio, and to greater abundance of naphtheno-aromatic units than are associated with asphaltenes.

As well as standing in good accord with qualitative macrostructure representations of "low"- and "high"-evolution kerogens (Tissot and Espitalié, 1975; Fig. 7.2.7), such inferences are consistent with three pieces of experimental evidence:

1. ^{13}C NMR spectra of kerogens show two pronounced bands peaking at 30 and 90–130 ppm and attributed, respectively, to *n*-alkyls and aromatics;
2. oxidative degradation of kerogens by RuO_4 or alkaline $KMnO_4$ generated an abundance of α,ω-dicyclic carboxy, isoprenoid, and cyclic acids; and
3. solvent extraction of kerogens or kerogen concentrates (see Robinson, 1969) delivered paraffins, methyl-substituted alkanoids, cycloalkanes, polynuclear aromatics, *N*-heterocycles, etc., that are virtually identical with species identified in component fractions of bitumens and in part also detected in coal.

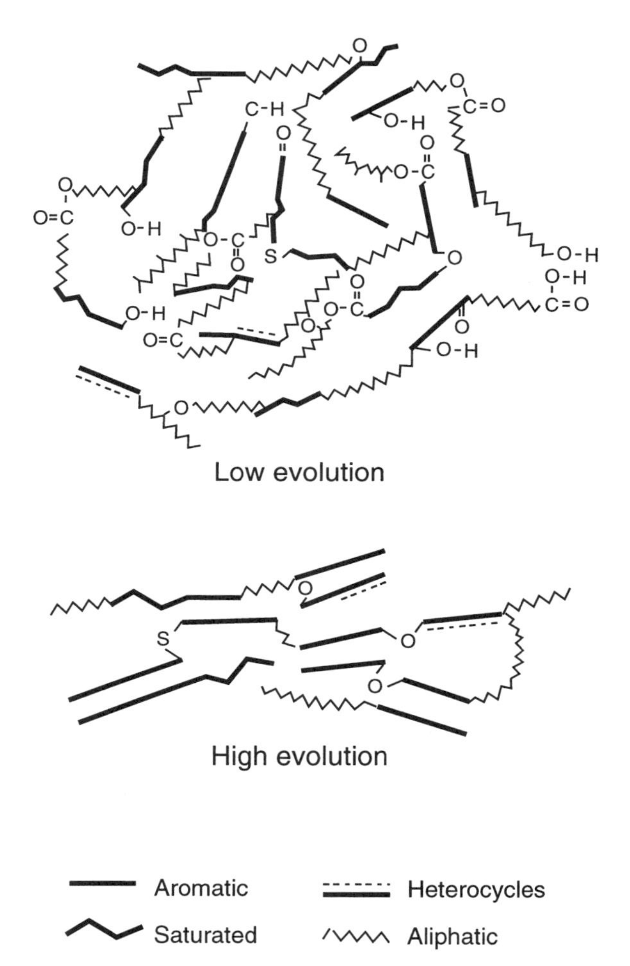

FIGURE 7.2.7 Proposed molecular structure of "low-" and "high-evolution" kerogens (Tissot and Espitalié, 1975).

Humic Substances

Humic acids extracted from Alberta oil sands with 0.5 *N* NaOH or 0.1 *N* NaOH + 0.1 *N* $Na_4P_2O_7$ have proved to be very similar to humic acids from subbituminous coals (Majid and Ripmeester, 1990). Positioned in an H/C vs O/C diagram, they overlapped, or fell entirely into, the area occupied by recent (or shallow) kerogens, and may consequently also resemble them structurally. Their origins are, however, by no means necessarily the same as those of the bitumens that contained them [23].

The Structure of Coal

Molecular Structure

Information on molecular structures in coal has accured from

(i) chemical degradation, principally by oxidation and thermolysis that defined the nature of aromatic entities
(ii) x-ray diffraction measurements, which provided additional data for the average size of aromatic "regions," as well as information about coal macrostructure and
(iii) infrared and NMR spectroscopy, which quantified structural parameters such as C_{ar}/C and H_{ar}/H (see Table 7.2.2).

The data now at hand are, however, limited to *humic* coals and, even so, relate almost exclusively to vitrinite concentrates or coals principally composed of vitrinites. H-richer liptinites and H-poorer inertinites are assumed to possess similar configurations that, supposedly, could be formulated by such addition or removal of CH_2 bridge units and peripheral substituents ($-CH_3$, $-COOH$, $-OH$, $-OCH_3$) as their respective elemental compositions and their C_{ar}/C and H/C ratios may require [24]. Metamorphic development, or maturation, which is broadly equated with increasing C_{ar}/C and falling H/C, is similarly accommodated, and so, presumably, could be *sapropelic* coals, which have received little, if any, attention in this context.

The "average molecular structure" models that began to be so formulated in the 1960s have not required significant revisions in light of data from later, more discriminating investigations, and set against hypothesized asphaltene configurations (see, for example, Figs. 7.2.1, 7.2.3, and 7.2.4), they show a direct structural progression—definable in terms of C_{ar}/C, H_{ar}/H, H/C, and derivative parameters—from heavy oils and bituminous substances to kerogens and coals. Table 7.2.2 shows how these parameters continue to change during maturation of coal *per se*.

Structural data for complex molecules can, of course, almost always be presented in acceptable alternative formats, and differences between "average structure" models for vitrinite-rich coals with 80–90% C, shown here as illustrative examples, therefore reflect personal preferences as well as innovative interpretations of the data at hand.

TABLE 7.2.2 Elemental Compositions and "Average" Structural Parameters for Coals and Coal Extracts[a,b]

Rank	%C	%H	H-distribution			f_a	s_r	H_{aru}/C_{ar}
			H_{ar}	H_a	H_o			
Subbit. A								
Vitrain	76.3	5.2						
Py-extr. (6.7%)	77.5	6.4	0.20	0.30	0.44	0.64	0.59	0.76
CS_2-extr. (1.5%)	83.4	9.4	0.15	0.31	0.52	0.44	0.57	1.00
hvAb								
Whole coal	82.6	5.9						
Py-extr. (21.4%)	83.9	6.2	0.35	0.26	0.36	0.73	0.37	0.68
CS_s-extr. (2.5%)	86.5	7.4	0.27	0.33	0.39	0.62	0.44	0.79
hvAb								
Vitrain	85.1	5.3						
Py-extr. (20.9%)	83.0	5.8	0.27	0.29	0.39	0.72	0.47	0.59
CS_2-extr. (5.7%)	86.9	7.2	0.29	0.35	0.34	0.61	0.45	0.86
mvb								
Vitrain	86.1	5.5						
Py-extr. (19.3%)	87.1	6.0	0.32	0.28	0.37	0.73	0.37	0.57
CS_2-extr. (9.7%)	88.1	6.9	0.29	0.36	0.33	0.64	0.44	0.76
lvb								
Whole coal	90.0	4.4						
Py-extr. (2.8%)	88.4	5.2	0.55	0.31	0.13	0.85	0.27	0.63
CS_2-extr. (0.8%)	89.9	5.8	0.53	0.36	0.11	0.81	0.29	0.71
lvb								
Vitrain	90.4	4.6						
Py-extr. (2.5%)	90.5	5.3	0.50	0.30	0.19	0.83	0.26	0.57
CS_2-extr. (1.1%)	90.7	5.7	0.50	0.36	0.14	0.78	0.32	0.70

[a] Retcofsky (1977).

[b] Elemental compositions expressed as % dmmf material. H distributions (expressed as fractions): H_{ar} = H in aromatic entities, including phenolic —OH; H_a = benzylic H; H_o = H in other aliphatic positions. f_a = fraction of aromatic carbon. s_r = degree of aromatic ring substitution. H_{aru}/C_{ar} = H/C ratio of hypothetical unsubstituted aromatic nuclei.

Given's formulation (Fig. 7.2.8) thus makes the basic aromatic unit a dihydro*phenanthrene* unit rather than, as was earlier customary, a dihydro*anthracene* moiety, and inserts a tryptycene-like element that, in addition to molecular buckling caused by the presence of hydroaromatic and alicyclic entities, endows the whole molecule with an extra spatial dimension that bears importantly on perceptions of coal *macro*structure (see later discussion).

Pitt's model (Fig. 7.2.9), although superficially resembling Given's, is *a priori* less attractive as a component of a macrostructure unit, but illustrates

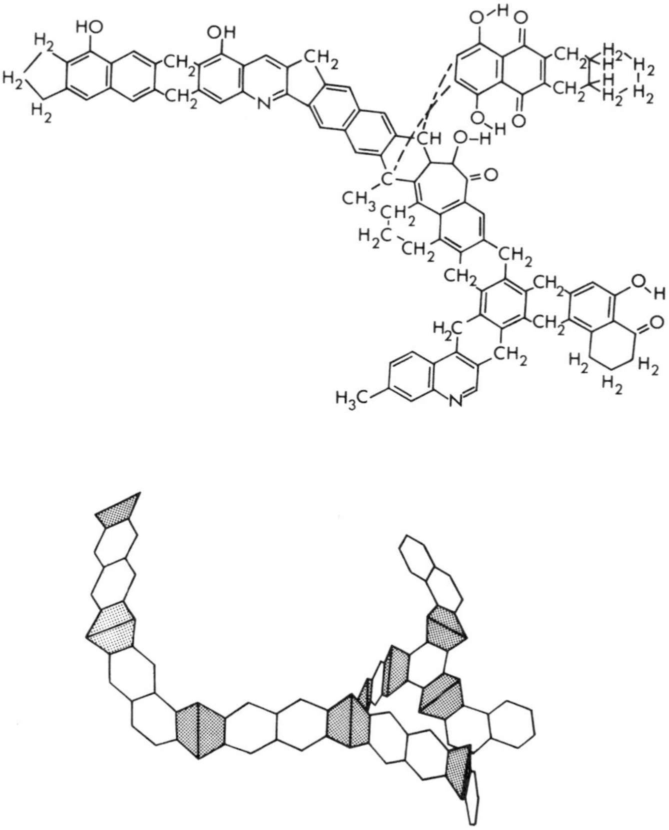

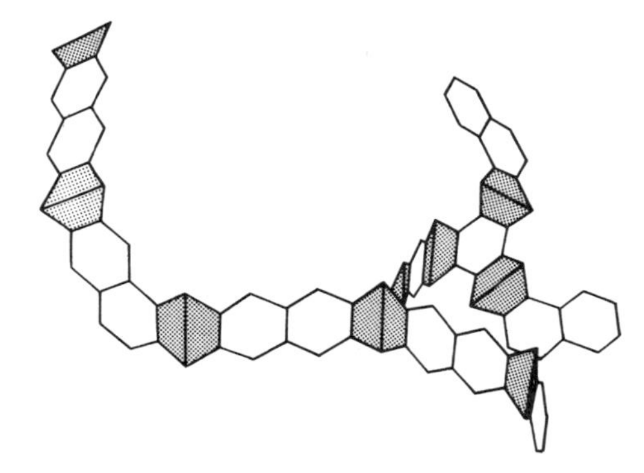

FIGURE 7.2.8 Hypothetical coal molecule at ~83% carbon (Given, 1960, 1961). The lower diagram, in which shaded areas represent *alicyclic* structural elements, shows the spatial configuration of the model.

how maturation from 80 to 90% C might be accommodated without appreciably changing the skeletal aspects of the molecule.

Heredy and Wender's (1980) configuration is noteworthy as a *monomeric* structure element that overtly recognizes the need to reconcile conjectured structures with *antecedents in their biosources* as well as with data for coal. The moieties incorporated in this model (Fig. 7.2.10) thus reflect

(i) that sugars (from degrading celluloses, hemicellulose, and glucosides in plant matter) can rearrange to polyhydric phenols by replacing —O— in

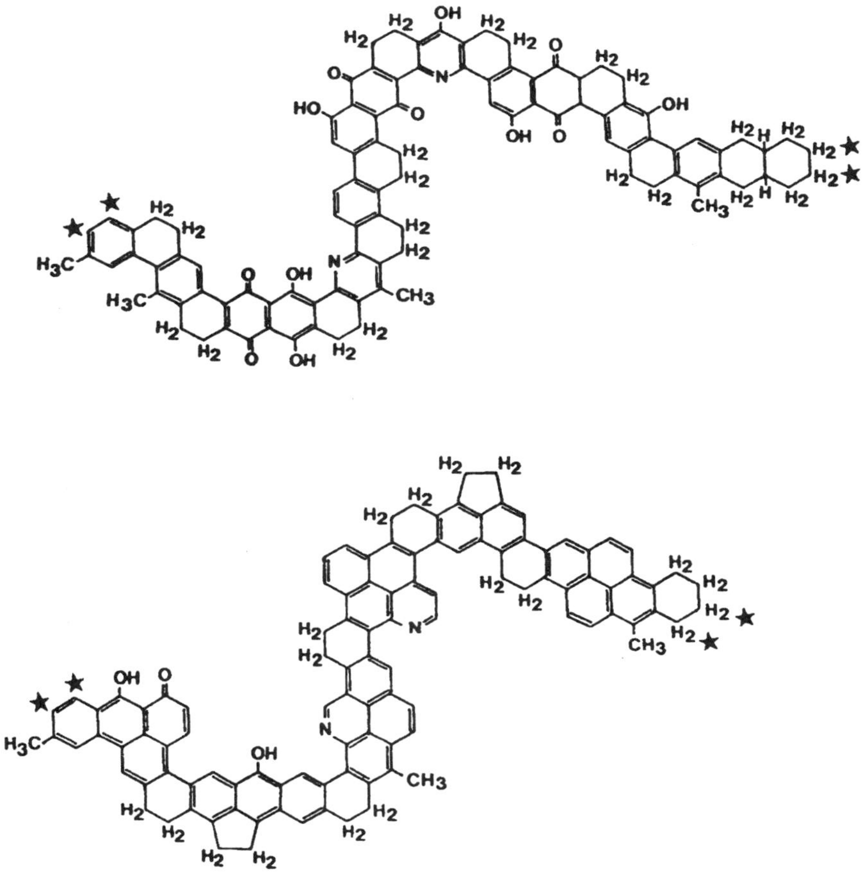

FIGURE 7.2.9 Hypothesized average molecular structures of vitrains at (a) 80%, and (b) 90% carbon (Pitt, 1979). Stars indicate where further molecular growth could occur.

their cyclic structures with C atoms, and then lose water as in

$$C_6H_{12}O\ [-3H_2O] \rightarrow C_6H_3(OH)_3.$$

and

(ii) that derivatives of phenylpropylene units, the building blocks of lignin and its primary breakdown products, represent a substantial component of the product slate accruing from coal oxidation with performic acid.

However despite their careful construction, the significance of these and other

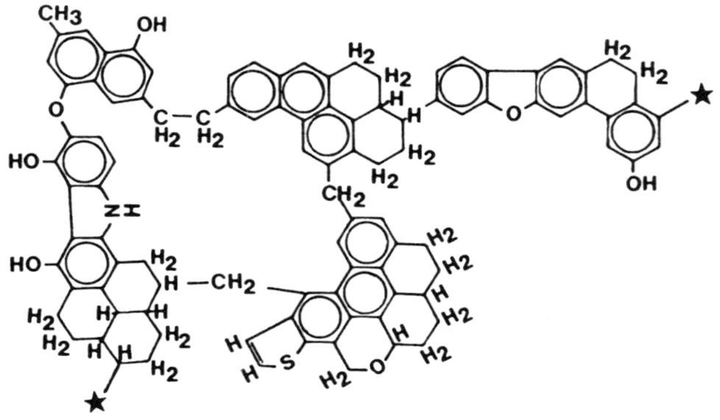

FIGURE 7.2.10 Proposed dominant structure element in vitrinite-rich coal at 83–84% carbon (Heredy and Wender, 1980). Stars indicate where further molecular growth could occur.

structure models tends to be jeopardized by the diversity of coal chemistries that began to be detected in the 1970s [25], and by uncertainties surrounding the *macromolecular* structure of coal.

The wide scatter of data points in comprehensive correlations of C_{ar} with

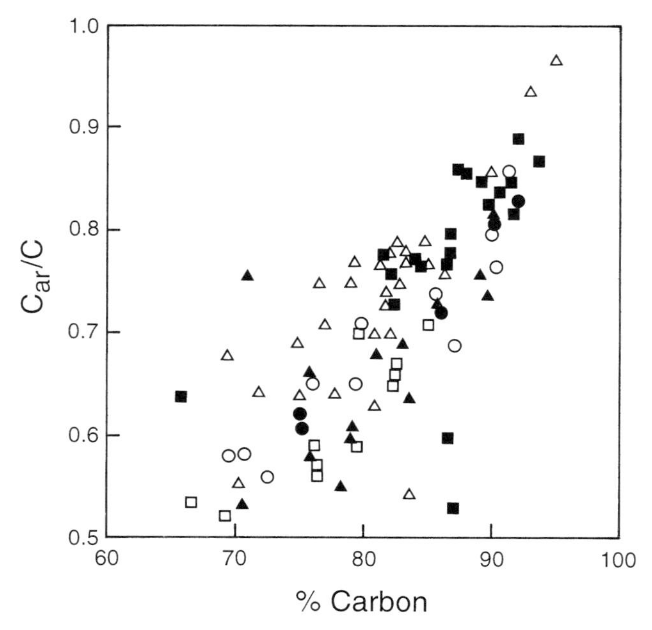

FIGURE 7.2.11 Variation of aromatic carbon [C_{ar}/C] with total carbon content (Sfihi *et al.* 1986; reproduced with permission).

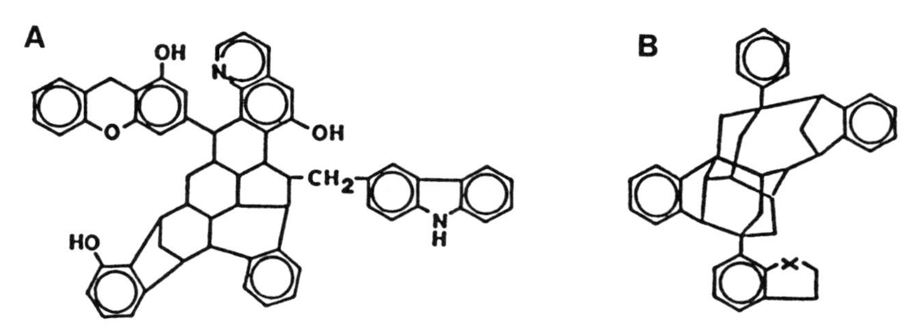

FIGURE 7.2.12 Proposed formal molecular structures in high-molecular-weight fractions from liquefaction of two similar coals (Whitehurst *et al.*, 1977). (a) Precursor: western Kentucky subbit. coal; fraction mol. wt. ~900, C_{ar} ~73%, H_{ar} ~ 60%. (b) Precursor: North Dakota subbit. coal; fraction mol. wt. ~600, C_{ar} ~26%, H_{ar} ~18%.

H/C and rank (see Sfihi *et al.*, 1986; Fig. 7.2.11), as well as "atypically" low or high values of C_{ar} and H_{ar} sometimes encountered in more limited data sets, appear to mirror regionally different depositional and/or developmental histories that have also been reported from other studies of coal [26]—and given such divergence, no model advanced as an "average molecular structure" can be accepted as such or, as asserted, "describe coal" unless it is related to a specific geographic area [27]. That ostensibly very similar coals should sometimes possess dramatically different chemical characteristics and structures (see Fig. 7.2.12) underscores this.

But questions about the validity of the models are also raised by the curious absence of $>C_2$ aliphatic moieties detected by oxidation (see Chapter 5)—i.e., by the absence of the precursors of aliphatic C_4–C_{20} diacids identified among products of HNO_3 oxidation; of straight-chain C_4–C_{22} diacids, branched-chain C_5–C_{10} diacids, and smaller amounts of C_6–C_8 triacids among products of oxidation with buffered $KMnO_4$; of phenyl derivatives with 3-C chains accruing from oxidation with HCO_3H; etc. [28]. That none of these are accommodated in the constructs indicates a persistent uncertainty about whether they derive from integral components of the coal macromolecule or exist as guest molecules in the host matrix. And that, in turn, effectively bars any definitive statement on whether the matrix is a three-dimensionally cross-linked polymer or made up of physically entangled, asymmetric molecules (see below).

Macromolecular Structure

X-ray diffraction measurements (see, among others, Hirsch, 1954; Peiser *et al.*, 1955; Diamond, 1957; Ergun and Tiensuu, 1959) suggest structures very similar to those of asphaltenes (Figs. 7.2.1 and 7.2.2) and kerogens (Fig.

TABLE 7.2.3 "Cluster"/"Crystallite" Dimensions (Å) in Heavy Hydrocarbons[a]

	1	2	3	4
Asphaltene cluster	3.6–3.8	6–15		1.0–1.6
Kerogen	3.4–3.7	<10		1.0–1.6
Coal crystallite				
at 80% carbon	3.9	~7.0	~1.39	1.2
at 92% carbon	3.5	~7.8	~1.40	1.8–2.0

[a] 1, Interlayer spacing; 2, layer diameter; 3, average C—C bond length; 4, average no. of aromatic platelets/stack.

7.2.7). The dominant feature of maturation is a continued change to greater aromaticity at the expense of aliphatic/naphthenic entities; however, crystallite dimensions, which change only slowly with increasing rank, do not differ greatly from those of asphaltene clusters and kerogens. Table 7.2.3 illustrates the crystallographic similarities between the three fossil hydrocarbon types [29].

But as already briefly noted, the shapes and spatial arrangements of macromolecules formed by associations of crystallites are still debatable.

The notion that metamorphic development of coal proceeds by progressive trifunctional polycondensation (Dormans and van Krevelen, 1960; van Krevelen, 1966) has led to viewing the coal matrix as composed of three-dimensionally cross-linked structures of variable (although uncertain) size, and to deeming this thesis to be confirmed by measurements of solvent imbibition and swelling of coal (Green *et al.*, 1982). However, imbibition and swelling could be equally well explained by supposing the matrix to be formed by highly *asymmetric, physically entangled* molecules (Hombach and Kölling, 1972; Hombach, 1975, 1979; Berkowitz, 1982). Such molecular association—which would lend *a priori* preference to Given's structure model (Fig. 7.2.8)—could, in fact, more readily explain the very high (>80%) extractability of mature coals by anthracene, phenanthrene, α-napthol, etc. at temperatures *more than 100°C below the onset of active decomposition* (Golumbic *et al.*, 1950; Orchin *et al.*, 1951); and because increasing overburden pressures promote progressive partial molecular disentanglement, and thereby allow some molecular alignment [30], it would also account for the increasingly pronounced optical anisotropy that begins to manifest itself in coals with 87–88% carbon.

NOTES

[1] Two examples may serve to exemplify the distortions of number-average molecular weights (M_n) by colloidally dispersed matter in extract solutions.

(a) Although pyridine and ethylenediamine extracts of coal in these solvents are usually reported with M_n values below 1200, microscopic examination of the extract solutions under high magnification showed much of the "dissolved" matter to exist as discrete particles whose diameters ran to several hundred angstroms, and that must therefore have possessed molecular weights on the order of 10^6 (Kann, 1951).
(b) After depolymerization of coal with phenol/BF_3 (Heredy *et al.*, 1963, 1964, 1965, 1966) as in

$$\text{Ar-(CH}_2)_n\text{-Ar}' + 2\phi\text{OH} \rightarrow \text{ArH} + \text{Ar}'\text{H} + \text{OH}\phi\text{-(CH}_2)_n\text{-}\phi\text{OH},$$

the pyridine-soluble fraction of an hvb coal in pyridine returned $M_n \approx 400$ (Ignasiak and Gawlak, 1977). However, this rose to ~1000 after colloidally dispersed matter was removed by centrifuging, and a molecular weight *distribution* showed <9 wt% with molecular weights <500 and slightly over 50 wt% with molecular weights above 3000 (Larsen *et al.*, 1978). Similar ranges, sometimes extending to $>2 \times 10^5$, have been reported by Sternberg *et al.* (1974) and Sun and Burk (1975).

[2] VPO is applicable to materials with average molecular weights in the range 200–3000, but adaptations have been used on polymers with molecular weights up to 4×10^4 (Kroschwitz, 1985).

[3] Where aggregation occurs, its extent will be affected by chain length, which can restrict solute aggregation by imposing steric hindrance.

[4] As noted in Chapter 5, derivations of molecular weights from GPC curves rely on correlations with molecular sizes, and as the proportionality factor that connects size and weight depends on the *shape* of the solute molecules, the choice of calibration compounds—which usually reflects what is known or assumed about the substance under study—is critical.

[5] Oxidation with air or oxygen, which has usually been conducted at elevated temperatures, has proved very much less useful.

[6] These are generated by action of H_2O_2 on an organic acid—e.g., $H_2O_2 + HCOOH \rightarrow HCO_3H$ (performic acid), or $H_2O_2 + CH_3COOH \rightarrow CH_3CO_3H$ (peracetic acid). In all cases, the only reaction by-product is H_2O.

[7] The core structures of porphyrins are similarly degraded and will, under mild oxidation conditions, furnish various derivatives of pyrrole-2,4-dicarboxylic acid. However, side chains other than —CHO and —CH=CH_2 (e.g., —CH_3, —CH_2CH_3, —$COCH_3$) can apparently sometimes survive into the mix of oxidation products.

[8] Unlike IR spectra, which arise from rotational–vibrational excitation, NMR spectra reflect interactions between the magnetic component (5–500 MHz) of electromagnetic radiation and the small magnetic moment of nuclei with nonzero spin numbers. In carbon-rich molecules, these are mainly associated with ^1H and ^{13}C.

[9] Fourier-transform NMR employs a succession of short (typically 10^{-5} s) high-power radiofrequency pulses that sample all nuclei simultaneously over a wide frequency range. Frequencies and amplitude responses are then accumulated during subsequent free induction decay (FID) and converted to slow-passage spectra by computer calculation of the Fourier transform of the FID.

[10] Pugmire *et al.* (1981) have, in fact, reported encountering nonquantitative behavior in studies of some hydrocarbon mixtures and polymers.

[11] For a detailed account of FTIR, the reader is referred to Griffiths (1975).

[12] Prior to the advent of FTIR, these could only be countered at the expense of spectral detail by attenuating the reference beam.

[13] Although the 1600 cm^{-1} band is always the most prominent absorption in IR spectra of coal, much uncertainty continues to surround assignments to it. It has been variously ascribed to polynuclear aromatics linked by predominantly aliphatic —CH_2— (Rao *et al.*, 1962), to H-

bonded or OH-chelated carbonyl (Roy, 1957; Fujii and Yokoyama, 1969), and to amorphous although not necessarily aromatic C—C moieties (Friedel *et al.*, 1971). It probably reflects *several* disparate structures.

[14] The packing distribution is the number of aromatic layers occurring singly or in crystallites with $n = 2, 3, 4, \ldots$.

[15] Interested readers can find the procedural details of statistical constitution analysis in van Krevelen's books.

[16] These observations and a consideration of structure-related properties of Athabasca asphaltenes and resins have subsequently led Selucky *et al.* (1981) to conclude that a definition of asphaltenes by solubility criteria is unsatisfactory, and that asphaltene behavior during chromatographic separation cannot be reconciled with structures that envisage polymer units mainly interconnected by σ-bonds. They suggested that asphaltenes represent an aggregational state that is more appropriately viewed as a stacked cluster or micelle—even though this cannot readily explain the gel permeation behavior of very dilute asphaltene solutions.

[17] Because VPO, low-resolution mass spectroscopy, and quantitative IR analyses showed average molecular weights of the precursor asphaltenes to lie between 500 and 800, this specimen mix was thought to represent the heaviest and most polar constituents of crude oil. Precipitation of asphaltenes on addition of *n*-pentane was ascribed to alteration of the solvent properties of the system.

[18] This suggests that asphaltene molecules can self-associate and form "solid-like" particles.

[19] An authoritative summary of relevant bitumen chemistry can be found in Hepler and Hsi (1989).

[20] The structure reproduced in Fig. 7.2.3 has led Strausz *et al.* to suggest that petroleum asphaltenes are mostly generated by catalytic cyclization, aromatization, and condensation of *n*-alkanoic (probably fatty acid) precursors. This view would accord with observations (Ignasiak *et al.*, 1977) that mild thermolysis of asphaltenes yields large amounts of mono- to pentacyclic condensed molecules, many in alkyl-substituted form and seemingly representing a homologous series.

[21] Aside from naphthenic moieties in this formulation appearing as part of volatile oily matter, at least half of the postulated aromatic units would do so: detailed studies of coal pyrolysis have shown that thermal splitting off of nonaromatic carbon will open immediately adjacent aromatic rings, and volatile material consequently will always take some C_{ar} with it.

[22] Some of the molecular species of which maltenes are composed have been unequivocally identified; however, that maltenes can be characterized by an "average structure" is even more unlikely than that asphaltenes can be so characterized in any meaningful sense.

[23] Humic acids form oxidatively in soils, bituminous substances, and coals, but whether their chemical structures are substantially identical or differ depending on their precursors is still unknown. Like bitumens and asphalts, they are essentially only defined by their dark-brown to black color and by solubility.

[24] Some details respecting structural variation among coal macerals have been reported by Pugmire *et al.* (1984).

[25] This diversity emerged as coal liquefaction research moved into high gear, and initially manifested itself in grossly divergent variations of liquefaction yields with rank (Gorin, 1981).

[26] A classic case presents itself in results of a study that sought to clarify the effects of coal rank on liquefaction yields (Yarzab *et al.*, 1980). A computer cluster analysis of more than 100 U.S. coals in terms of 15 coal characteristics discriminated between three populations—but a useful predictive relationship between coal properties and percent conversion required a *different set of properties for each population.* Two of these populations represented the U.S. Eastern (Appalachian) and Interior (Midwest) provinces and were comprised of Carbonifer-

ous coals with very different time-temperature-pressure histories; Yarzab *et al.*, 1979), and the third was composed of Cretaceous coals from the Rocky Mountain region.

[27] As more data accumulate, *asphaltene* structures may well come to need similar reappraisal.

[28] Also given unwarranted short shrift are bridged tricycloalkanes (analogous to variously extended polyamantanes) that were detected among the products of NaOCl oxidation (Chakrabartty and Kretschmer, 1972, 1974; Chakrabartty and Berkowitz, 1974). Reports of these substances were vigorously challenged, but were later implicitly confirmed by Mayo and Kirshen (1979) and have also been found in deep petroleum reservoirs (Lin and Wilk, 1995).

[29] Data for asphaltene clusters and kerogens are based, respectively, on Yen (1979) and Tissot and Welte (1984).

[30] A well-known laboratory demonstration provides an analogy: if molten wax in which mica flakes are randomly suspended is cooled and resolidified, and unidirectional pressure is applied to it, the mica flakes will tend to orient themselves at 90° to the direction of the pressure.

REFERENCES

Andrew, E. R. *Progr. Nucl. Magn. Reson. Spectrosc.* **8**, 1 (1971).

Baldwin, R. M., K. J. Voorhees, and S. L. Durfee. *Fuel Process. Technol.* **15**, 281 (1987).

Bartle, K. D., and D. W. Jones. In *Analytical Methods for Coal and Coal Products* (C. Karr, ed.), Vol. 2, p. 104, 1979. New York: Academic Press.

Bartle, K. D., T. G. Martin, and D. F. Williams. *Fuel* **54**, 226 (1975).

Bell, C. L. M., R. E. Richards, and R. W. Yorke. *Brennst. Chem.* **39**, 530 (1958).

Berkowitz, N. *Proc. 11th Biennial Lignite Symp., San Antonio, TX, June 1982.*

Berkowitz, N. *The Chemistry of Coal,* 1985. Amsterdam: Elsevier, 1985.

Blayden, H., J. Gibson, and H. L. Riley. *Proc. Conf. Ultrafine Struct. Coals and Cokes,* p. 176. BCURA, London, 1944.

Blom, L. Ph.D. Thesis, University of Delft, The Netherlands (1960).

Blom, L., L. Edelhausen, and D. W. van Krevelen. *Fuel* **36**, 135 (1957).

Brown, J. K., and W. R. Ladner. *Fuel* **39**, 87 (1960).

Bunger, J. W. *Proc. Symp. Anal. Chem. Tar Sands & Oil Shale, New Orleans, Louisiana,* **22**(2) 695 (1977).

Chakrabartty, S. K., and N. Berkowitz. *Fuel* **53**, 240 (1974).

Chakrabartty, S. K., and H. O. Kretschmer. *Fuel* **51**, 160 (1972); **53**, 132 (1974).

Diamond, R. *Acta Crystallogr.* **10**, 359 (1957); **11**, 129 (1958).

Dickie, J. P. and T. F. Yen. *Anal. Chem.* **39**, 1847 (1967).

Dormans, H. N. M., and D. W. van Krevelen. *Fuel* **39**, 273 (1960).

Durfee, S. L., and K. J. Voorhees. *Anal. Chem.* **57**, 2378 (1985).

Ergun, S. *Fuel* **37**, 365 (1958).

Ergun, S., and V. H. Tiensuu. *Fuel* **38**, 64 (1959); *Acta Crystallogr.* **12**, 1050, 1959.

Franklin, R. E. *Acta Crystallogr.* **10**, 105 (1950).

Friedel, R. A. *J. Chem. Phys.* **31**, 280 (1959).

Friedel, R. A., J. A. Queiser, and G. L. Carlson. *ACS Div. Fuel Chem. Prepr.* **15**(1), 123 (1971).

Fujii, S., and F. Yokoyama. *Nenryo Kyokai-shi* **48**, 694 (1969).

Gerstein, B. C., C. Chow, R. G. Pembleton, and R. C. Wilson. *J. Phys. Chem.* **81**, 565 (1977).

Given, P. H. *Fuel* **39**, 147 (1960); **40**, 427 (1961).

Golumbic, C., J. R. Anderson, M. Orchin and H. H. Storch. *US Bur. Mines Rept. Invest. No. 4662, Washington, D.C.* (1950).

Gorin, E. In *Chemistry of Coal Utilization,* 2nd Suppl. Vol. (M. A. Elliott, ed.), 1981. New York: Wiley-Interscience.

Green, T., J. Kovac, D. Brenner, and J. W. Larsen. In *Coal Structure* (R. A. Meyers, ed.), Chapter 6, 1982. San Diego: Academic Press.

Griffiths, P. R. *Chemical Infrared Fourier Transform Spectroscopy,* 1975. New York: Wiley.

Hepler, L. G., and C. Shi. *AOSTRA Tech. Publ. Ser. #6,* 1989. Edmonton, AB: Alberta Oil Sands Technology & Research Authority.

Heredy, L. A., and I. Wender. *ACS Div. Fuel Chem., Prepr.* **25**(4), 38 (1980).

Heredy, L. A., A. E. Kostyo, and M. B. Neuworth. *Fuel* **42**, 182 (1963); **43**, 414 (1964); **44**, 125 (1965); *ACS Adv. Chem. Ser.* **55**, 493 (1966).

Hirsch, P. B. *Proc. Roy. Soc. London A* **226**, 143 (1954); *Phil. Trans Roy. Soc. (London) A* **252**, 68 (1960).

Holmes, S. A. *AOSTRA J. Res.* **2**, 167 (1986).

Hombach, H. P. *Erdol Kohle Erdgas Petrochem.* **28**, 90 (1975); **32**, 85 (1979).

Hombach, H. P., and G. Kölling. *Erdol, Kohle Erdgas Petrochem.* **25**, 644 (1972).

Ignasiak, B. S., and M. Gawlak. *Fuel* **56**, 216 (1977).

Ignasiask, B. S., T. M. Ignasiak, and N. Berkowitz. *Rev. Anal. Chem.* **2**, 265 (1975).

Ignasiak, T. M., A. V. Kemp-Jones, and O. P. Strausz. *J. Org. Chem.* **42**, 312 (1977).

Ihnatowicz, A. *Pr. Gl. Inst. Gorn.*, 125 (1952).

Kann, L. *Fuel* **30**, 47 (1951).

Kotlyar, L. S., J. A. Ripmeester, B. D. Sparks, and J. Woods. *Fuel* **67**, 1529 (1988).

Kotlyar, L. S., C. Morat, and J. A. Ripmeester. *Fuel* **70**, 90 (1991).

Kröger, C., K. Fuhr, and G. Darsow. *Erdöl Kohle* **18**, 36, 701 (1965).

Kroschwitz, J. I. *Encycl. Polymer Sci. and Technol.*, 1985. New York: Wiley & Sons.

Kukharenko, T. A., and V. I. Belikova. *Khim. Tverd. Topl. #13* (1968).

Larsen, J. W., P. Choudhoury, and L. O. Urban. *ACS Div. Fuel Chem., Prepr.* **23**(4), 181 (1978).

Lin, R., and Z. A. Wilk. *Fuel* **74**, 1512 (1995).

Majid, A. and J. A. Ripmeester. *Fuel* **69**, 1527 (1990).

Mayo, F. R. and N. A. Kirshen. *Fuel* **57**, 405 (1978); **58**, 698 (1979).

McIntyre, D. D., D. S. Montgomery, and O. P. Strausz. *AOSTRA J. Res.* **2**, 251 (1986).

McKay, J. F., P. J. Amend, T. E. Cogswell, P. M. Harnsberger, R. B. Erickson, and D. R. Latham. *ACS Adv. Chem. Ser.* **170**, 128 (1978).

Moschopedis, S. E., J. F. Fryer, and J. G. Speight. *Fuel* **55**, 227 (1976).

Nelson, J. B. *Fuel* **32**, 153, 381 (1954).

Orchin, M., C. Golumbic, J. R. Anderson, and H. H. Storch. *U.S. Bur. Mines Bull. No. 505, Washington, D.C.* (1951).

Osjord, E. H., and D. Malthe-Sorenssen. *J. Chromatogr.* **297**, 219 (1983).

Osjord, E. H., H. P. Ronningsen, and L. Tau. *J. High Res. Chrom. & Chrom. Comm.* **8**, 683 (1985).

Pederson, K. S., Aa. Fredenslund, and P. Thomassen. *Properties of Oils and Natural Gas,* Chapter 2, 1989. Houston: Gulf Publ. Co.

Peiser, H. S., M. P. Rooksby, and A. J. C. Wilson. *X-Ray Diffraction of Polycrystalline Materials,* pp. 409–429, 438–453, 1955. London: Institute of Physics.

Pembleton, R. G., L. M. Ryan, and B. C. Gerstein. *Rev. Sci. Instrum.* **48**, 1286 (1977).

Pines, A., M. G. Gibby, and J. S. Waugh. *J. Chem. Phys.* **56**, 1776 (1972); **59**, 569 (1973).

Pitt, J. G. In *Coal and Modern Coal Processing* (J. G. Pitt and G. R. Millward, eds.), 1979. London: Academic Press.

Pugmire, R. J., W. R. Woolfenden, C. L. Mayne and D. M. Grant. *Proc. Basic Coal Sci. Workshop, Houston TX, Dec. 1981.*

Pugmire, R. J., W. R. Woolfenden, C. L. Mayne, J. Karas, and D. M. Grant. In *Chemistry and*

Characterization of Coal Macerals (R. E. Winans and J. C. Crelling, eds.), Chapter 6, 1984. Washington, D.C.: ACS Symp. Ser.
Rao, H. S., P. L. Gupta, F. Kaiser, and A. Lahiri. *Fuel* **41**, 417 (1962).
Ravey, J. C., G. Doucouret, and D. Espinat. *Fuel* **67**, 1560 (1988).
Retcofsky, H. L. *Appl. Spectrosc.* **31**, 116 (1977).
Robinson, W. E. In *Organic Geochemistry* (G. Eglinton and M. T. J. Murphy, eds.), p. 619, 1969. Berlin: Springer.
Roy, M. M. *Fuel* **36**, 249 (1957).
Selucky, M., T. Ruo, Y. Chu, and O. F. Strausz. *ACS Adv. Chem. Ser.* **170**, 117 (1978).
Selucky, M. L., S. S. Kim, F. Skinner, and O. P. Strausz. *ACS Adv. Chem. Ser.* **195**, 84 (1981).
Sfihi, H., M. F. Quinton, A. Legard, S. Pregermain, D. Carson, and P. Chiche. *Fuel* **65**, 1007 (1986).
Shaw, D. *Fourier Transform NMR Spectroscopy*, 1976. Amsterdam: Elsevier.
Solomon, P. R. *ACS Div. Fuel Chem., Prep.* **24**(2), 184 (1979).
Speight, J. G. *Proc. Natl. Sci. Found. Symp. Fund. Org. Chem. Coal, Knoxville, TN, 1975,* p. 125.
Speight, J. G. *Info. Ser. No. 81,* 1978. Edmonton, AB: Alberta Research Council.
Speight, J. G. *The Chemistry and Technology of Petroleum,* 2nd ed., 1991. New York: Marcel Dekker.
Speight, J. G., D. L. Wernick, K. A. Gould, R. E. Overfield, B. M. L. Rao, and D. W. Savage. *Inst. Français Petr. Rev.* **40**, 51 (1985).
Sternberg, H. W., and C. L. Delle Donne. *Fuel* **53**, 172 (1974).
Storm, D. A., R. J. Barresi, and S. J. DeCanio. *Fuel* **70**, 779 (1991).
Strausz, O. P. In *Fundamentals of Resid. Upgrading* (R. H. Heck and T. F. Degnan, eds.), 1989. New York: AlCHemE.
Strausz, O. P., T. W. Mojelsky, and E. M. Lown. *Fuel* **71**, 1355 (1992).
Sun, J. Y., and E. H. Burk. *Proc. Workshop on Fund. Org. Chem. of Coal, U. of Tennessee, PB-264119 1975,* p. 80.
Tissot, B. P. & J. Espitalié. *Inst. Français Petr. Rev.* **30**, 743 (1975).
Tissot, B. P., and D. H. Welte. *Petroleum Formation and Occurrence,* p. 361, 1978; 2nd ed., 1984. Berlin: Springer.
van Krevelen, D. W. *Brennst. Chem.* **35**, 289 (1954).
van Krevelen, D. W. *Coal,* 1961. Amsterdam: Elsevier.
van Krevelen, D. W. *Fuel* **45**, 99, 229 (1966).
Vaughan, R. W., L. S. Schreiber, and J. A. Schwartz. *ACS Symp, Ser.* **34**, 275 (1976).
Vluchter, J. C., H. I. Waterman, and H. A. van Westen. *J. Inst. Petrol. Technol.* **21**, 661, 701 (1935).
Whitehurst, D. D., M. Farcasiu, T. O. Mitchell, and J. K. Dickert. *EPRI-AF-480, Research Project 410-1, Final Rept.* (1977).
Winniford, R. S. *ASTM Spec. Tech. Publ. No. 294,* 31 (1960).
Witherspoon, R. S., and P. A. Winniford. *Proc. Conf. Chem. and Chem. Process. Petrol. Nat. Gas, Budapest 1968,* 967.
Witherspoon, R. S., and P. A. Winniford. In *Fundamental Aspects of Petroleum Geochemistry,* (B. Nagy and U. Colombo, eds.), p. 261, 1967. Amsterdam: Elsevier.
Yarzab *et al.* (1979); Yarzab *et al.* (1980) p. 19.
Yen, T. F. *ACS Div. Petrol. Chem., Prepr.* **24**(4), 901 (1979).
Yen, T. F. *Proc. 2nd Internatl. Conf. Heavy Crude & Tar Sands, Caracas, Venezuela, 1982.*
Zilm, K. W., R. J. Pugmire, D. M. Grant, R. E. Wood, and W. H. Wiser. *Fuel* **58**, 11 (1979).

CHAPTER 8

Preparation

Regardless of how they are produced, crude fossil hydrocarbons usually contain environmentally and/or economically unacceptable contaminants—H_2O, condensates and H_2S in natural gas; saline formation waters, waxy matter, and asphaltics in crude oils; excessive amounts of H_2O, dust, and mineral matter in coal—that require removal in order to meet market specifications or criteria for downstream processing. It is therefore almost always mandatory to take them through a sequence of preparation steps [1] that depend on *what is to be prepared* and *what removed.*

1. NATURAL GAS

Raw natural gas contains substantial concentrations of C_2–C_6 hydrocarbons as well as H_2S and other inorganic contaminants (see Chapter 5) and is, in fact, designated at the wellhead as *sweet* or *sour* if holding $<0.5\%$ or $>2.0\%$ H_2S, and *wet* or *dry* if containing >4.01 or <1.34 liters of condensible liquids per 100 m^3 [2].

Preparation consequently entails drying, extraction of C_2+ hydrocarbons, desulfurization, etc., and the techniques used for these purposes are virtually identical with the procedures used to clean other gas streams for their respective destinations.

Drying

Since condensation of moisture impedes pipeline transport and/or poses downstream problems by corroding equipment, preparation of natural gas usually begins with removal of water. This can be accomplished by cryogenic drying, but under field conditions is more efficiently and cost-effectively achieved by contacting the raw gas with a hygroscopic liquid (normally ethylene glycol) and running continuously with two units in series—one operating while the other regenerates the spent sorbent. Regeneration can be effected with bauxite (which will remove ~6 wt% H_2O from the sorbent), SiO_2 or Al_2O_3 (which can extract up to ~8% H_2O), or a molecular sieve (which can accommodate as much as 15% H_2O); if the gas is sour, however, which it frequently is, a molecular sieve must be preceded by SiO_2 in order to prevent its premature plugging by elemental sulfur from dissolved H_2S.

Recovery of Condensates

Gas liquids, which are typically comprised of C_2–C_6 hydrocarbons, are routinely extracted by compressing the dry gas and sending it through cold *n*-hexane. This furnishes a socalled "rich light oil" stream that can be fractionated by distillation into ethane, ethylene, and C_3–C_6 alkanes designated as *liquefied petroleum gases* (LPGs).

Removal of Acid Gases

Extraction of H_2S

If natural gas is destined for use as primary fuel, it is only necessary to substantially free it of associated H_2S, and if concentrations of H_2S are relatively modest (<10%), this can be efficiently done by recourse to the Stretford or the Giammarco–Vetrocoke process (U.S. Dept. of Commerce, 1976; Riesenfeld and Kohl, 1985). Both are capable of lowering H_2S levels in the gas to ~1 ppm, and yield elemental sulfur without requiring tail-gas cleanup; but neither delivers pure sulfur, and neither is therefore a method of choice if processing economics dictate recovery of *saleable* sulfur.

The Stretford process uses an aqueous solution of Na_2CO_3, Na_2VO_3, and a proprietary oxidation catalyst to extract H_2S by formation of a sodium acid sulfide that generates elemental sulfur by reduction of V^{5+} to V^{4+}. Air-blowing then restores V^{4+} to the higher valency state.

In the Giammarco–Vetrocoke process, the feed gas is contacted by aqueous potassium arsenite (K_3AsO_3), which reacts with H_2S to form the thioarsenite (K_3AsS_3). This then reacts with potassium arsenate to produce a monothioarsenate, and K_2AsO_3 is finally regenerated by air blowing, which also furnishes elemental S.

However, if H_2S concentrations in the raw gas exceed ~10% and/or pure elemental sulfur is sought as by-product, desulfurization mandates use of the Claus process, which converts H_2S to elemental S by a net reaction such as

$$2H_2S + O_2 \rightarrow 2S + 2H_2O.$$

In the first stage of this process, the gas is combusted in limited air to yield

SO_2, H_2S, and elemental S corresponding to ~65% conversion, and in the second, that mix is then made to generate more free S via

$$2H_2S + SO_2 \rightarrow 3S + 2H_2O$$

by passing it over bauxite or an Al_2O_3 catalyst at ~215°C. With two or three catalytic converters in the second stage, and gas reheating between the first and second, ~96% of H_2S in the feed gas can in this manner be drawn off as pure liquid sulfur.

To improve the thermal efficiency of gas desulfurization, some attention has also been given to desulfurization of *hot* gases. In one such process, developed by Conoco (Curran *et al.*, 1973), the hot feed is moved through fluidized dolomite at 915°C/1.5 MPa in order to initiate

$$H_2S + CaCO_3 \cdot MgO \rightarrow CaS \cdot MgO + H_2O + CO_2,$$

and the steam + CO_2 produced by this reaction is then used to transfer $CaS \cdot MgO$ to a regenerator where the reverse occurs at ~700°C. The more concentrated H_2S streams thereby generated are sent to Claus processing.

A similar scheme, which operates with much the same chemistry, employs a $Li_2CO_3/CaCO_3/K_2CO_3/Na_2CO_3$ melt which is thought to react mainly through $CaCO_3$ (Moore 1977). Desulfurization proceeds optimally at ~800°C, and as in the dolomite process, spent melt is regenerated by reversing the reaction.

Complete Acid-Gas Extraction

If virtually *complete* removal of H_2S, CO_2, and other acid gas components is mandated—as is, for example, the case when a syngas (see Chapter 10) has to be prepared for processing—recourse can be had to a solid sorbent or to absorption in a liquid (or solution) with which acid gases will react reversibly. Both methods are technically identical with procedures for stripping undesirable components from other raw gas streams, and in both, the spent sorbent is regenerated by raising the temperature and/or flashing it to a lower pressure. Figure 8.1.1 exemplifies this mode of operation.

How CO_2- and H_2S-rich wastes from sorbent regeneration are finally disposed of does, however, depend on the waste-gas volumes and compositions.

As a rule, "rich" gas streams with H_2S concentrations greater than ~8–10% are directly taken to a Claus plant for sulfur recovery, and small volumes of "lean" gas with relatively low H_2S contents are either incinerated and SO_2 vented or, where that is disallowed by emission standards, H_2S and other noxious impurities are fixed by sorption on silica gel (which also abstracts any residual light hydrocarbons), activated carbon, or a molecular sieve such as Linde Type 4A or 5A (which selectively sorb H_2S, SO_2, COS, CO_2, and mercaptans) [3]. Spent sorbents are then regenerated by heating in dry air or

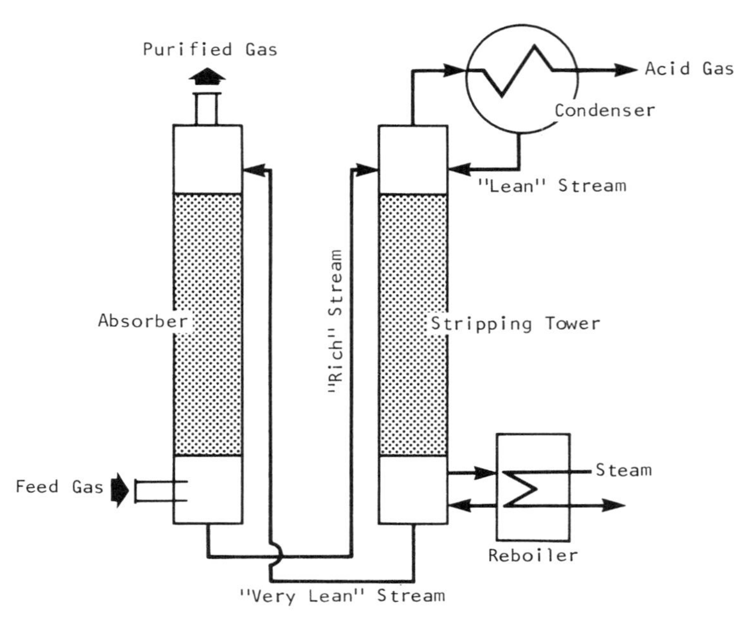

FIGURE 8.1.1 A flowsheet for hot carbonate gas purification.

N_2 at 120–175°C, and the released, concentrated gas is finally disposed of by Claus processing.

The same approach is taken when acid gases are abstracted by absorption in liquids—usually aqueous alkanoamines, methanol, *N*-methyl-2-pyrrolidone, etc., or, in some cases, solutions of copper–ammonium salts (preferably formates, carbonates, or acetates) [4]. Table 8.1.1 lists the most common sorbents for such extraction, and shows the levels to which they can reduce H_2S and CO_2 concentrations in the gas.

The use of alkanolamines was pioneered by Bottoms (1930), but the triethanolamine (TEA) initially employed was quickly replaced by monoethanolamine (MEA) and diethanolamine (DEA), which offered greater stability and higher absorption capacities. Diisopropanolamine (DIPA) is also sometimes used (Bally, 1960). The basic process chemistry is exemplified by the equilibrium reactions

$$2\,RNH_2 + H_2S \rightarrow (RNH_3)_2S$$

$$(RNH_3)_2S + H_2S \rightarrow 2\,RNH_3SH$$

$$2\,RNH_2 + CO_2 + H_2O \rightarrow (RNH_3)_2CO_3$$

$$(RNH_3)_2CO_3 + CO_2 + H_2O \rightarrow 2\,RNH_3HCO_3$$

$$2\,RNH_2 + CO_2 \rightarrow RNHCOONH_3R,$$

TABLE 8.1.1 Acid Gas Removal Processes[a]

Process	Scrubbing fluid	Attainable purity (ppm)	
		CO_2	H_2S
Chemical absorption:			
Alkazid	25% aq. soln. of M[b] or DIK[c]	<0.2%	15
Amine	15% aq. monoethanolamine [5]	20	1
	or		
	25% aq. diethanolamine	200	1
Econamine	50–70% aq. diglycolamine	100	4
Hot carbonate	20–30% aq. K_2CO_3	0.2%	15
Shell	diisopropanolamine	100	4
Physical absorption:			
Purisol	*N*-methyl-2-pyrrolidone	10	2
Rectisol	methanol	10	<1
Selexol	Me_2-poly(ethylene)glycol ether	0.5%	<4
Sulfinol	sulfolane[d]/diisopropanolamine	200	1

[a] Riesenfeld and Kohl (1985).
[b] Potassium salt of methyl-aminopropionic acid.
[c] Potassium salt of dimethyl-aminoacetic acid; this is more selective for H_2S in presence of CO_2.
[d] Tetrahydrothiophene dioxide.

and compositions of equilibrium solutions vary with the partial pressures of the acid gases over them.

Aqueous monoethanolamine is the preferred sorbent when maximum removal of H_2S and CO_2 at relatively low pressures is sought, and mixtures of MEA and di- or triethylene glycol are widely used (as in the glycolamine process) to remove acid gases *and simultaneously dry* natural gas. Aqueous diethanolamine is the absorbent of choice for treating gases with high concentrations of COS and/or CS_2.

Methanol, used in Lurgi's Rectisol process between −18 and −62°C, can remove organic sulfur compounds and HCN as well as CO_2 and H_2S, but is particularly advantaged by its ability to *selectively* dissolve CO_2 and H_2S at low temperatures and pressures above ~1 MPa (see Table 8.1.2). Table 8.1.3

TABLE 8.1.2 Equilibrium Solubilities of CO_2 and H_2S in Methanol

Temperature	Solubility (v.v)		Selectivity H_2S/CO_2
	CO_2	H_2S	
−10°C	9	41	5.1
−30°C	15	92	6.1

TABLE 8.1.3 Rectisol Purification of a Syngas[a,b]

Component	Concentration (mol %)	
	Inlet gas	Outlet gas
$H_2S + CO_2$	30.2	0.9
CO	19.8	28.5
H_2	38.6	55.6
CH_4	11.0	14.2
N_2	0.4	0.8

[a] Hoogendorn and Solomon (1957).
[b] At SASOL I (see Chapter 10).

illustrates this selectivity with typical data for Rectisol purification of a syngas. However, *N*-methyl-2-pyrrolidone, the absorbent in the Purisol process, is claimed to be equally effective for selectively extracting H_2S in the presence of CO_2, and reportedly delivers a clean gas with <0.1% CO_2, 2–3 ppm H_2S (Hochgesang, 1968).

The dimethyl ether of poly(ethylene) glycol, used in the Selexol process, dissolves 9 times more H_2S than CO_2, and therefore also tends to remove H_2S selectively, as well as almost all COS and methyl mercaptans, but it does not quite achieve the exit-gas purity attainable by the Rectisol or Purisol processes.

The mixture of diisopropanolamine, sulfolane $(C_2H_4)_2 \cdot SO_2$, and water used in Shell's Sulfinol process will efficiently extract H_2S, but suffers from leaving relatively high CO_2 concentrations in the cleaned gas.

Choices among these technologies are therefore ultimately determined by cleaning requirements and process economics.

Aqueous solutions of copper–ammonium salts, to which reference was made earlier, are mainly used for removal of CO and CO_2 from other raw gas streams—notably from H_2 required for synthesis of ammonia (see Chapter 10). The sorption chemistry apparently involves

$$Cu(NH_3)^{2+} + CO + NH_3 \rightarrow Cu(NH_3)_3(CO)^+,$$

and the secondary reactions resulting in absorption of CO_2 are probably

$$2\,NH_4OH + CO_2 \rightarrow (NH_4)_2CO_3 + H_2O$$
$$(NH_4)_2CO_3 + CO_2 + H_2O \rightarrow 2\,NH_4HCO_3.$$

A typical CO-absorption system reduces CO from ~3.5% in the inlet gas stream to less than 25 ppm (Yeandle and Klein, 1952).

An alternative for extraction of CO offers itself in Tenneco Chemicals' COsorb process (Walker, 1975), in which the sorbent is a solution of $CuAlCl_4$

in toluene or another suitable aromatic compound. This selectively takes up CO and releases it again when heated to ~135°C in a stripper unit. The volatile solvent is recovered by compression, refrigeration, or sorption on active carbon [6].

2. REFINERY GASES

C_1–C_4 hydrocarbons accruing as by-products of refinery operations (and therefore termed "refinery gases") are compressed, cooled to liquefy the bulk of C_3 and C_4 hydrocarbons, and transferred to separation drums from which CH_4 and C_2H_6 is withdrawn overhead for use as fuels. C_3/C_4 streams are disposed of as LPGs. Figure 8.2.1 shows a flowsheet for such operations.

Refinery gases with comparatively high concentrations of unsaturates are similarly processed. But instead of being used as fuels, as are the more or less olefin-free gases, they are disposed of as petrochemical feedstocks—with C_2H_4 going to polyethylene manufacture, and propylene/butylenes streams being

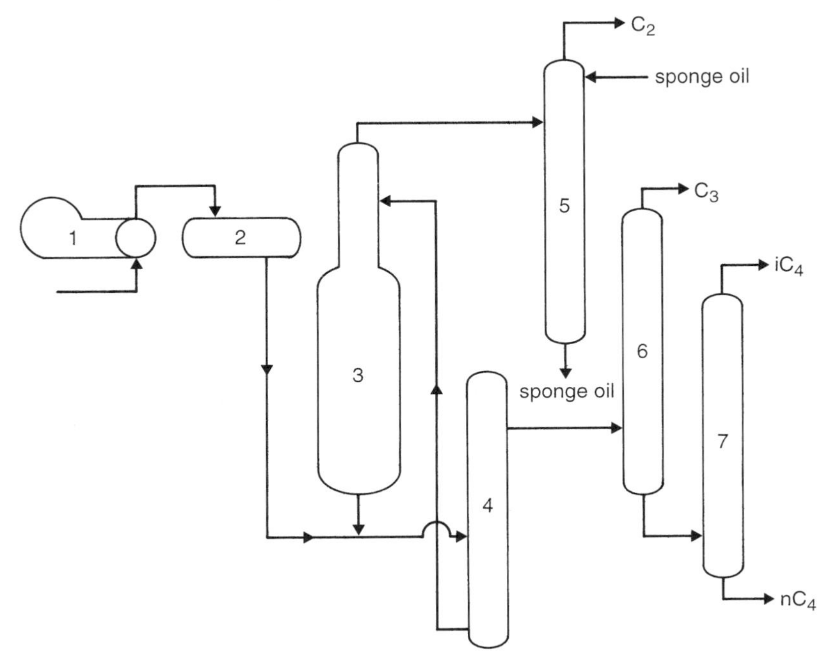

FIGURE 8.2.1 A flowsheet for refinery gas separation. 1, compressor; 2, phase separator; 3, absorber; 4, debutanizer; 5, sponge absorber; 6, depropanizer; 7, de*iso*butanizer.

alkylated at 4.5°C over HF or H_2SO_4 and thereby transformed into propane, *n*-butane, and high-octane blend components for gasoline.

3. CRUDE OILS

Desalting

Analogous to drying of raw natural gas, processing of crude oil begins with separation of formation waters that are coproduced with oil and usually contain enough Cl^- to damage refinery equipment by forming HCl at elevated temperatures. These fluids are removed by *desalting,* which can be accomplished thermally or by emulsion breaking.

Thermal desalting is achieved by heating the crude to 90–150°C/0.35–1.75 MPa, at which little vapor loss is incurred while oil and water form two easily separated phases: in most cases rapid separation and settling requires no more than sending the hot stream through a sand or gravel bed.

For phase separation by emulsion breaking, a fatty acid, sulfonate, soap, or long-chain alcohol is added to the crude, and formation of two phases is accelerated by applying a high-potential electric gradient across the oil/H_2O mix.

In both cases, nearly all dissolved mineral matter in the crude moves into the aqueous phase and is removed with it.

Primary Fractionation

Separation of crude oils into primary components is rooted in European processing of coal tars and shale oils, and was initially conducted in retort-like kettles from which oil began to distill at ~82°C and furnished three fractions in particular demand at the time—a 72–74 API gasoline, a 62–65 API naphtha, and a 40–50 API kerosine. Retorts were, however, quickly replaced by more efficient shell-stills, and these, in turn, soon gave way to the progressively improved fractionation towers now in use.

The fundamentals of modern fractionation practices and the major product streams generated by them are shown in Fig. 8.3.1 and identified in Table 8.3.1. But because crudes from different pools can yield markedly dissimilar product slates, the minutiae of fractionation and downstream processing sequences (see Chapter 10) are dictated by feedstock composition and quality.

Modern towers (Figure 8.3.2) accept preheated crude, and operate in much the same manner as the pipe stills that serve to fractionate coal tars (see Chapter 9). Effective fractionation is sought by fitting the towers with bubble-

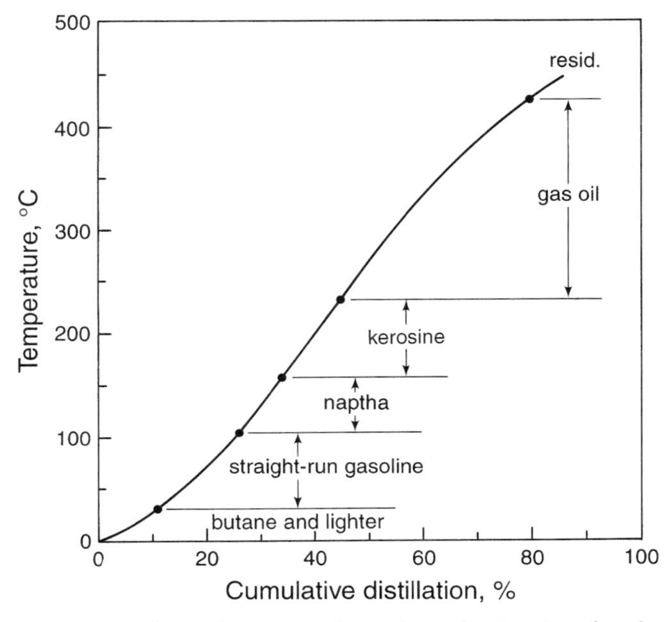

FIGURE 8.3.1 The product streams from primary fractionation of crude oil.

cap distillation plates, and separation of gas oils from lighter fractions is maximized by injecting the crude oils at ~300°C [7]. Even so, however, most product streams must be freed of more volatile components that may have misreported to them; unacceptably wide-cut naphthas must be resolved into light and heavy naphthas in "rerun" towers, which are also used to strip chemically treated fractions of process chemicals and waste products generated by them; and stripped straight-run gasolines as well as light naphtha fractions

TABLE 8.3.1 **Product Streams from Primary Fractionation of Crude Oil**

Designation	Boiling range [°C (°F)]		Disposition
Propane/butanes	<32	(<90)	gas processing
Straight-run gasoline	32–105	(90–220)	motor gasoline blending
Naphtha	105–160	(220–310)	catalytic reforming
Kerosine	160–230	(315–450)	hydrotreating
Light gas oil	230–340	(450–650)	distillate fuel blending
Heavy gas oil	340–425	(650–800)	catalytic cracking
Straight-run resid.	>425	(>800)	flashing

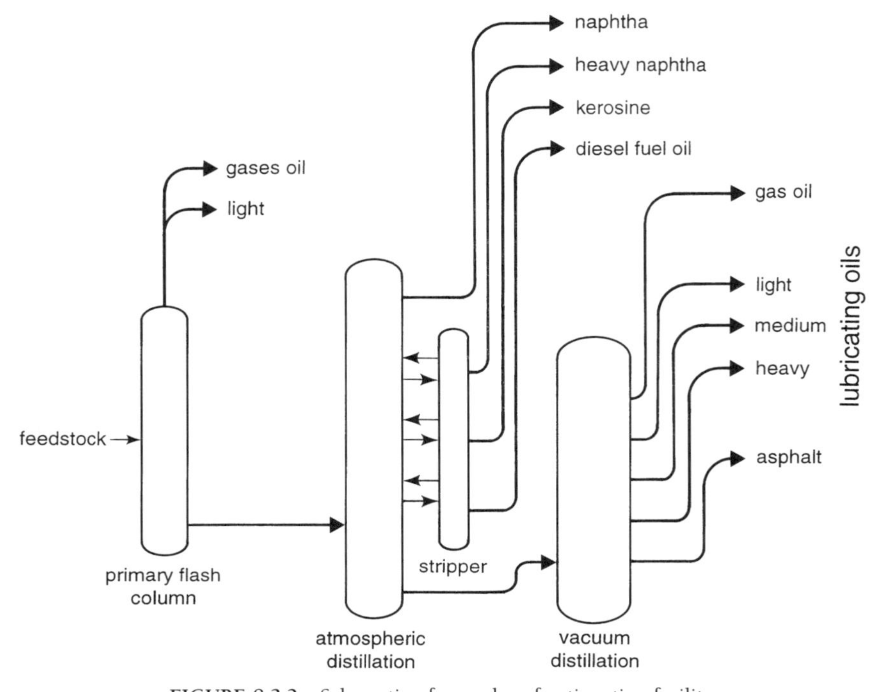

FIGURE 8.3.2 Schematic of a modern fractionating facility.

must be stabilized by removing "light ends" (see Table 8.3.2) and thereby lowering their vapor pressure for safe storage and minimal vaporization.

On occasion, fractionation is augmented by azeotropic and/or extractive distillation to recover particular components of a crude oil or oil fraction. Separation of butadiene from butylene is thus accomplished by liquid NH_3, with which it forms an azeotrope; extractive distillation with an H_2O/acetone mixture serves to separate butanes from butylene; and other solvents are similarly used to separate heavier aromatics, e.g., anthracene and phenanthrene from carbazole (see Chapter 9).

Deasphalting

Formation of "coke"—a residual carbonaceous solid with very low H contents—during thermal processing [8] of middle distillates and heavier fractions (such as gas oils and residua) can be minimized by stripping them of asphaltic and resinous matter with fluids that quickly dissolve or precipitate such matter.

TABLE 8.3.2 The "Light Hydrocarbon Ends"

Hydrocarbon	Mol. wt.	b.p. (°C)	Disposition/use
Methane	16	−182	fuel gas
Ethylene	28	−104	fuel gas, petrochem. feedstock
Ethane	30	−89	fuel gas
Propylene	42	−48	fuel gas, petrochem. feedstock
Propane	44	−42	fuel gas (LPG)
Butylene-1[a]	56	−6	motor gasoline
Butylene-2[a]	56	1	synthetic rubber
Isobutylene	56	−7	alkylate, chemical feedstock
n-Butane	58	−1	motor gasoline
Isobutane	58	−12	motor gasoline, alkylate
Pentylenes	70	30	motor gasoline
n-Pentane	72	36	motor gasoline, aviation fuel
Isopentane	72	28	motor gasoline, aviation fuel
n-Hexane	86	69	motor gasoline, aviation fuel
Isohexane	86	61	motor gasoline, aviation fuel

[a] Differentiated by the double-bond position: butylene-1 (≡ butene-1 or but-1-ene) is $CH_3CH_2CH{=}CH_2$; butylene-2 (≡ butene-2 or but-2-ene) is $CH_3CH{=}CHCH_3$.

These include C_3H_8 and other low-molecular-weight hydrocarbons, as well as aliphatic ethers and some alcohols [9], and enable deasphalting by either of two procedures.

In one, the oil is diluted with several times its volume of liquified propane, moved into a settling tank, pressurized to ~140 kPa, and heated to 50 ± 20°C. The slight pressure maintains propane in its liquid state, and precipitation can be accelerated by using the higher of the indicated temperatures. After recovery of asphaltics for later processing (see Chapter 10), the oil/propane mix can then also be dewaxed (see below) by cooling to ~40°C and extracting wax crystals by filtration. Residual propane is recovered by steam-heating the filtrate. Alternatively, the oil can be dewaxed by injecting liquefied propane near the bottom of a countercurrent extractor tower that the oil enters near the top. The diluted deasphalted oil then exits with propane at the top of the tower, and the precipitated asphaltics are withdrawn near its bottom.

The utility of these procedures is increased by the fact that propane similarly serves to segregate residua from, for example, vacuum distillation into different grades of lubricating oils.

Dewaxing

Instead of by solvent extraction (as noted above), crude oil fractions are now sometimes dewaxed by more recently introduced catalytic hydrocracking at

290–400°C/2-10 MPa. Examples are a Mobil Oil process that selectively cracks *n*-paraffins and isoparaffins with short side chains over a proprietary catalyst, and a British Petroleum process that selectively cracks *n*-paraffins over a platinized molecular sieve and minimizes secondary cracking by concurrent hydrogenation of reactive intermediates.

Refining

As well as treating oil fractions with clays to remove unwanted color and undesirable components (such as acidic and asphaltic matter, diolefins, and resins [10]) by sorption, several other methods are used in suitable formats to improve oil quality. Although often loosely referred to as *refining* and considered to be processing procedures, they are noted here because they are essentially *preparative* techniques. Examples include:

1. Transformation of thiols (or mercaptans) into more acceptable disulfides by

(a) controlled air oxidation, as in

$$4\ RSH + O_2 \rightarrow 2\ RS{-}SR + H_2O;$$

(b) oxidative reaction with Cu^{2+} salts, as in

$$4\ RSH + 2\ CuCl_2 \rightarrow RS{-}SR + 2\ CuSR + 4\ HCl$$

in which case Cu^{2+} is regenerated by

$$2\ CuSR + 2\ CuCl_2 \rightarrow RS{-}SR + 4\ CuCl$$

$$4\ CuCl + 4\ HCl + O_2 \rightarrow 4\ CuCl_2 + H_2O;$$

or (c) interaction with sodium plumbite, as in

$$2\ RSH + Na_2PbO_2 \rightarrow Pb(RS)_2 + 2\ NaOH$$

$$Pb(RS)_2 + S \rightarrow PbS + RS{-}SR$$

and regenerating Na_2PbO_2 by air blowing:

$$PbS + 4\ NaOH + 2O_2 \rightarrow Na_2PbO_2 + Na_2SO_4 + H_2O.$$

2. Removal of unwanted naphthenic or aromatic components from an oil fraction by

(a) Extraction with liquid SO_2 at 0°C to generate aromatics-free solutions from which nonaromatics are obtained by addition of kerosine, which is recovered with SO_2 by stripping and fractionation or (b) extraction of aromatics from mixed hydrocarbons with diethylene glycol and H_2O at ~120°C (as in the UDEX process).

3. Extraction of catalytic reformates with sulfolane to free them of lower aromatics.

4. Extraction of naphthas, middle distillates, and gas oils with HF, usually at 50°C and slightly elevated pressure, to remove sulfur and high-molecular-weight species given to coke formation.

4. HEAVY HYDROCARBONS

Bitumen

Preparation of *mined* bituminous sands [11] centers on separation of bitumen from the massive amounts of inorganic matter with which it is intimately associated [12]. In principle, this can be accomplished by solvent extraction or retorting. But these routes are deemed to be insufficiently well developed for high-volume processing [13], and the two currently active commercial plants [14] therefore use a hot water extraction process (HWEP), which began to be studied by K. A. Clark and co-workers at the Alberta Research Council in the 1920s. This entails mixing the raw sand with hot (~85°C) water and sufficient NaOH to maintain the pH at 8.5–9.0, and moving the slurry to a separation chamber from which a bitumen froth with ~6% solids and 30–40% water exits. The froth is then diluted with naphtha and centrifuged to less than ~1% solids; the naphtha is recovered by distillation; and the bitumen is sent to processing. Tailings from extraction and froth treatment are ponded (Houlihan, 1980; ERCB, 1984).

From sands with >10% bitumen, HWEP recovers up to 90% bitumen as an 8–9° API hydrocarbon melange, of which half can be distilled without concurrent cracking. But as this efficiency falls to ~80 or 60%, respectively if the sands contain only 8 or 6% bitumen [15], and the tailings tend to form colloidal suspensions that disallow substantial recycling of process water [16], some attention has been given to alternative extraction techniques—notably to direct retorting [17] and to solvent-extraction methods such as the SESA and Magna processes [18]. The latter, like others of this genre, are technically less developed than retorting, but are considered likely to play important roles in future large-scale processing (Houlihan and Williams, 1987).

The SESA process (Sparks and Meadus, 1979, 1981; [19]) mills and extracts the raw oil sand with naphtha, and thereby generates two separately handled process streams. The extracted sand is agglomerated by addition of water and rapid stirring, washed once more with naphtha to remove residual bitumen, and then brought to 110°C to recover the solvent and generate easily disposable, dry tailings. From the bitumen/solvent stream, which is combined with the naphtha washings of the extracted sand, the solvent is recovered by distillation, which produces a virtually solvent-free feed with >99.4% bitumen for downstream processing by hydrotreating or coking (see Chapter 9).

Systematic tests of SESA technology with Athabasca oil sands indicate a capability of recovering 95% bitumen, regardless of ore quality, at significantly lower energy expenditures than are associated with HWEP.

The Magna process, which recovers ~90% bitumen regardless of ore grade, extracts the bitumen with cold water and kerosine in two successive stages, and separates the resultant hydrocarbon froth at ~85°C into three zones—an upper zone, which contains ~94.8% hydrocarbons, 5% water, and 0.2% solids and is taken directly to solvent recovery; a middle aqueous layer, which is recycled to feed conditioning; and a bottom layer, comprising ~60% water, 20% organics, and 20% solids, which is centrifuged to remove solids and recover hydrocarbon material. After solvent recovery by two-stage distillation and stripping, the process is reported to deliver a bitumen stream with ≤1% solids, but energy expenditures are still undetermined.

Other solvent extraction methods for which some technical details have been disclosed (ERCB, 1984), but which are still in early stages of development and face a more uncertain future, include

1. the Dravo process, which employs a percolating moving-bed reactor to extract bitumen from the ore with a hot naphtha-solution of bitumen, and
2. Standard Oil's SOHIO process, which conditions the ore with aqueous Na_2CO_3 at 70°C/pH ~8.5 before extracting it with countercurrent kerosine

Bitumen in strata *beyond* surface-mineable depths are recoverable by steam stimulation or partial *in-situ* combustion methods that closely resemble techniques used for producing heavy oils at acceptable rates—i.e., by heating the formation in order to lower the viscosity of the bitumen to produceable levels, and in such cases, preparation centers primarily, as in preparation of crude oils, on stripping the bitumen of associated salty formation fluids.

Oil Shales

Since retorting of oil shale, which is usually conducted at temperatures between 480 and 510°C (Chapter 9), is disadvantaged by extensive coking, and consequently by appreciably lower yields of hydrocarbon liquids than potentially obtainable, some interest has been expressed in *thermal dissolution,* which proceeds during pyrolysis at 300–400°C in the presence of a solvent that

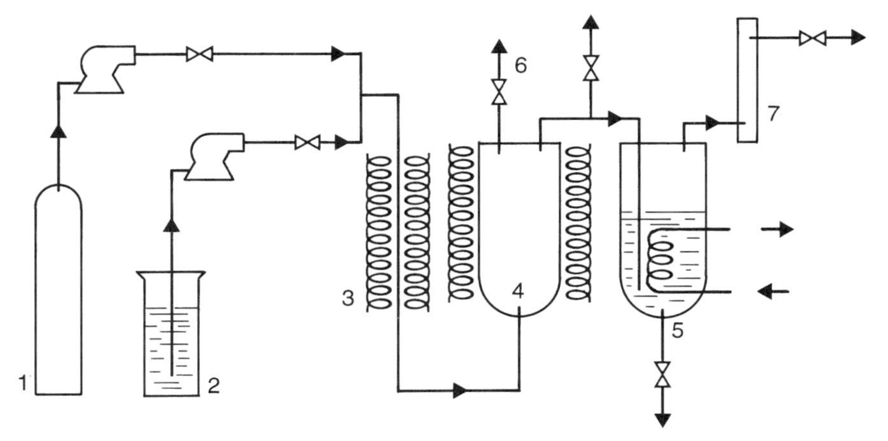

FIGURE 8.4.1 A laboratory assembly for supercritical extraction (Berkowitz and Calderon, 1990). 1, High-pressure gas cylinder (nitrogen, CO); 2, solvent reservoir; 3, preheater; 4, stirred autoclave; 5, pressure letdown vessel; 6, sampling valve; 7, activated carbon trap.

isolates soluble matter as soon as it forms. This solvent can be, but is not necessarily, a hydrogen donor.

First studied in the 1920s (Day, 1922; Ryan, 1920, 1928; Hampton, 1928, 1929), the technique was only sporadically pursued during the 1940s and 1950s as means for oil shale processing (Dulhunty, 1942; Buchan, 1949; Jensen *et al.*, 1953), but gained renewed attention in the 1970s and 1980s as a promising method for recovering oils or oil precursors when operated under (nominally) *supercritical* conditions (Cummins *et al.*, 1976, 1978, 1980; Kesavan *et al.*, 1988; Kramer and Levy, 1989; Ogunsola and Berkowitz, 1995). The fluids used for such extraction included toluene, methanol, water, and mixtures thereof, and extraction was typically conducted at 350–425°C under pressures of 7–14 MPa. In some cases up to 90% of the organic matter in the shale could be recovered. Laboratory studies (see Fig. 8.4.1) indicate that

1. supercritical fluid extraction with H_2O in the presence of an H donor or of H_2 generated by an *in-situ* shift reaction may prove superior to retorting,
2. comminution of the raw shale before processing may be obviated by observations that particles as large as several centimeters respond well to extraction, and
3. uncontrolled (and undesirably extensive) thermal degradation of organic matter in the shale increases in the sequence solvent extraction < supercritical extraction < pyrolysis [21]

Coal

Preparation of coal—often referred to as *beneficiation* or upgrading, and designed to meet end-use specifications and/or reduce transportation costs—centers on drying, sizing, and cleaning and parallels preparation of crude oils and bitumen.

Drying

Inherent moisture contents greater than 12–15% (see Chapter 5) are only encountered in low-rank coals, which are currently mainly used for on-site power generation [22]. Dewatering and drying [23] is therefore, as a rule, restricted to washed bituminous coals, which must meet market-dictated specifications, and to lignites or subbituminous coals used for the production of briquettes and other specialty products, such as absorbent or activated carbons.

Sufficiently coarse (>6-mm) washed bituminous coals and anthracites are routinely dewatered by drainage on inclined vibrating screens that avoid packing and/or stratification of small particles by forcing the bed to periodically cross a set of dams. Coal < 6 mm and cleaned fine coal from flotation cells is usually dewatered by cyclones, centrifuges, or pressure filters that are sometimes preceded by conventional thickeners (Sandy and Matoney, 1979).

For drying, the equipment of choice are rotary kilns, cascade dryers or, where coal size allows it, entrainment dryers operating with a fluidized coal bed. But unless low-rank coals are intended for immediate further processing (as in briquetting), reducing their moisture contents to less than their *air-dried* moisture contents (see Chapter 5) is unproductive, as moisture will be quickly reabsorbed from the air. This may, indeed, present a serious fire hazard from autogenous heating when coal is stored in the open [24].

Sizing

Screening of superficially dry or dried coal to specified size ranges is now only mandatory if the coal is to be used in combustion devices that demand specific fuel sizes for satisfactory performance. Table 8.4.1 lists the standard commercial size designations [25].

For pulverized fuel combustion, which is the preferred operating mode in central power stations, the coal is comminuted to the required nominal size (<74 μm) in ball mills from which it is pneumatically conveyed to the boilers by combustion air through internal screens.

TABLE 8.4.1 Coal Size Designations and Size Limits[a,b]

Size designation	Top size (in.)	Bottom size (in.)
Run-of-mine	variable	0
Large lump	variable	4 (10)
Lump	variable	1 (2.5)
Cobble, egg, stove[c]	6 (15.2)	2 (5)
Nut	2 (5)	$\frac{3}{4}$ (2)
Prepared stoker		
Large	2 (5)	$\frac{1}{4}$ (0.6)
Intermediate	1 (2.5)	$\frac{1}{8}$ (0.3)
Small	$\frac{3}{4}$ (2)	$\frac{1}{16}$ (0.2)
Nut slack	2 (5)	0
Slack	1.25 (3)	0
Fines	0.5 (1.3)	0

[a] Determined with round hole screens.
[b] Sizes in parentheses are approximate metric equivalents (cm).
[c] Alternative designations.

Cleaning

Some reduction of inorganic material is often imperative if the coal is to be environmentally acceptable, and when it must be transported to distant markets, partial removal of mineral matter can also bring significant economic benefits from lower transportation costs per unit weight of *useable* carbon. But selection of optimum cleaning methods depends on the distribution of mineral matter in the coal [26], and this must be specifically assessed in each case.

Preliminary information is generally provided by "float-and-sink" tests, which quantify the yields and ash contents of cleaned coal (or "floats") at specific gravities between 1.20 and 2.20, and also show the proportions of "middlings," i.e., coal likely to remain in suspension at those gravities. Figure 8.4.2 illustrates such data.

However, because separation of clean coal from high-ash discard material is affected by the size distribution, plant design and operating parameters require more definitive information, and that can be obtained from float–sink data for a set of closely sized samples from which a prorated composite for the projected feed can be computed, or from distribution curves (Tromp, 1937) for suitably sized fractions of other coals. The latter (see Fig. 8.4.3) connect the specific gravities of cleaning media to distribution coefficients that define the weight-percentages of the feed that will separate as clean coal within a defined range of gravities. An example is

$$C_{1.45} = 100F/(F + S),$$

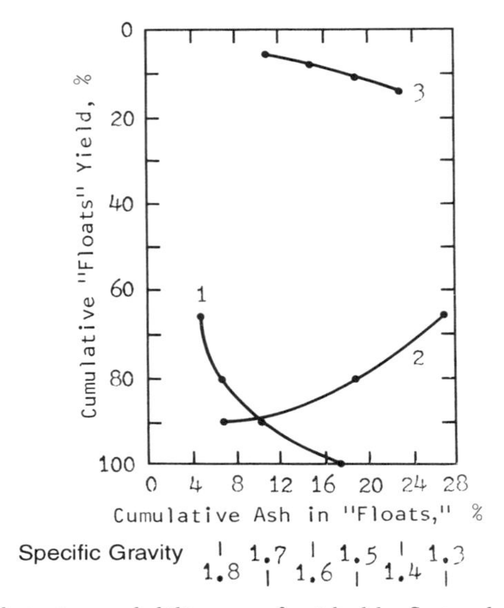

FIGURE 8.4.2 An illustrative washability curve furnished by float-and-sink tests. At s.g. 1.5, ~80% cleaned coal (2) with ~6.5% ash (1) could be recovered, and middlings (3) would run to ~11%. At s.g. 1.3, ~65% cleaned coal with ~4.5% ash would be obtained, and middlings would approach ~20%.

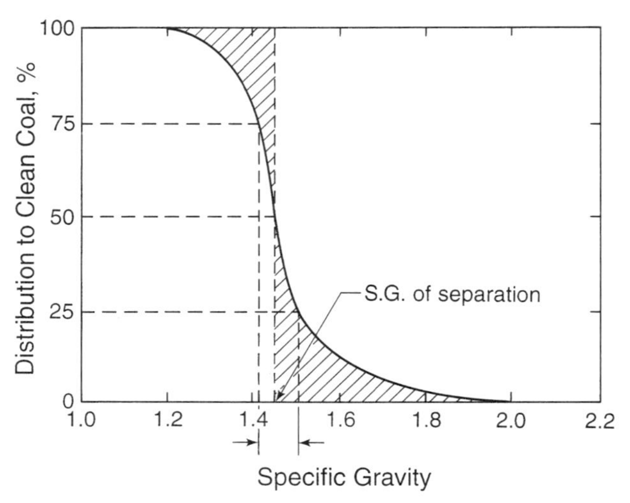

FIGURE 8.4.3 An illustrative distribution curve.

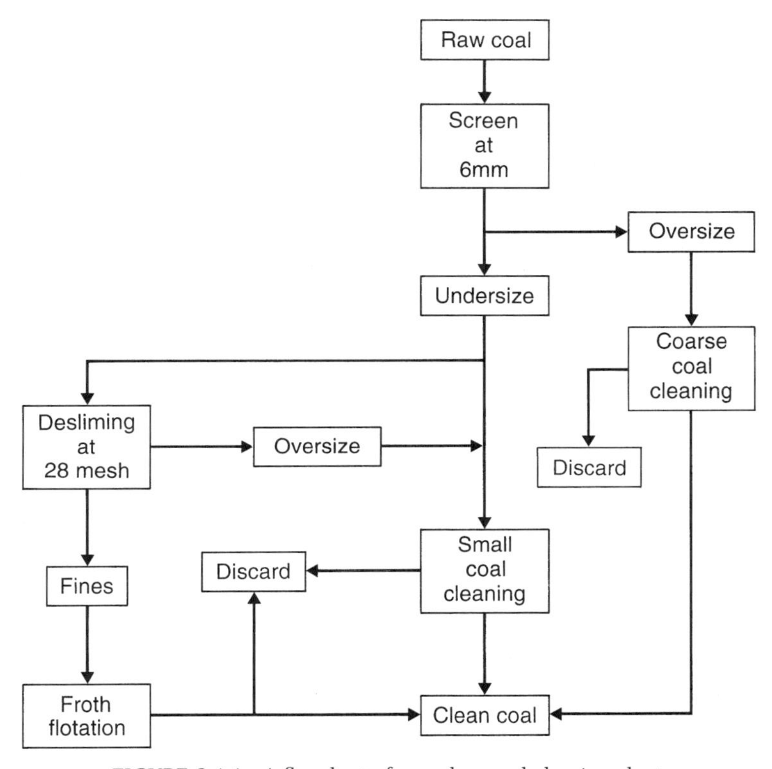

FIGURE 8.4.4 A flowsheet of a modern coal cleaning plant.

where F and S are the weight percentages of raw coal appearing as floats and sinks, respectively, between 1.40 and 1.50 s.g., and $C_{1.45}$ is the distribution coefficient. The specific gravity at which $F = S$ is defined by C = 50% [27].

To optimize recovery of clean coal, the raw feed is usually divided into at least three separately processed size fractions. The largest sizes are cleaned in jigs, dense media cells, or cleaning tables; <12-mm coal is processed in cyclones; and flotation is, as in mineral dressing, used to clean <0.5-mm fines. Figure 8.4.4 illustrates this mode of operation.

In jigs, the feed is subjected to an alternating up-and-down flow of water while moving across a slightly inclined perforated screen, and a consequent rapidly alternating compression and expansion of the bed shifts the relatively lighter clean coal particles to the surface where they are recovered (Citron, 1958).

Cleaning tables (Deurbrouck and Palowitch, 1979) make use of the fact that heavy particles carried by flowing water settle faster than light ones. The equipment is therefore designed to separate clean coal from discard while the feed travels horizontally from one end of the table to the other, and discard material is retained by riffles that divert it to the side of the table.

Cleaning by a dense medium (Whitmore, 1959; Palowitch and Deurbrouck, 1979) is accomplished by passing the feed through a drum in which it encounters a suitably concentrated, circulating suspension of sand or magnetite in water [28]. Clean coal and discard are then extracted at the top and bottom of the drum, respectively; sand or magnetite entrained in the two streams is removed on rinsing screens; and magnetite, if used, is stripped of residual fine coal in magnetic separators before being recycled. Specific gravities between ~1.30 and 1.90 can be attained by setting the concentrations of sand or magnetite in the circulating suspension between 25 and 45%.

In cyclones, which are routinely employed for cleaning <12-mm coal, separation of mineral matter is enhanced by centrifugal forces (Fig. 8.4.5). But because cut points can be changed by adjusting the vortex finders, cyclones could also process coal as fine as 75 μm (Sokaski *et al.*, 1979), and this flexibility makes them useful for preliminary cleaning, which reduces the load on other downstream equipment (Zimmerman, 1979).

Fine (<0.5-mm) coal is, however, in practice usually processed by froth flotation (see Fuerstenau, 1962) or agglomeration (Sirianni *et al.*, 1969; Capes *et al.*, 1974).

In flotation, which is preferred because of its simplicity, air is passed upward through an aqueous suspension of coal to which a frothing agent—commonly methyl isobutyl carbinol at ~50 g/tonne coal—and sometimes an oily collector (Sun, 1954) has been added. The lighter clean particles then attach themselves to the air bubbles and rise with them to the top where they are skimmed off.

Agglomeration is based on the same principle, but offers advantages by delivering a *consolidated* clean product from coal as fine as 75 μm. This accrues when the raw particles are suspended and rapidly agitated in water to which an immiscible hydrocarbon that preferentially wets the coal has been added [29]. The clean flocs can thereafter be more firmly agglomerated by briefly stirring them less rapidly in a holding tank [30].

More recently, attention has been focused on high-gradient magnetic separation (HGMS) as a means of removing pyrite from coal (Beddow, 1981). Small-scale tests have achieved separations that are claimed to equal or surpass those achieved by flotation (Hucko and Miller, 1980; Hise *et al.*, 1981), and further improvements are thought likely to accrue from prior fragmentation of coal by anhydrous NH_3 or HF. NH_3 tends to expose mineral matter by rupturing the coal along bedding planes and interfacial boundaries between organic and inorganic material (Howard and Datta, 1977), whereas HF causes almost instantaneous crumbling of the coal (Jensen, 1977).

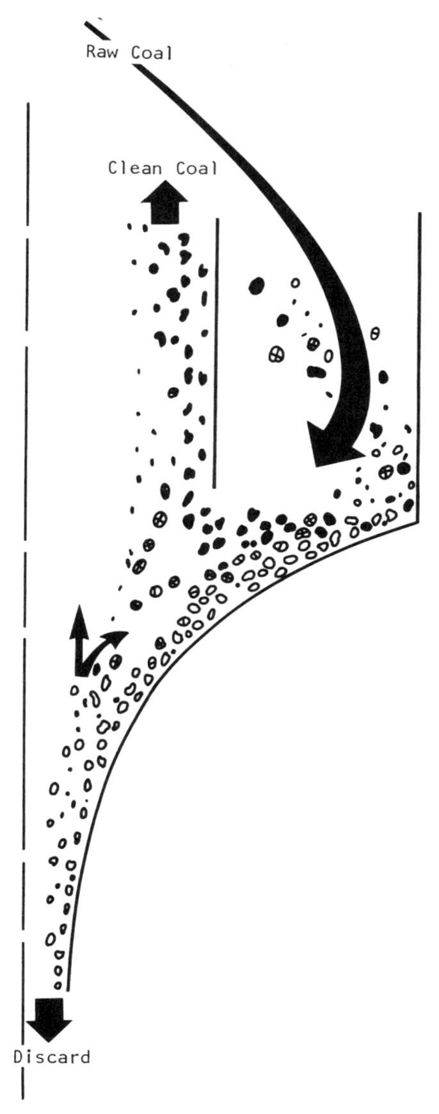

FIGURE 8.4.5 Schematic illustration of the operating principles of a hydrocyclone.

But since physical cleaning techniques cannot appreciably reduce organic sulfur which normally represent 40–50% of total [S], environmental concerns about emission of SO_2/SO_3 from combustion systems have served to focus attention on potential chemical and microbiological cleaning processes.

Chemical methods are specifically directed to removal of organic sulfur, and some also help to eliminate inorganic forms; but each is beset by problems that may preclude their large-scale use.

1. Chlorination of moist coal in CCl_4 with Cl_2 at 65–75°C, and subsequent hydrolysis with steam at 300–500°C, thus converts FeS into $FeCl_3$, removes organic sulfur as H_2SO_4, as in

$$R{-}S{-}R' + 4\,Cl_2 + 4\,H_2O \rightarrow RCl + R'Cl + H_2SO_4 + 6\,HCl$$

and generates more HCl by dechlorination [31]:

$$RCl + H_2O \rightarrow ROH + HCl.$$

The procedure reportedly lowers total sulfur contents by as much as 50–60% (Chatterjee *et al.*, 1990) and allows recovery of Cl_2 from HCl by, e.g., reaction with pyrolusite ($Mn^{4+}O_2$), but poses potential difficulties from HCl.

2. Oxidation of bituminous coals with NaOCl (Brubaker and Stoicos, 1985) is claimed to remove organic and sulfate S by a reaction sequence of the form

$$RSH \rightarrow RS{-}SR \rightarrow RSO_3R + H_2SO_4,$$

which is facilitated by pretreating the coal with NH_3, but must be followed by high-temperature hydrolysis to remove the up to 2.5% Cl_2 that NaOCl incorporates into the coal.

3. Direct oxidation by O_2 eliminates inorganic sulfur by transforming FeS into Fe^{3+} oxide and/or sulfate, and at 200–225°C, also removes organic S in alkyl sulfides and disulfides as SO_2. However, because $Fe_2(SO_4)_3$, which is strongly favored at relatively low processing temperatures, tends to be retained in the coal, excess O_2 is needed to promote formation of Fe_2O_3, and substantial excision of S from aromatic compounds such as thiophenes and aryl sulfides would require processing at 450–475°C (Stephenson *et al.*, 1985).

4. Mild hydrogenation of catalyst-impregnated high-S lignites at 275–325°C can remove >50% of total [S]. But hydrogenation is precluded by high hydrogen costs, and an alternative, although more complex, *convert/remove* hydrodesulfurization strategy (Rogoff *et al.*, 1962) could therefore conceivably prove more attractive. This entails

$$RSH + H_2 \rightarrow RH + H_2S,$$

followed by

(a) interaction of H_2S with an *in situ* scavenger such as

$$FeO + H_2S \rightarrow FeS + H_2O;$$

(b) steam-driven regeneration of the scavenger by

$$FeS + H_2O \rightarrow FeO + H_2S;$$

(c) conversion of H_2S to elemental sulfur on activated carbon or in a Claus plant.

However, unfavorable economics, handling difficulties, and/or undesirable side effects—e.g., of oxidation on caking properties of metallurgical coal—have so far militated against implementation of chemical S-removal schemes, and rather greater interest seems to have been expressed in *microbial* desulfurization [32]. This has mostly centered on species such as *Thiobacillus ferrooxidans* and *Pseudomonas putida,* which can lower pyritic-S contents by as much as 80% (Rai, 1985; Kargi, 1986), and on a *Thiobacillus*-like strain (Th1) that appears capable of removing organic as well as inorganic S from lignites (Gockay and Yuteri, 1983). However, only one such microbial method has so far been demonstrated in a continuous 4.5 kg/day bench-scale unit—and even that with only mixed success (El Sawy and Gray, 1991). Progress depends apparently on the development of more active and faster-growing microorganisms than are now at hand.

NOTES

[1] With few exceptions, *preparation* entails physical upgrading (or *beneficiation*) of the resource. In contrast, *processing,* discussed in Chapter 9, involves upgrading by chemical modification.

[2] The term "condensate gas" is synonymous with wet gas containing substantial proportions of C_4–C_7 hydrocarbons.

[3] Because of pore-size limitations, Linde Types 4A and 5A can only sorb the lightest mercaptans.

[4] That Cu salts can "dissolve" CO was discovered by Leblanc (1850).

[5] Union Carbide has developed a corrosion inhibitor that is claimed to allow use of much higher concentations of monoethanolamine (Hawkes and Mago, 1971). Dailey (1970) has also reported safe commercial use of 35–50% diethanolamine.

[6] $CuAlCl_4$ in toluene (or another light aromatic solvent) is inert toward H_2, CH_4, CO_2, and N_2, but moisture, if present, will react to form HCl. The feed gas must therefore be dry, preferably with <1 ppm H_2O.

[7] This also avoids unacceptable thermal cracking of asphaltic residua that are routinely discharged near the tower bottom at ~280°C.

[8] Such processing, which is exemplified by thermal and catalytic cracking, is discussed in Chapter 9.

[9] Propane and LPGs that precipitate asphalts and resins can also be used to dewax heavy lubricating oils. The choice of solvent is therefore usually dictated by solvent costs (which tend to favor propane).

[10] The diolefins tend to generate gums in motor gasolines.

[11] The formations in which bitumens and related heavy hydrocarbons occur are referred to as bituminous sands or tar sands, and in Alberta are designated as oil sands. To avoid confusion, the term "bituminous sands" is used here.

[12] All available data show that bitumen concentrations rarely exceed ~15–16%.

[13] In different formats, direct retorting is, however, being operated with other feedstocks. An example is the Lurgi–Ruhrgas process, which was developed in the late 1940s for LT coal carbonization (Peters, 1960; Sommers, 1974) and was later also used for cracking crude oils, naphtha, and fuel oil to olefins. Like the UMATAC/Taciuk process (see Chapter 9), the L-R process has been successfully tested with Athabasca oil sands as well as with sands from Nigeria, California, and Venezuela.

[14] Both plants operate in Alberta and process surface-mined sands from the Athabasca deposit. (One (*Suncor*) currently produces some 19,100 m^3 synthetic sweet crude per day, and the other (*Syncrude Canada*) produces ~31,800 m^3/day. The combined output represents ~18% of Canada's present total oil production.

[15] A modified hot-water extraction method, the RTR/Gulf process noted in the ERCB study (1984), seems better able to handle low-grade ores and, coincidentally, also has lower water and energy requirements than the HWEP.

[16] One consequence of this, aside from the need for large volumes of make-up water, is a need to progressively expand the voluminous tailing ponds.

[17] Since retorting involves extensive thermal cracking, it is in the context of this survey deemed to be a form of processing and is discussed in Chapter 9.

[18] Much of the earlier work on solvent extraction centered on *in-situ* extraction of bitumen and was abandoned because of unacceptable solvent losses to the strata.

[19] SESA: an acronym for solvent extraction/spherical agglomeration. The underlying principles of the process have been described by Farnand *et al.* (1961).

[20] This can amount to as much as 80% of the produced fluids.

[21] The molecular weights of oils generated by supercritical fluid extraction and pyrolysis thus averaged ~700 and ~300, respectively.

[22] For the potential of such low-rank coals as sources of hydrocarbon liquids, fuel gas or syngas, see Chapter 10.

[23] In this context, *dewatering* means almost total removal of surficial bulk water, whereas *drying* implies partial or complete removal of connate and/or sorbed water. The latter contributes to the capacity moisture content of a coal.

[24] Mechanisms of autogenous heating have been discussed by Berkowitz (1994). Overdrying *bituminous* coals will rarely pose significant hazards from such heating, but will enhance dustiness and consequently compound handling difficulties, as well as increase windage losses during haulage by truck or train.

[25] These are now infrequently used. In most cases, size requirements are directly indicated in metric units (i.e., centimeters or millimeters).

[26] From *autochthonous* coals, in which inorganic material tends to be concentrated along bedding planes and major fractures, it can usually be fairly easily removed. But in *allochthonous* coals, in which colloidally dispersed clays represent a large proportion of mineral matter, physical means commonly cannot achieve effective cleaning.

[27] Even then, a *definitive* assessment of cleaning efficiency requires comparing actual yields and ash contents of cleaned coal with predictions from distribution curves. An appropriate formula for this purpose is

$$E = R/R_t(A_1 - A_2)/(A_1 - A_3),$$

where R and R_t are the actual and predicted yields of cleaned coal, and A_1, A_2, and A_3, respectively, denote the ash contents of the feed, of the actually produced clean coal, and of the predicted yield of clean coal.

[28] Dense media are, however, not suitable for preparing coals that contain high proportions of clay minerals (notably bentonite); these minerals tend to disperse in such media.

[29] A suitable liquid is diesel oil that, depending on the nature of the coal, can be used alone or with a surface conditioning agent.

[30] Agglomeration has also attracted interest as a technique for collecting coal particles from accidental spills into water reservoirs and natural water courses, and may thus gain importance as an environmental management tool in coal mines and preparation plants.

[31] Dechlorination is imperative in order to prevent release of HCl, which could seriously corrode and crack reaction vessels.

[32] An excellent review of the action of microbiota on coal has been presented by Hsu-Chou *et al.* (1989).

REFERENCES

Bally, A. F. *Erdöl Kohle* **14**, 921 (1960).

Beddow, J. K., *Chem. Eng.* **88**, 70 (1981).

Berkowitz, N. *An Introduction to Coal Technology*, 2nd ed., 1994. New York: Academic Press.

Berkowitz, N., and J. Calderon. *Fuel Process. Technol.* **25**, 33 (1990).

Bottoms, R. R. U.S. Pat. No. 1,783,901 (1930).

Brubaker, I. M., and T. Stoicos. *Processing High-S Coals* (Y. A. Attia, ed.), 1985. Amsterdam: Elsevier.

Buchan, F. E. U.S. Pat. No. 2,487,788 (Nov. 1949).

Capes, C. E., A. E. Smith, and I. E. Puddington. *Bull. Can. Inst. Min. Met.*, July (1974).

Chatterjee, K., R. Wolny, and L. M. Stock. *Energy & Fuels* **4**, 402 (1990).

Citron, E. H. *Min. Eng.* **10**, 488 (1958).

Cummins, J. J., and W. E. Robinson; *ACS Div. Fuel Chem. Prepr.* **21**(6), 94 (1976); *DOE Laramie Energy Research Center, LERC/RI-78/1* (1978).

Cummins, J. J., D. A. Sanchez, and W. E. Robinson. *Energy Comm.* **6**(2), 117 (1980).

Curran, G. P., J. T. Clancy, B. Pasek, M. Pell, G. D. Rutledge, and E. Gorin. *NTIS PB 232-695/AS* (1973).

Dailey, L. W. *Oil Gas J.* **136**, May 4 (1970).

Day, D. T. U.S. Pat. Nos. 1,447,296, 1,447,297 (March 1922).

Deurbrouck, A. W., and E. R. Palowitch. *Coal Preparation*, 4th ed. (J. W. Leonard, ed.), 1979. New York: Amer. Inst. Min. Met. Pet. Engrs.

Dulhunty, J. A. *Linnean Soc. NSW* **67**, 238 (1942).

El Sawy, A., and D. Gray. *Fuel* **70**, 591 (1991).

ERCB (Energy Resources Conservation Board, Alberta); *Rept. CA2AL-ER-2-84037*, **2**, Calgary, Alberta (1984).

Farnand, J. R., H. M. Smith, and I. E. Puddington. *Can. J. Chem. Eng.* **39**, 94 (1961).

Fuerstenau, D. W. *Froth Flotation*, 1962. New York: AIME.

Gockay, C. F., and R. N. Yuteri. *Fuel* **62**, 1223 (1983).

Hampton, H. U.S. Pat. No. 1,668,898 (May 1928); U.S. Pat. No. 1,707,759 (April 1929).

Hawkes, E. N., and B. F. Mago. *Hydrocarbon Process.* **8**, 109 (1971).

Hise, E. C., I. Wechsler, and J. M. Doulin. *ORNL-5571, Oak Ridge Natl. Lab., Oak Ridge, TN* (1981).

Hochgesang, G. *Chem. Ing. Techn.* **40**, 432 (1968).

Hoogendoorn, J. C., and J. M. Solomon. *Brit. Chem. Engng.* **40**, 432 (1957).

Houlihan, R. *Proc. UNITAR II, Caracas, Venezuela* (1980).

Houlihan, R., and K. H. Williams. *J. Can. Petroleum Technol.* **26**, 91 (1987).
Howard, P. H., and R. S. Datta. *ACS Symp. Ser.* **64**, 58 (1977).
Hsu-Chou, R. S., H. C. C. G. do Nascimento, and T. F. Yen. In *Sample Selection, Aging and Reactivity of Coal* (R. Klein and R. Wellek, eds.), Chapter 9, 1989. New York: Wiley.
Hucko, R. E., and K. J. Miller. *Rept. RI-PMTC-10(80)*, 1980. Washington, D.C.: U.S. Dept. of Energy.
Jensen, H. P. U.S. Pat. No. 4,169,710 (1977).
Jensen, H. B., W. I. Barnet, and W. I. R. Murphy. *U.S. Bur. Mines Bull. No. 533* (1953).
Kargi, F. *Trends Biotechnol.* **4**, 293 (1986).
Kesavan, S. K., A. Ghosh, M. E. Polasky, V. Parameswaran, and S. Lee. *Fuel Sci. Technol. Internatl.* **6**(5), 505 (1988).
Kramer, R., and M. Levy. *Fuel* **68**, 702 (1989).
Leblanc, F. *Compt. Rend.* **30**, 483 (1850).
Lovell, H. L. In *Coal Preparation,* 4th ed. (J. W. Leonard, ed.), 1979. New York: Amer. Inst. Min. Met. Pet. Engrs.
Ogunsola, O. M., and N. Berkowitz. *Fuel Process. Technol.* **45**, 95 (1995).
Palowitch, E. R., and A. W. Deurbrouck. In Coal Preparation, 4th ed. (J. W. Leonard, ed.), 1979. New York: Amer. Inst. Min. Met. Pet. Engrs.
Peters, W. *Chem. Ing. Technol.* **32**(3), 178 (1960).
Rai, C. *Biotechnol. Progr.* **1**(3), 200 (1985).
Riesenfeld, F. C., and A. L. Kohl. *Gas Purification,* 4th ed., 1985. Houston: Gulf Publ.
Rogoff, M. H., I. Wender, and R. B. Anderson. *U.S. Bur. Mines Inf. Circ. 8075* (1962).
Ryan, H. D. U.S. Pat. No. 1,327,572 (Jan. 1920); U.S. Pat. No. 1,672,231 (June 1928).
Sandy, E. J., and J. P. Matoney. In *Coal Preparation,* 4th ed. (J. W. Leonard, ed.), 1979. New York: Amer. Inst. Min. Met. Pet. Engrs.
Sirianni, A. F., C. E. Capes, and I. E. Puddington. *Can. J. Chem. Eng.* **47**, 166 (1969).
Sokaski, M., M. R. Geer, and W. L. Morris. In *Coal Preparation,* 4th ed. (J. W. Leonard, ed.), 1979. New York: Amer. Inst. Min. Met. Pet. Engrs.
Sommers, H. *VGB Kraftwerkstechnik* **5**, 306 (1974).
Sparks, B. D., and F. W. Meadus. *Energy Processing/Can.* **72**, 55 (1979); *Fuel Process. Technol.* **4**, 251 (1981).
Stephenson, M. D., M. Rostam-Abadi, L. A. Johnson, and C. W. Kruse. In *Processing High-S Coals* (Y. A. Attia, ed.), Amsterdam: Elsevier.
Sun, S. C. *Trans. AIME* **199**, 306 (1954).
Tromp, K. F. *Gluckauf* **73**, 121 (1937).
U.S. Dept. of Commerce. *NTIS Publ. No. FE-1772-11* (1976).
Walker, D. G. *Chem. Tech.* 303 (May) 1975.
Whitmore, R. L. *Colliery Eng.* **36**, 151 (1959).
Yeandle, W. W., and G. F. Klein. *Chem. Eng. Progr.* **48**, 349 (1952).
Zimmerman, R. E. In *Coal Preparation,* 4th ed. (J. W. Leonard, ed.), 1979. New York: Amer. Inst. Min. Met. Pet. Engrs.

CHAPTER 9

Processing

1. THE CHEMICAL FOUNDATIONS

Unlike preparation, which concerns itself mainly with removal of unwanted matter in fossil hydrocarbons by physical means [1], processing effects chemical modifications designed to enhance the economic value of the feedstock or meet specialized market demands, and for the most part, these objectives are achieved by some form of thermal cracking or hydrogenation. Cracking—in essence, a controlled LT pyrolysis—will, if sufficiently mild, reduce the number of C atoms in a molecule without appreciably changing the H/C ratio of the feed or, if more severe, yield an H-enriched fraction by H-disproportionation and rejection of an equivalent amount of excess carbon to high-molecular-weight polynuclear aromatics and/or coke. The other method, hydrogenation, inserts H atoms into the host molecule, and thereby increases the H/C ratio without necessarily changing the molecular size unless the required hydrogenation depth—i.e., the extent of hydrogenation—demands reaction conditions sufficiently severe to cause concurrent thermal cracking.

Both approaches offer wide operational freedoms and are used in diverse formats- and if appropriately adapted to feedstock characteristics, both are technically capable of accepting *any* fossil hydrocarbon feedstock [2].

Thermal Cracking

Although limited decomposition and molecular rearrangement can occur at low temperatures [3], significant thermal cracking can only proceed above ~300–350°C, where all hydrocarbons other than CH_4 and C_2H_6 are less stable than their constituent elements. But even at these temperatures, it follows well-defined patterns: *n*-paraffins, isoparaffins, and olefins are invariably less stable than naphthenes; aromatics, more stable than naphthenes, frequently do not crack significantly below ~450–500°C; and formation of tars or coke, both promoted by secondary cracking and random recombination of the free radicals thereby generated, is the more pronounced the lower the atomic

H/C ratio of the feed, the higher the temperature, and/or the longer the duration of processing.

As a general rule, *n*-paraffins degrade by random scission of C—C bonds, and consequently either generate lower *n*-paraffins and olefins, as in

$$CH_3—(CH_2)_6—CH_3 \rightarrow CH_3CH_2CH_2CH_3 + CH_2{=}CH—CH_2—CH_3,$$

or yield only lower *n*-paraffins if $CH_3—(CH_2)_3^*$ radicals are immediately stabilized by H-addition (as in hydrocracking). Isoparaffins will crack similarly after losing the carbon branch, as in

$$\begin{array}{l} \quad CH_3 \\ \quad | \\ CH_3—CH—(CH_2)_n—CH_3 \rightarrow CH_2{=}CH—(CH_2)_n—CH_3 + CH_4 \\ \downarrow \\ [CH_3—CH^*—(CH_2)_n—CH_3] \rightarrow CH_3—(CH_2)_{n+1}—CH_3, \\ \qquad \uparrow \\ \qquad [H] \end{array}$$

where [H] indicates H-capping of the bracketed radical. And substituted naphthenes and aromatics will either be dealkylated through loss of peripheral functions such as $—CH_3$ and $—C_2H_6$, or suffer staged breakdown by sequential ring opening, chain scission, and H-disproportionation.

However, the diverse hydrocarbon mixtures that are feedstocks for thermal cracking operations make product slates dependent on feed composition as well as on the minutiae of cracking conditions; and although the chemical pathways of cracking can be modified by cracking catalysts, very severe reaction conditions will always cause extinction of the precursor molecules—i.e., will only yield gas and coke.

The simplest example of thermal cracking offers itself in the decomposition of *n*-butane to ethane, ethylene, and propylene. This reaction can be formally written as

$$CH_3CH_2CH_2CH_3 \rightarrow CH_4 + CH_3—CH{=}CH_2 \rightarrow H_3C—CH_3 + H_2C{=}CH_2$$

and, if not arrested, will continue with secondary interactions and/or rearrangement of the initial reaction products, as in

$$2\ H_2C{=}CH_2 \rightarrow H_3CCH_2CH{=}CH_2.$$

Heavier *n*-paraffins crack similarly with formation of lower *n*-paraffins and olefins by random scission of C—C bonds, as in

$$H_3C—(CH_2)_6—CH_3 \rightarrow H_3C—CH_2CH_2—CH_3 + H_2C{=}CH—CH_2—CH_3,$$

or only form lower paraffins if $CH_3—(CH_2)_3^*$ radicals are rapidly H-capped.

Isoparaffins can decompose like *n*-paraffins, as in

$$R{-}CH_2{-}CH_2{-}\underset{\displaystyle CH_3}{\underset{|}{CH}}{-}CH_3 \rightarrow R{-}CH_3 + H_2C{=}\underset{\displaystyle CH_3}{\underset{|}{C}}{-}CH_3,$$

but can also furnish straight-chain olefins by internally rearranging, as in

$$CH_3{-}\overset{\displaystyle CH_3}{\overset{|}{CH}}{-}(CH_2)_n{-}CH_3 \rightarrow CH_2{=}CH{-}(CH_2)_n{-}CH_3 + CH_4,$$

or, as already noted, yield *n*-paraffins by rapidly H-capping radicals, as in

$$CH_3{-}\overset{\displaystyle CH_3}{\overset{|}{CH}}{-}(CH_2)_n{-}CH_3 \rightarrow [CH_3{-}\underset{\displaystyle [H]}{\underset{\uparrow}{CH^*}}{-}(CH_2)_n{-}CH_3]$$

$$\rightarrow CH_3{-}(CH_2)_{n+1}{-}CH_3.$$

Cycloparaffins with relatively long *n*-alkyl substituent functions can undergo bond scissions that eliminate these functions, and thereby form an olefin and the corresponding CH_3- or C_2H_5-cycloparaffins, as in

$$C_6H_{11}{-}CH_2{-}(CH_2)_n{-}CH_2R \rightarrow C_6H_{11}{-}CH_3 + (CH_2)_n{=}CHR,$$

and a similar decomposition pattern is followed by alkylated aromatics, in which rings are generally stable up to ~500°C.

However, very different reaction paths are taken in the presence of acid catalysts such as SiO_2/Al_2O_3, which can donate a proton to an olefin or abstract a hydride ion (H^+) from a hydrocarbon. In that case, instead of proceeding by free radical reactions, cracking involves a sequence of ionic interactions that generate carbonium ions, such as

$$H_3C{-}\underset{\displaystyle CH_3}{\underset{|}{C^+}}{-}CH_3,$$

and consequently allow reactions to proceed at much lower activation energies. Such sequences yield mainly isoparaffins and aromatics, but promote simultaneous abstraction of organic sulfur as H_2S. Examples are

1. incorporation of carbonium ions into $>C_6$ hydrocarbons, eventually followed by aromatization;
2. formation of olefins from *n*-paraffins that would otherwise mainly crack at γ-C positions; and
3. formation of olefins from isoparaffins, which usually rupture between the β- and γ-positions of a tertiary C-atom.

Dehydrogenation

An alternative to thermal cracking, which sometimes offers advantages, presents itself in abstraction of H-atoms, as in

$$H_3C—CH_2—CH_2—CH_3 \begin{cases} \rightarrow H_3C—CH_2—CH{=}CH_2 \\ \rightarrow H_3C—CH{=}CH—CH_3. \end{cases}$$

Internal H-disproportionations of this type—controlled by catalysts such as Al_2O_3-supported Cr_2O_3, which also minimizes formation of high-molecular-weight species—reduce the H/C ratio without appreciably affecting the number of C atoms/molecule.

The variety of useful dehydrogenation reactions is exemplified by

1. dehydrogenation of cyclopentane derivatives to aromatics via a substituted cyclohexane, as in

$$C_5H_9—CH_2—R \rightarrow [C_6H_{11}—R] \rightarrow C_6H_5—R\ (+3\ H_2);$$

2. transformation of cyclohexane and alkyl-substituted cyclohexanes into benzene, as in

$$C_6H_{12} \rightarrow C_6H_6\ (+3\ H_2)$$
$$C_6H_{11}—CH_3 \rightarrow C_6H_5—CH_3\ (+3\ H_2);$$

3. conversion of polycyclic naphthenes into the corresponding aromatic compounds—e.g., $C_{10}H_{18}$ (decalin) $\rightarrow$ $C_{10}H_8$ (naphthalene); and
4. dehydrogenation of alkyl benzenes, as in

$$\underset{\text{ethylbenzene}}{\phi—CH_2CH_3} \rightarrow \underset{\text{styrene}}{\phi—CH{=}CH_2}\ (+H_2)$$

or

$$\underset{\text{isopropylbenzene}}{\phi—\underset{|\atop CH_3}{CH}—CH_3} \rightarrow \underset{\alpha\text{-methylstyrene}}{\phi—\underset{|\atop CH_3}{C}{=}CH_2}$$

In some cases—for instance, in the conversion of *n*-hexane to benzene, *n*-heptane to toluene, or *n*-octane to *o*-xylene + ethylbenzene—dehydrogenation will also promote cyclization and subsequent aromatization.

Hydrogenation Reactions

Hydrocracking

When control of thermal cracking in an inert atmosphere proves difficult, the operation can be conducted under H_2 at $>350°C/1–7$ MPa, and is then termed hydrocracking (or *hydropyrolysis*; [4]). This offers major advantages by suppressing unwanted secondary decomposition through H-capping of transient radical species, and allows close control over product slates because its flexibility makes for an exceptionally wide range of operating conditions.

n-Paraffins will typically react as in

$$\mathrm{R{-}(CH_2)_n{-}CH_3 \rightarrow RCH_3\ [+CH_3{-}(CH_2)_{n-3}{-}CH{=}CH_2]}$$
$$\downarrow$$
$$\mathrm{CH_3{-}CH_2)_{n-2}{-}CH_3.}$$

Isoparaffins tend to be demethanated, as in

$$\mathrm{R{-}(CH_2)_n{-}\underset{\substack{|\\ CH_3}}{CH}{-}R' + H_2 \rightarrow R{-}(CH_2)_{n+1}{-}R'.}$$

And naphthenes respond with ring opening and addition of H atoms at each end of the resultant chain. But, as illustrated by

$$\text{cyclo-}[\mathrm{CH_2{-}CH_2{-}CH(CH_3){-}CH_2{-}CH_2}]$$
$$\downarrow$$
$$\mathrm{CH_3{-}\underset{\substack{|\\ CH_3}}{CH}{-}CH_2{-}CH_2{-}CH_3}$$
$$\mathrm{+\ CH_3{-}CH_2{-}\underset{\substack{|\\ CH_3}}{CH}{-}CH_2{-}CH_3 + CH_3{-}(CH_2)_4{-}CH_3,}$$

which occurs over carbon-supported Pt catalysts, alkylated cyclopentanes can undergo more indiscriminate scission of C—C bonds, and furnish more diverse products.

In contrast, aromatics will generally only hydrocrack at 350–500°C/10–11 MPa in the presence of a sulfided molybdenum catalyst, and then respond sequentially with (i) loss of alkyl substituents (if present), (ii) transformation into corresponding naphthenes, and, in sufficiently prolonged reaction, (iii) formation of paraffins.

Polycyclic aromatics follow much the same paths by saturating and opening one ring at a time.

Hydrotreating

A milder hydrogenation procedure, unaccompanied by significant cracking, presents itself in *hydrotreating*, which is generally conducted over a Co/Mo/

Al_2O_3 catalyst [5] and mainly used to remove sulfur as H_2S and/or nitrogen as NH_3. Illustrative reactions are

$$\begin{aligned} R—CH_2—SH + H_2 &\rightarrow R—CH_3 + H_2S \\ R—NH—R + 2\,H_2 &\rightarrow 2\,RH + NH_3. \end{aligned}$$

Aromatics are first saturated, so that thiophene reacts as in

$$\begin{array}{l} CH{=}CH \\ \vert \qquad \backslash \\ \vert \qquad\;\; S \\ \vert \qquad / \\ CH{=}CH \end{array} \rightarrow \begin{array}{l} CH_2—CH_2 \\ \vert \qquad \backslash \\ \vert \qquad\;\; S \\ \vert \qquad / \\ CH{=}CH \end{array} \rightarrow CH_3CH_2CH{=}CH_2 + H_2S,$$

and quinoline (C_9H_7N) would furnish n-propylbenzene + NH_3.

However, hydrotreating can also saturate aromatics and olefins, as in

$$\begin{aligned} C_6H_5—R &\rightarrow C_6H_{11}—R \\ R—CH{=}CH—R' &\rightarrow R—CH_2—CH_2—R', \end{aligned}$$

where R, R′ denote peripheral substituents.

Other Processing Techniques

Although in their present forms only suitable for processing gaseous and liquid feedstocks, hydrocarbon structures can also be variously modified by some other techniques that merit noting here.

Reforming

Rearrangement of molecular structure *without* concurrent changes in elemental composition can be induced over an $Al_2O_3/SiO_2/Pt$ catalyst at 480–525°C/ 1.5–3.5 MPa. Such *reforming* is procedurally shown in Fig. 9.1.1 and is particularly useful for aromatizing n-paraffins and naphthenes in naphtha fractions.

Isomerization

Isomerization driven by carbonium ions and proceeding over Pt or HCl-promoted $AlCl_3$ will stereospecifically convert n-butane to isobutane and effect similar, if less specific, transformations of other paraffins. It can also isomerize olefins by shifting an H atom, as in

$$CH_3—(CH_2)_n—CH{=}CH_2 \rightarrow CH_3—(CH_2)_{n-1}—CH{=}CH—CH_3,$$

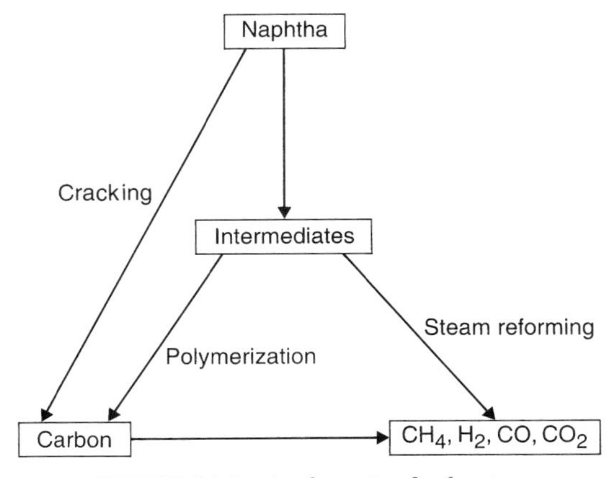

FIGURE 9.1.1 A schematic of *reforming*.

or bring about skeletal changes, as in

$$CH_3{-}(CH_2)_n{-}CH_2{-}CH_3 \rightarrow CH_3{-}(CH_2)_{n-1}{-}CH{=}\underset{\displaystyle CH_3}{\underset{|}{C}}{-}CH_3,$$

and in some cases promote *cis* → *trans* isomerization, as in

$$\underset{\text{1-butylene}}{CH_2{=}CH{-}CH_2{-}CH_3} \rightarrow \underset{\textit{cis}\text{-butylene}}{\begin{matrix} HC{-}CH_3 \\ \| \\ HC{-}CH_3 \end{matrix}} + \underset{\textit{trans}\text{-butylene}}{\begin{matrix} CH_3{-}CH. \\ \| \\ HC{-}CH_3 \end{matrix}}$$

Isomerization of napthalenes generates olefins by ring scission, as in

$$\begin{matrix} CH_2{-}CH_2 \\ | \quad\quad | \\ CH_2{-}CH_2 \end{matrix} \rightarrow CH_3{-}CH_2{-}CH{=}CH_2,$$

and on alkyl aromatics it operates by disproportionating side chains or shifting them to other positions on the molecular core. An example of position shifting is the isomerization of ethylbenzene to *o*-, *m*-, and *p*-xylene.

Alkylation

A reaction over $AlCl_3$, which serves to produce high-octane gasolines from a mix of paraffins and olefins, is exemplified by an alkylation sequence such as

$$CH_2{=}CH_2 + H^+ \rightarrow CH_3CH_2^+$$

$$\begin{array}{cc} & CH_3 \\ & | \\ CH_3{=} & CH \end{array} + \begin{array}{cc} CH_3 & {-}CH_2^+ \\ | & \\ CH_3 & \end{array} \rightarrow \begin{array}{cc} & CH_3 \\ & | \\ CH_3{-} & C^+ \\ & | \\ & CH_3 \end{array} + C_2H_6$$

$$\begin{array}{cc} & CH_3 \\ & | \\ CH_3{-} & C^+ \\ & | \\ & CH_3 \end{array} + CH_2{=}CH_2 \rightarrow \begin{array}{ccc} & CH_3 & \\ & | & \\ CH_3{-} & C & {-}CH_2{-}CH_2^+ \\ & | & \\ & CH_3 & \end{array}$$

$$\begin{array}{ccc} & CH_3 & \\ & | & \\ CH_3{-} & C & {-}CH_2{-}CH_2^+ \\ & | & \\ & CH_3 & \end{array} + \begin{array}{cc} & CH_3 \\ & | \\ CH_3{-} & CH \\ & | \\ & CH_3 \end{array} \rightarrow \begin{array}{ccc} & CH_3 & \\ & | & \\ CH_3{-} & C & {-}CH_2{-}CH_3 \\ & | & \\ & CH_3 & \end{array} + \begin{array}{cc} & CH_3 \\ & | \\ CH_3{-} & C^+ \\ & | \\ & CH_3 \end{array},$$

in which isobutane reacts with $CH_3CH_2^+$ and eventually yields 2,2-dimethylbutane as the "alkylate." A similar series of reactions proceeds with, e.g., cyclohexane or benzene plus $CH_3{-}CH{=}CH_2$ and furnishes 1,3,5-trimethylcyclohexane or isopropylbenzene. Reactions with $>C_3$ olefins can, however, also incur H exchanges that ultimately generate more uncertain product slates.

2. CRUDE OIL PROCESSING

The cracking techniques used in processing of crude oils and oil fractions differ mainly in process severity, but occasionally also employ specific (usually proprietary) catalysts.

Visbreaking

As its designation implies, visbreaking reduces the viscosity of a crude oil or heavy oil fraction, mainly to facilitate its pipeline transportation and handling, and is achieved by briefly cracking the feed under mild conditions, usually ~90 seconds at 450–510°C/0.35–2 MPa. The effect on viscosity—and hence on the pour point of the oil and on its volatility (as indicated by its flash point)—is due to partial removal of peripheral substituent functions on molecular cores and to consequent physical "untangling" of molecules after excission of aliphatic linkages between aromatic units. No other significant compositional changes occur.

TABLE 9.2.1 Effects of Visbreaking[a]

	Feed	Processing temperature (°C)			
		435	450	465	480
Specific gravity					
15/15°C	1.0190	1.0132	1.0102	1.0053	0.9995
Pour point (°C)	33	1	12	6.4	−2
Flash point (Pensky) (°C)	199	75	61.5	42.5	27.5
Viscosity					
cSt at 50°C	12,362	4123	2470	1374	709
cSt at 100°C	384	174	134	91	76
Asphaltenes (wt%)	9.5	9.8	10.1	10.5	10.8
Conradson carbon (wt%)	14.76	15.61	16.30	16.59	17.22
Total sulfur (wt%)	6.68	6.68	6.64	6.61	6.59

[a] After Al-Soufi *et al.* (1988).

However, as illustrated by data relating to a heavy residuum with b.p. > 350°C (see Table 9.2.1), overall effects of visbreaking are governed by process severity, and although the method is procedurally simple, some care is required: improperly conducted, visbreaking can generate unstable olefins that polymerize into gums, and in heavy oil fractions, it may also generate solids that settle out as sediments. Both effects depend on the composition of the feedstock and can usually be avoided by limiting viscosity reductions to 12–30% (Decroocq, 1984; Gray, 1994).

Thermal Cracking

Uncatalyzed Cracking

More extensive chemical changes, which maximize gasoline yields from crude oils and heavy oil fractions, accrue from thermal cracking reactions which, among other effects, convert long-chain paraffins with octane numbers near zero into a mix of lower paraffins and olefins whose octane numbers approach 100. An example is

$$CH_3(CH_2)_{10}CH_3 \rightarrow CH_3(CH_4)_4CH_3 + CH_2{=}CH(CH_2)_3CH_3.$$

Such primary cracking is, however, always followed by secondary scission of hydrocarbon chains, and consequent formation of hydrocarbon gases and olefin-rich liquids that promote formation of unsaturated diolefins. These can effectively reverse cracking by generating heavy hydrocarbons by oligomerization and Diels–Alder cycloadditions—i.e., by addition of olefins to conjugated diolefins, dehydrogenation of the resultant six-membered naphthenic rings to alkylated aromatics, and eventual dealkylation to polycondensed tars and/or cokes with 2–10% H_2. At temperatures above ~300°C, secondary cracking and coke formation can indeed only be minimized by carefully selecting operating parameters, keeping reaction rates low, and recycling any substantially unchanged feed exiting from the reactor.

Early forms of thermal cracking are illustrated by the Burton process, operated between 1913 and 1920: this degraded heavy oils in boiler-type stills in a 40- 50 to 50-h reaction at ~400°C/0.5–0.7 MPa, and thereby accomplished 60–70% conversion to distillates of which nearly 50% vaporized in the gasoline range. However, by the mid-1920s such refining was superseded by continuous cracking techniques exemplified by the Cross, Dubbs, and Holmes processes; and since the late 1920s, a veritable plethora of cracking procedures, many using specially formulated catalysts, has come to hand.

In contemporary practice, direct thermal cracking has come to be progressively replaced by catalytic cracking (see below). But where still used, it is conducted at 450–550°C/0.7–7 MPa, and reaction minutiae depend on feed composition as well as on the desired product slate. As a rule, relatively light feedstocks such as gas oils require longer in-reactor residence times than residua, and cracking above 500°C at pressures below 3 MPa yields hydrocarbons of lower molecular weight than otherwise form.

Crude oils and middle distillates such as kerosine are generally cracked in liquid-phase operations in which the feedstock is quickly brought to 400–480°C/2.4 MPa, held for the necessary period, and then cooled in a flash chamber. The overhead vapors are taken to fractionation into gasoline and recycle oil, while bottoms are withdrawn as heavy fuel oil. The relatively high reaction pressure serves to minimize coke formation at the reaction temperature.

An alternative vapor-phase method, which cracks feedstocks at 550–590°C and pressures below ~0.35 MPa, and thereby promotes dehydrogenation that favors conversion of light hydrocarbons into olefins and aromatics, is frequently bedevilled by unacceptable coke deposition.

Catalytic Cracking

The advantages of catalytic cracking spring from different reaction pathways as well as from much higher reaction rates. For example:

1. Instead of random scission of *n*-paraffins, C—C bonds tend to rupture more toward the middle of the molecule, e.g., between the δ- and ε-carbon atoms of the chain:

$$—CH_2—CH_2—CH_2—\overset{\varepsilon}{C}H_2—/—\overset{\delta}{C}H_2—CH_2—CH_2—\overset{\alpha}{C}H_3$$

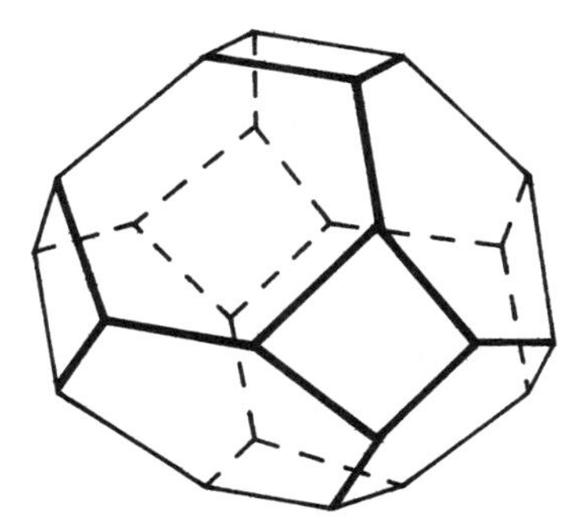

FIGURE 9.2.1 The sodalite structure, made up of tetrahedra of aluminum and silicon oxides.

2. $>C_5$ *n*-paraffins can rearrange and undergo limited dehydrocyclization to aromatics, as in

$$CH_3{-}(CH_2)_5{-}CH_3 \rightarrow \begin{array}{c} CH{=}CH \\ H_2C \qquad\quad CH_2 \\ CH{=}C \\ \quad | \\ \quad CH_3 \end{array} \; (+\,3\,H_2)$$

3. In isoparaffins, bond scission is favored between C atoms β and γ to a tertiary C, as in

$$\begin{array}{c} \qquad CH_3 \\ \qquad | \\ H_3C{-}C{-}CH{-}\underset{\beta}{C}{-}/{-}\underset{\gamma}{C}{-}CH_3 \end{array}$$

4. Cracking of olefins will usually proceed 10^3–10^4 times faster than in the absence of catalysts.

The essential roles in the chemistry of catalytic cracking are played by carbocation intermediates that must be generated by protonating a C–C double bond or promoting heterolytic bond scission, as in

$$CH_3{-}CH_2{-}CH_3 \rightarrow CH_3{-}CH^+{-}CH_3 + H^-$$

and this is conveniently achieved in the presence of strong Bronsted and Lewis acids—i.e., when a proton donor as well as an acceptor for an electron pair on the hydride ion are at hand. In such situations, both reactions can proceed simultaneously; and because optimal cracking requires formation of carbocations from alkenes as well as from alkanes, the most suitable catalysts are zeolites, i.e., aluminosilicates in which the AlO_4 and SiO_4 tetrahedra are three-dimensionally linked in a so-called "sodalite structure" (see Fig. 9.2.1).

TABLE 9.2.2 Yields (vol %) from Fluid Catalytic Cracking of a Gas Oil[a]

	SiO_2/Al_2O_3	Zeolite
C_1 and C_2 gases	3.8	2.1
C_3 gases	17.6	13.1
C_4 gases	20.8	15.4
Gasoline to 200°C	55.5	62.0
Light fuel oil	4.2	6.1
Heavy fuel oil	15.8	13.9
Coke	5.6	4.1

[a] Schobert (1990).

Some 30 such zeolites occur in nature, most commonly as faujasite and mordenite. But several, notably Type A zeolites with Na^+ or K^+ ions and pore entrance diameters of 0.4 and 0.3 nm, respectively, have also been synthesized, and because they offer higher concentrations of active surface sites, these have almost entirely replaced the earlier SiO_2/Al_2O_3 and SiO_2/MgO preparations. By effectively stabilizing carbocations via H transfer, they also arrest cracking, and therefore promote formation of C_5–C_{10} hydrocarbons, rather than the large amounts of C_3–C_4 moieties that accrue from use of SiO_2/Al_2O_3 preparations. Table 9.2.2 illustrates this with data for catalytic cracking of a gas oil.

The catalysts are usually prepared by embedding 3–25% 1-mm diameter zeolite crystals in SiO_2/Al_2O_3, and if to be used for fluid catalytic cracking, this preparation is formed into 20- to 60-mm particles. Cracking itself is usually conducted in moving or fluidized beds at 480–540°C/70–140 kPa, but maintenance of acceptable catalyst activity and selectivity requires periodic burn-off at ~600°C of carbon and metal compounds, mainly Cu, Fe, Ni, and V deposited from the feedstock.

The major advances in reactor design and catalysis since the 1950s are reflected in widespread replacement of fixed bed techniques by:

1. continuous moving-bed processes such as the Houdriflow and Socony Airlift Thermofor systems, which differ mainly in materials handling and catalyst regeneration, and
2. fluidized-bed techniques that, characteristically, employ powdered rather than pelletized catalysts, but otherwise operate, like the UOP and Orthoflow processes, with process configurations much like those adopted for moving-bed catalytic cracking (see Fig. 9.2.2).

Preferred feedstocks for catalytic cracking are the relatively heavy (340–425°C) gas oils [6], and cracking at 480–500°C/75–140 kPa is usually com-

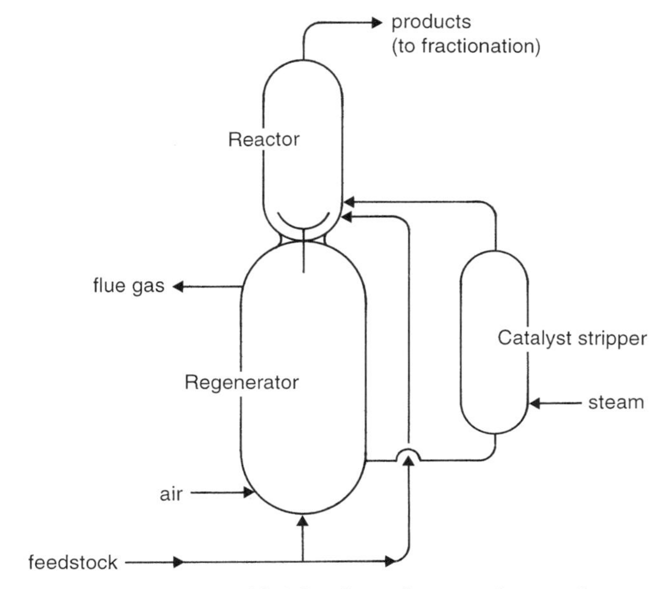

FIGURE 9.2.2 Simplified flowsheet of UOP catalytic cracking.

pleted within a few seconds. Fractionation of product streams then recovers C_1–C_4 gases, "catalytic" gasolines for gasoline blending, "catalytic" light (LGO) and heavy (HGO) gas oils disposed as fuel oils, and a residual heavy oil that is usually recycled and cracked to extinction.

Figure 9.2.3 shows a schematic flowsheet of a modern catalytic cracking facility for which typical inputs and outputs are identified in Table 9.2.3.

Hydroprocessing

To facilitate processing of aromatics, which tend to be fairly abundant in the heavier fractions of a crude, and to minimize deposition of coke on hydrocracking catalysts, recourse is commonly had to prior low-temperature hydrogenation of the feedstock at <390°C. Because this transforms aromatics into naphthenic species that respond to hydrocracking with ring opening and H addition at chain ends, substantial advantages accrue from decoupling hydrogenation from hydrocracking and conducting the two operations successively.

Similar advantages accrue also from relatively milder *hydrotreating*, which is primarily designed to remove objectionable heteroatoms (notably N, O,

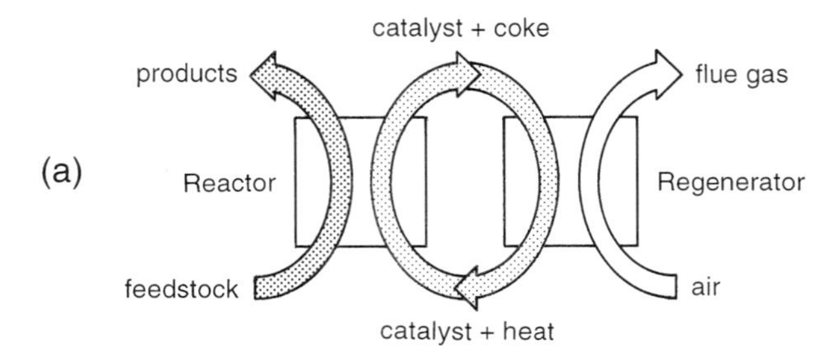

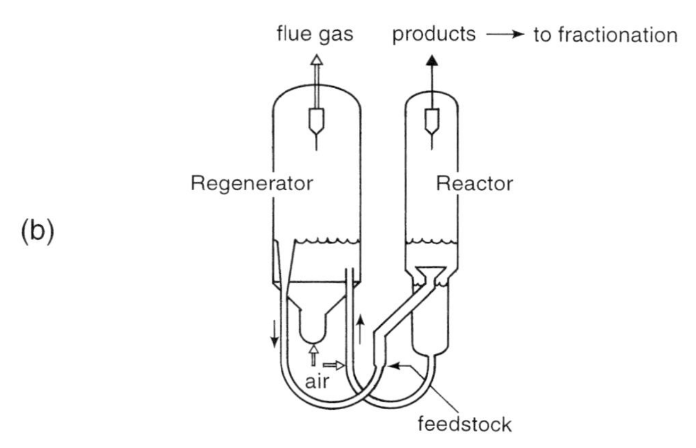

FIGURE 9.2.3 A schematic flowsheet of a modern catalytic cracking facility: Diagram (a) illustrates the principles of such a facility, diagram (b) shows a typical application. (after Decroocq, 1984; reproduced with permission)

TABLE 9.2.3 Input/Ouput of Typical Catalytic Cracking[a]

Input	
Heavy gas oil	40
Flasher tops	60
Output	
Coke	8
C_1–C_4 hydrocarbons	35
Gasoline	55
Light gas oil	12
Heavy gas oil	8

[a] Vol %; excludes ~10% recycle oil returned to cracker.

TABLE 9.2.4 Objectives in Hydroprocessing of Crude Oil Fractions

			Hydrotreat to remove[a]					
Fraction	Hydrocrack to	↓	C_{ar}	S	N	M	*n*-CH	Olefins
Naphtha	x	reformer feed		x	x			x
		LPGs						
Gas oil								
from atm. dist.		diesel fuel	x				x	
		jet fuel	x					
		petrochem. feed	x					
	x	naphtha						
Gas oil								
from vac. dist.		cat. cracker feed		x	x			
	x	kerosine		x	x	x		
	x	diesel fuel	x	x				
	x	jet fuel	x	x				
		naphtha	x					
		LPGs						
	x	lubricating oil	x					
Residua		cat. crack. feed[b]		x	x	x		
		coker feed[b]		x		x		
	x	diesel fuel						

[a] C_{ar} = aromatics; *n*-CH = *n*-paraffins.
[b] In these cases, processing also seeks to reduce the liability to coke formation in secondary processing.

and S) and/or unwanted organic constituents from crude oil fractions. Table 9.2.4 identifies the objectives of such processing.

Coking

For heavy fractions from a crude oil, which are characterized by greater aromaticity and a more compact molecular structure, acceptable upgrading requires more aggressive pyrolytic processing, and straight-run or cracked resids from fractionation or catalytic cracking of crude oils are therefore processed by *coking*. A form of severe thermal cracking, this delivers C_1–C_4 gases, naphtha, fuel oil, gas oils, and a carbonaceous solid conventionally (but misleadingly) termed "coke." The gas oils are then often further degraded by catalytic cracking.

The simplest form of coking, and the method of choice when coke can be profitably disposed, is "delayed" coking (see Fig. 9.2.4), which optimizes yields

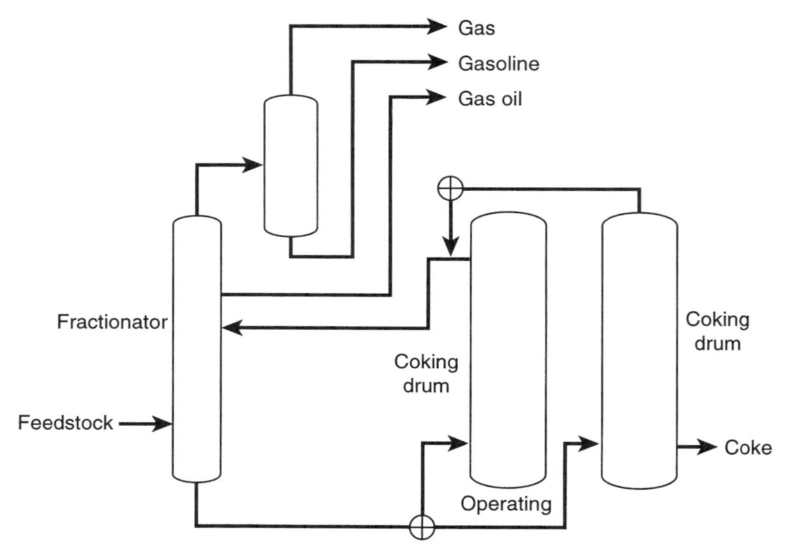

FIGURE 9.2.4 Schematic flowsheet for delayed coking.

of naphtha and middle distillates at the expense of gas oils. Such cracking is carried out in drumlike reactors in which the feedstock is heated to temperatures between 435 and 480°C. Light product vapors are continuously taken off at 430–440°C and quenched, while coke is allowed to slowly accumulate until the drum is full. At that point, the feed is diverted to a second drum, and the first is emptied. The entire cycle usually takes ~48 hr.

However, because delayed coking generates highly aromatic cokes that retain most of the [N], [S], and [O] of the precursor feed, preference is often given to *fluid coking,* which was introduced in the mid-1950s (Fig. 9.2.5). This entails spraying the feed into a fluidized bed of 1- to 6- mm coke particles on which cracking proceeds at 500–530°C. The resultant light hydrocarbon vapors are then sent through cyclones that remove entrained coke, and cooled by heat exchange against incoming fresh feed. Surplus coke is burned as fuel or, as in flexicoking in which the coke is gasified with O_2, converted into a syngas (see Chapter 10).

Advantages claimed for fluid coking include improved heat transfer, less product polymerization, and, because of shorter in-reactor residence of the cracked vapors, improved yields of distillate oils as well as significantly lower losses to coke (see Table 9.2.5).

A variant of fluid coking, particularly suitable for converting gas oils into naphtha and middle distillate oils, presents itself in *fluid catalytic cracking,* for

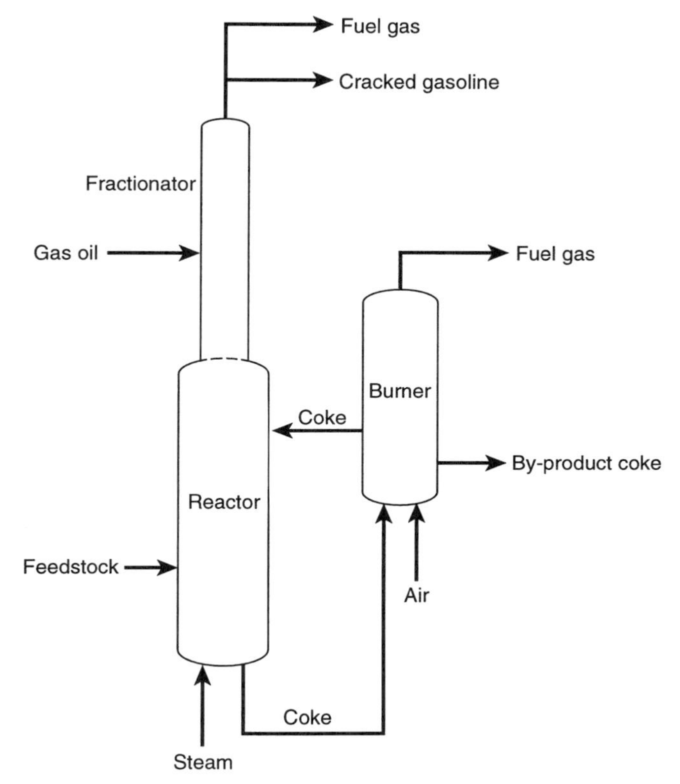

FIGURE 9.2.5 Schematic flowsheet for fluid coking.

which a suitable catalyst consists of 5–10% zeolite in an amorphous SiO_2/Al_2O_3 matrix. The zeolite particles are usually 2–10 mm in size and formulated from a synthetic Na/Al silicate (Zeolite Y) by substituting H^+ for Na^+ and reducing the aluminum content by steam treatment. Catalysts so prepared (and sometimes referred to as "ultrastable Y") can protonate *n*-alkanes and initiate carbonium ion sequences that

1. crack and isomerize paraffins to C_3-paraffins and olefins,
2. transform olefins into branched olefins, diolefins, and paraffins by H transfer and,
3. open naphthenic rings, cleave peripheral carbon chains, and thereafter form paraffins, olefins, and aromatics by dehydrogenation.

Fluid catalytic cracking also serves to open hydroaromatic rings, which then react like naphthenes and to oligomerize diolefins, which generate polynuclear

TABLE 9.2.5 Coke Yields from Delayed and Fluid Coking as Functions of Feed Gravity and Carbon Residue

Carbon residue (wt%)	Gravity (°API)	Coke yield (wt%)	
		Delayed coking	Fluid coking
5	26	8.5	3
10	16	18	11.5
15	10	27.5	17
20	3.5	42	29

aromatics and coke by cycloaddition. Aromatic rings will, however, merely lose side chains (which appear as olefins, diolefins, and paraffins) and otherwise survive substantially intact.

Industrially, fluid catalytic cracking is accomplished by contacting the feed with the catalyst at between 470 and 510°C, and sending the cracked vapors through a series of cyclones before withdrawing them from the reactor system. The spent catalyst is steam stripped to free it of residual hydrocarbons, and regenerated by fluidizing it with air and burning coke off at 670–720°C.

3. THE PROCESSING OF HEAVY HYDROCARBONS

BITUMENS

As noted in Chapter 8, environmental problems associated with the hot-water extraction process (HWEP) have directed attention to alternative techniques for extracting bitumen from oil sands, and particular interest has been shown in retorting—a method that mimics coking of petroleum residua. At atmospheric pressure, total oil plus gas yields from such pyrolysis approach 95% on bitumen (Rammler, 1970) and increase with temperature to a maximum at ~550°C, but fall slightly if the pressure exceeds ~1 atm (Stangeby and Sears, 1978).

The status of retorting technology for large-scale processing is illustrated by the UMATAC/Taciuk (Taciuk, 1977; Taciuk *et al.*, 1989) and Lurgi–Ruhrgas (Peters, 1960; Sommers, 1974) processes, both of which can deliver high oil yields regardless of the bitumen contents of the sands, and both of which leave dry tailings for final disposal.

Extensive field trials, in which the UMATAC/Taciuk retort was operated at 520°C, and >525°C material was recycled to extinction, delivered C_4^+ hydrocarbons that accounted for ~74 wt% of the bitumen in the feed, and generated net yields of synthetic crude that reached 70–71 wt% at an overall

TABLE 9.3.1 Oil Qualities from Two Bitumen Recovery Processes (wt%)

	UMATAC/Taciuk	HWEP/Fluid coking
C_1–C_4	3.03	2.04
Naphtha (C_5/220°C)	20.43	19.26
Distillate (220–345°C)	33.25	26.29
Gas oil (345–525°C)	43.29	52.41

energy efficiency of ~82%. The quality of the syncrude (Table 9.3.1) compared favorably with the quality of the oil produced by fluid coking of bitumen extracted from HWEP (Chapter 8).

The Lurgi–Ruhrgas process—which evolved from LT coal carbonization technology, successfully cracked heavy fuel oils and crudes to naphtha, and was more recently also operated with oil shales (see below)—has similarly undergone testing with oil sands and appears to offer a satisfactory alternative to UMATAC/Taciuk retorting. From rich sands with >10 wt% bitumen, which were processed at 500°C, it recovered up to 85 wt% of the feed bitumen as a good-quality synthetic crude, and operated with low-grade sands that contained 6–8 wt% bitumen, it furnished >70 wt% in C_{4^+} hydrocarbons.

Both operations recycle heavy bottom oils and thereby generate a stream that can be directly sent to hydrocracking over fixed or fluidized ("ebullated") catalyst beds (Gray, 1994).

In fixed-bed operations, processing is usually conducted in downflow systems (or trickle beds), and the catalyst—commonly 10–14% molybdenum/2–3% cobalt or 10–14% molybdenum/2–3% nickel on γ-alumina—is deployed in the form of small (~2-mm) pellets, rings, or short cylinders. But because metals and coke deposited on the catalyst from the feed will seriously impair its activity, acceptable operation requires gradually raising the processing temperatures.

"Ebullated"-bed operations are illustrated by Hydrocarbon Research Inc.'s H-Oil process [7] and LC-fining, two very similar technologies that differ primarily in equipment design details. Reaction conditions are typically 420–450°C/10–15 MPa at a hydrogen recycle rate that ensures excess H_2 in the liquid as it moves through the catalyst bed; the catalyst size is chosen to allow substantially unhindered flow through the ebullated bed; and spent catalyst is replaced at rates—commonly 0.5–1.5 kg/m^3 fresh feed—that maintain the required conversion level.

The ebullated-bed reactor gains major advantages from an intimate contact between feedstock and catalyst and from temperature gradients as low as 2°C.

TABLE 9.3.2 LC-Fining of Athabasca Bitumen[a]

	Yield (% feed)	
	wt%	vol%
H_2S and NH_3	3.39	
C_1–C_3 fuel gas	3.81	
C_4 hydrocarbons	1.49	3.4
light naphtha	2.29	4.0
heavy naphtha	8.12	10.7
light gas oil	35.38	41.2
heavy gas oil	9.24	9.9
resid.	37.20	34.8

[a] Bishop (1990).

LC-fining of Athabasca bitumen (see Table 9.3.2) illustrate the efficacy of such processing with a one-reactor system. More extensive conversion, in some cases running as high as 80% of a >525°C resid, has been reported from deploying two ebullated bed reactors in series (van Driesen *et al.*, 1987).

Alternatives to fixed- or ebullated-bed processing present themselves in *additive-based* methods that control coke formation, maximize conversion of residua, and furnish low-N and -S liquids that can be satisfactorily hydrotreated. Examples are the Veba Combi-Cracking technology (Wenzel, 1992) and the CANMET process (Pruden *et al.*, 1989). The additives include iron oxides or sulfates and colloidally dispersed organometallics—in particular, naphthenates that furnish metal sulfides when decomposing during processing. As a rule, these once-through additives are slurried into the feedstock, which then interacts with H_2 in an upflow reactor. Table 9.3.3 shows data from the CANMET process for four levels of pitch conversion.

TABLE 9.3.3 CANMET Conversion of a Cold Lake (Alberta) Resid[a]

	Pitch converstion (%)			
	70	77	79	86
C_1–C_4 (wt%)	5.7	6.9	7.3	8.6
Naphtha (vol%)	18.8	22.6	23.7	27.6
Distillate (vol%)	30.1	33.9	34.9	37.9
Gas oil (vol%)	33.1	33.1	33.1	33.1
>524°C pitch (vol%)	25.6	19.1	17.4	11.3

[a] Feedstock: 80% Cold Lake vacuum bottoms + 20% IPL vacuum bottoms, 3.3° API.

Coke formation can reportedly also be controlled by diluting the feed with hydrogenated middle oils that concurrently enhance hydroconversion by donating [H] to the residuum (Carlson *et al.*, 1958; [8]).

Oil Shales

At temperatures between ~450 and 475°C, organic matter (kerogen) in oil shales begins to break down by C–C bond scission in aliphatic and heteroatom linkages, undergoes H disproportionation, and forms H-rich liquid hydrocarbons comparable to those in crude oils. The C-rich solids thereby generated become increasingly aromatic as the pyrolysis progresses, and finally assume pseudo-graphitic character.

Kinetically, this breakdown of kerogen has been represented as a two-step sequence of the form

$$\text{kerogen} \xrightarrow{k_1} \text{bitumen} \xrightarrow{k_2} \text{oil} + \text{gas} + \text{coke},$$

in which the formation of a bitumen intermediate proceeds at lower temperatures and faster than secondary degradation of that intermediate (see Franks and Goodier, 1922; Maier and Zimmerly, 1924). However, later studies (Cane, 1951) indicate formation of *two* prior intermediates—a heat-altered kerogen II and, as a second step, generation of an insoluble "rubberoid" material that develops by rupture of cross-linkages and forms a more nearly linear elastomeric polymer (Johnson *et al.*, 1975). For pyrolysis in an inert atmosphere, the mechanism is therefore better written as

$$\begin{array}{ccccccccl} & & & & & & & & \rightarrow \text{gas} \\ \text{kerogen I} & \rightarrow & \text{kerogen II} & \rightarrow & \text{“rubberoid”} & \rightarrow & \text{bitumen} & & \rightarrow \text{oil} \\ & & & & & & & & \rightarrow \text{coke.} \\ & & \downarrow & & & & \downarrow & & \\ & & H_2S,\ CO_2 & & & & H_2S & & \end{array}$$

In either formulation, however, the kerogen → oil transition closely resembles the pyrolysis of a caking coal (see Fitzgerald and van Krevelen, 1959), which has been represented by

$$\begin{array}{lllll} \text{coal} & \rightarrow & \text{plastic coal} & \rightarrow & \text{semicoke.} \\ \downarrow & & \downarrow & & \\ \text{volatiles} & & \text{volatiles} & & \end{array}$$

Virtually all hydrocarbon material [9] and some H_2 are emitted between ~450 and 500°C; weight losses below 200°C are primarily caused by dehydration; and emissions between 700 and 800°C reflect decomposition and silication of carbonate minerals.

Oil yields are directly related to the fraction of aliphatic carbon (C_{al}) in the kerogen, and what yields can be expected from processing can be estimated by standard Fischer assays or from the empirical

$$y = 2.216\ w_k,$$

where y is the anticipated oil yield in U.S. gal/ton), and w_k the weight percentage of kerogen in the shale (Cook, 1976).

Because kerogens, except for a small (<10%) proportion of occluded bitumen, are virtually insoluble in common solvents, the only method seriously considered for hydrocarbon recovery is retorting [10], and this has been explored as means for *in situ* recovery as well as for processing of mined shales. Suitable methods have been identified by the U.S. Office of Technology Assessment (1980) as true *in situ* (TIS), modified *in situ* (MIS), and above-ground retorting (AGR). None of these is considered ready for immediate commercialization, but MIS and AGR are thought to be promising options for future large-scale application.

Current concepts of TIS envisage fracturing the formation much as in the preparation of oil sand strata for *in situ* bitumen recovery, pyrolyzing the organic matter by partial shale combustion or injection of hot combustion gas, and extracting the resultant hydrocarbon vapors through production holes. However, this was found to deliver poor oil yields [12] and has therefore, at least for the time being, been virtually abandoned in favor of a modified procedure (MIS) in which a substantially empty underground (u/g) retort is developed by mining. Fresh shale rubble is then caved into the retort and pyrolyzed by partial combustion or hot gas injection. Satisfactory recovery of oil depends on uniform distribution of broken shale within the retort; and even given that, yields of hydrocarbon liquids have not yet exceeded ~60% of potential. Substantial improvements are only expected from turning the retort through 90°—and so in effect conducting MIS much like a u/g coal gastification scheme (see Chapter 10). Table 9.3.4 offers some indication of the composition of oil from two u/g retorting trials, one conducted by Occidental Petroleum Corporation and the other by the U.S. Bureau of Mines.

Processing of *mined* oil shale is commonly conducted in retorts or rotary kilns with established performance histories in other similar operations—e.g., drying, ore roasting, and calcination. Some, like the Lurgi gasifier (Chapter 10), are heated internally by partial combustion of the charge, and exemplified by the Nevada–Texas–Utah (NTU) downdraft retort, which resembles the producer gas generators employed for manufacture of a low-Btu fuel gas from coal. Others, illustrated by the Paraho retort, which differs from the Lurgi reactor by injecting a hot gas into the middle zone and cold air above the bottom before discharging the spent shale, are heated by circulating externally generated hot combustion gas. Still others are operated like the Tosco II and

TABLE 9.3.4 Oil Quality from *in-Situ* Oil Shale Retorting[a]

	Retort	
	Occidental	USBM
Elem. comp. (%)		
C	84.9	84.6
H	11.8	12.1
O	1.1	1.2
N	1.5	1.6, 1.4
S	0.7	0.6, 0.7
°API	25	28.4, 28.4
Distillation [vol% at (°F)]		
50	700, 680	700
70	700, 765	768
80	825	809
90	920, 910	867

[a] U.S. Office of Technology Assessment (1980).

Lurgi–Ruhrgas reactors and, like these units, are heated by externally prepared solid heat carriers [13].

Other than the NTU retort, which is now deemed to be obsolescent, different versions of these retorts—in particular, the designs used in Union Oil's retorting process (see U.S. Pat. No. 4,010,092), and in Brazil's Petrosix process [14] developed for Irati oil shale (Matar, 1982)—have been and in part still are operated in a number of small commercial facilities and continuing pilot-scale trials [15]. Table 9.3.5 summarizes the compositions of oils from some of these operations.

However, R&D studies have also explored recovery of liquid hydrocarbons from oil shales by extraction with supercritical organic solvents (Nowacki, 1981; Das, 1989) and supercritical H_2O (Ogunsola and Berkowitz, 1995). Significantly improved oil yields have in some cases accrued from hydropyrolysis (Matthews and Feldkirchner, 1983; [16]). And hydroretorting of Alabama and Indiana oil shales in a fluidized-bed reactor at 515°C/4–7 MPa H_2 has furnished 170–200% of the Fischer assay yields (Roberts *et al.*, 1989).

Methods for *refining* raw shale oils (Whitcomb & Vawter, 1976; Yen & Chilingarian, 1976) replicate techniques for processing crude oil and include visbreaking, fluid coking, flexicoking, and catalytic hydrogenation (for example, as in the H-Oil process [17]). Figure 9.3.1 illustrates three possible refining schemes; a configuration resembling Fig. 9.3.1c, but replacing initial hydrotreating by fluidized catalytic cracking, has in fact been used by SOHIO to refine 13 500 m^3 of shale oil from the Paraho project (Robinson, 1979).

Like other oils, shale oils will, however, slowly deteriorate when exposed

TABLE 9.3.5 Composition of Oils from Retorting of Mined Oil Shales[a]

	Retort				
	Fischer	NTU	Tosco II	Union Oil	Paraho
Elem. comp. (wt%)					
C	84.6	84.6	85.1	84.0 84.8	84.9
H	11.5	11.4	11.6	12.0 11.6	11.5
O	1.1	0.8	0.9	0.9	1.4
N	2.0	2.1	1.9	2.0 1.7	2.2
S	0.6	0.8	0.9	0.9 0.8	0.6
API gravity		20.3	21.2	18.6 22.7	19.3
Distillation [vol% at (°F)]					
50	655	670	700	775 731	810
70	705		850	980	
80			920	960	980
90				1040	

[a] U.S. Office of Technology Assessment (1980).

to air. A study in which such oxidative deterioration at 45°C was followed over 4000 hr (Fookes and Walters, 1990) identified the dominant reactions—all of which could be accelerated by visible light—as involving oxidation of alkenes, phenols and pyrroles. Fookes *et al.* (1990) have also shown that when exposed to heat (e.g., through contact with hot shale solids), the oils (i) crack and form additional alk-1-enes, (ii) respond to acidic sites in the shale with dealkylation of aromatics, isomerization of alkenes, and formation of coke, and (iii) suffer catalytic dehydrogenation of aliphatics with consequent formation of, mainly, *O*-substituted benzenes.

Coals

Depending on its rank, coal contains between ~65 and 90% aromatic carbon, and processing is therefore now, much like processing of bitumens and kerogens, limited to severe thermal cracking by retorting (or coking; [18]). This delivers hydrocarbon gases and H-rich tars that are sometimes (unfortunately) referred to as "oils" (Eddinger *et al.*, 1968; Squires, 1975) and leaves correspondingly carbon-enriched solids, which represent the *bulk* of the product slate. The by-products in this case are the tars rather than the "coke", and how carbonization is conducted is determined by what properties the coke is required to possess.

Low-temperature (LT) carbonization at temperatures between 550 and 700°C thus produces reactive chars that are, for the most part, used as a

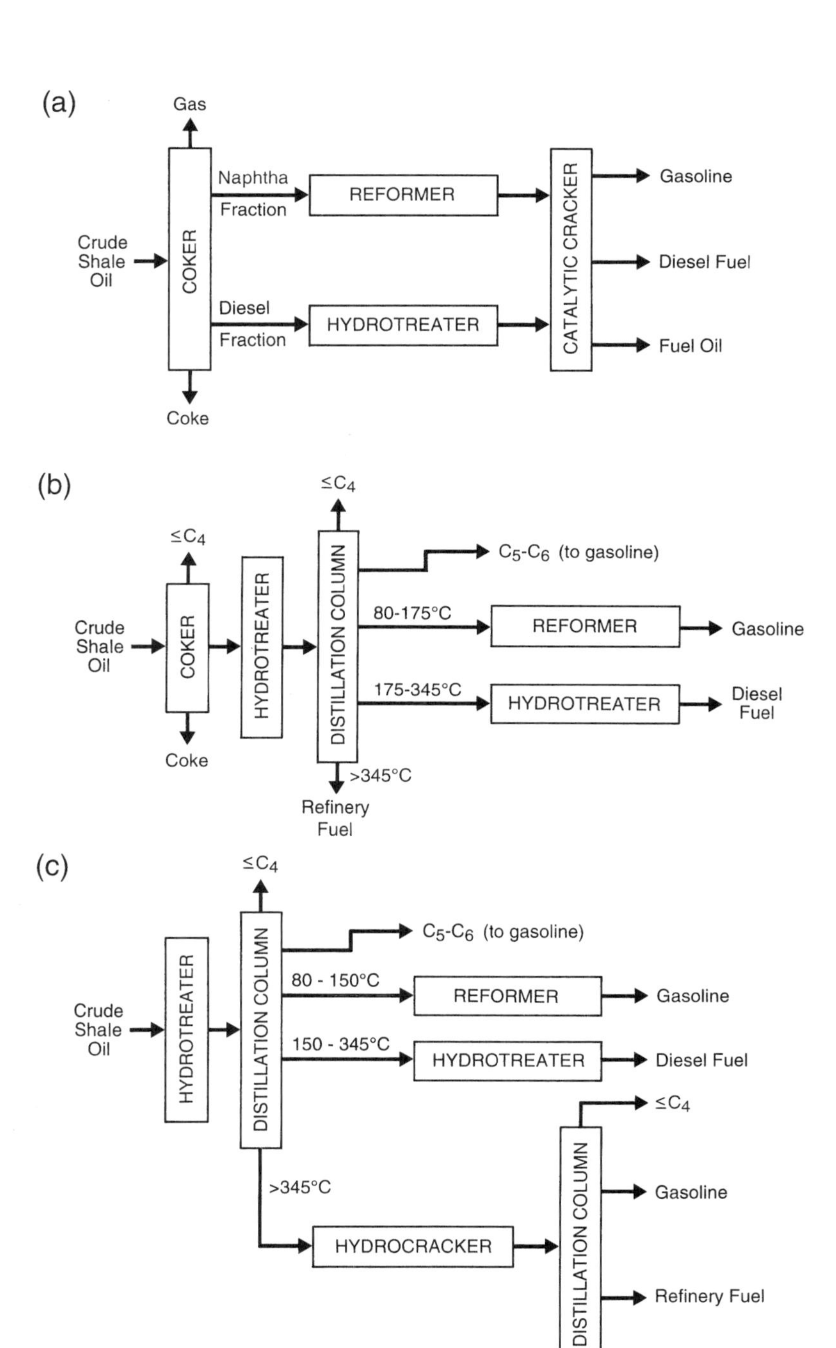

FIGURE 9.3.1 Refining options for raw shale oil: (a) refining for maximum gasoline production; (b) refining of coked raw oil; (c) refining of hydrotreated raw oil.

TABLE 9.3.6 500°C Fischer Assays of Coals[a,b]

Rank	No.	Tar	Water	Gas
Subbit. B	7	0.055–0.070	0.090–0.085	57.1–86.1
Subbit. A	5	0.076–0.101		
hvCb	7	0.076–0.159	0.046–0.073	48.7–64.6
hvBb	11	0.099–0.179	0.047–0.050	51.8–75.5
hvAb	134	0.094–0.160	0.012–0.035	52.7–73.6
mvb[c]	30	0.041–0.107	0.011–0.027	
lvb[c]	17	0.023–0.055	0.004–0.025	49.9–61.1
Cannel	7	0.221–0.445	0.007–0.018	46.8–66.1

[a] Adapted from Selvig and Ode (1957).
[b] Yields in m^3/tonne, tar includes light oils.
[c] Included for comparison, but not suitable for LT carbonization

"smokeless" domestic fuel [19], whereas high-temperature (HT) processing at ~900°C, almost always centered on blends of suitable mvb and/or lvb coal [20], delivers metallurgical cokes that are essential raw materials for iron making and certain nonferrous metallurgical operations.

LT Carbonization

Although details of alternative low-temperature carbonization methods are beyond the scope of this survey, it is germane to note their wide operational freedoms [21], and to observe that LT *tar* processing techniques are virtually identical with procedures employed for upgrading heavy oils and bitumens [22].

The preferred feedstocks for LT carbonization are subbituminous and hvb coals, which yield porous chars and, depending on rank, can deliver up to 0.18 m^3 tar and 75 m^3 gas per tonne (Table 9.3.6). However, LT tars are much more heterogeneous than heavy oils. They contain a wide spectrum of paraffins and olefins as well as variously substituted aromatics and heterocycles; and because none of these components occurs in concentrations high enough to warrant its extraction, processing has centered on transforming the tars into synthetic gasolines, diesel fuels, heating oils, and waxes [23].

The major processing sequences, used in several full-scale industrial plants between 1927 and 1945, involved high-pressure hydrogenation or hydrocracking. Maximum gasoline yields were generally obtained by hydrogenation over a molybdenum or tungsten sulfide catalyst at 420–500°C/14–42 MPa, but product streams with substantial proportions of light diesel fuels were also generated over 10% molybdena on γ-alumina at 425–450°C/1.5–7 MPa. Motor gasolines and diesel fuels from such processing represented ~80% of the tar for overall average hydrogen consumptions of 34–42 m^3 per product m^3.

Production of gasoline and diesel fuel by *hydrocracking* involved similar processing of the tar (or <400°C tar fractions), then freeing the cracked stream of tar acids and bases, and finally aromatizing it over a molybdena–alumina or chromia–molybdena–alumina catalyst as ~425°C/5–10 MPa.

Thermal or catalytic cracking at 430–450°C/4–6 MPa yielded mainly diesel fuels and heating oils, so that a typical product stream would contain ~20% diesel fuel, 47% heating oils, and 4% gasoline (Chilton, 1958), but substantially higher gasoline yields at the expense of heating oils could be secured by delayed coking of the tar (Pursglove, 1957; Dell, 1959).

In some instances LT tars were also, much like petroleum hydrocarbons, refined by selective solvent extraction. Successive extraction with SO_2, naphtha, and $(CH_2)_2Cl_2$ thus recovered gasoline, diesel fuel, heating oils, and various soft and hard waxes (Terres, 1946), and mixed polyhydric phenols were obtained by continuous extraction with H_2O at 165°C/1.65 MPa (Bahmüller, 1957) or by countercurrent extraction with C_2H_6 and aqueous CH_3OH (Batchelder *et al.*, 1959).

The precipitous decline of domestic demand for solid fuels in the late 1940s forced almost total abandonment of such operations. But by the mid-1970s, LT carbonization began to attract renewed interest as possible means for meeting a projected continuing high-volume demand for *industrial* solid fuels [24] and augmenting supplies of *synthetic liquid* fuels [25]—and this has prompted development of carbonization techniques that maximize tar generation by utilizing observations that (nominal) volatile matter contents of coal increase rapidly with heating rates [26]. Figure 9.3.2 illustrates this dependence [27]. Although most of these techniques still require further testing and scale-up before being available for commercial use, several are thought to hold much promise for future coal processing.

The Lurgi–Ruhrgas process [28] (Fig. 9.3.3), which cracks <6-mm coal at 600–650°C by 25- to 30-second contact with hot process-generated char, exemplifies this genre of LT techniques. Hydrocarbon vapors are drawn off and quenched as fast as they form, and char particles are stripped of fines (which are used as a boiler fuel), reheated by partial combustion in off-gas, and recycled to the carbonizer. Operated with hvAb or hvBb coals, tar yields run at 0.20–0.22 m^3 (1.25–1.40 bbl) per tonne (calculated on dry, ash-free coal).

The Toscoal process (Carlson *et al.*, 1975), which was adapted for coal processing from Tosco II technology developed for oil shale retorting, resembles the Lurgi–Ruhrgas system, but employs hot ceramic balls as a heat carrier for cracking the coal at temperatures between 430 and 520°C. This avoids

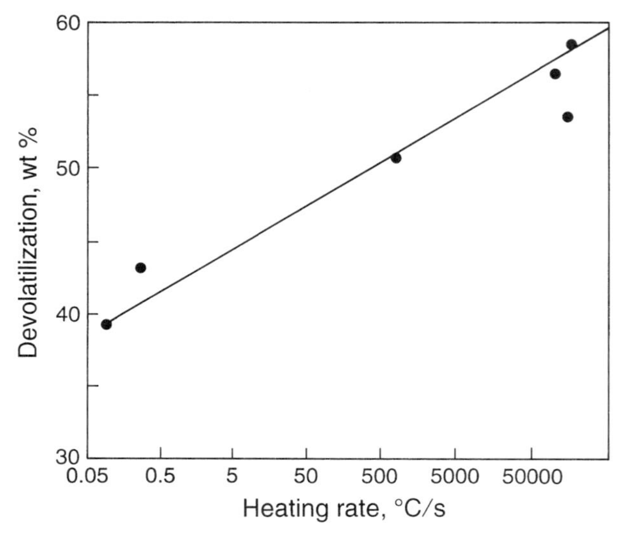

FIGURE 9.3.2 The variation of volatile matter yields with heating rates (Eddinger *et al.*, 1968). Amounts incremental to volatile matter contents determined by prescribed standard analytical methods accrue almost entirely to *tarry* matter.

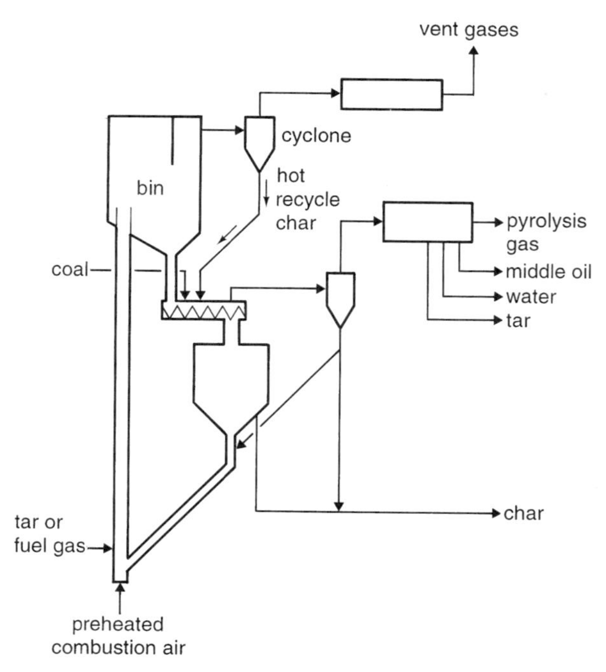

FIGURE 9.3.3 A schematic of the Lurgi–Ruhrgas process.

diluting the off-gas with N_2, CO and CO_2, and thereby delivers an H-rich medium-Btu fuel gas as by-product.

A more dramatic departure from classic approaches is seen in the COED process (FMC, 1966, 1967, 1975), which cracks fluidized <3-mm coal while it cascades against a hot gas stream through a set of reactors at ~450, 540, and 870°C (Fig. 9.3.4). With hvAb and hvBb coals, this also delivers 0.20–0.22 m^3 tar/tonne at ~40% carbon conversion, but can be taken further by close-coupling it with gasification of the hot char. In that mode, the scheme is known as the COGas process (Paige, 1975). Typical COED/COGas yields from three hvb coals are listed in Table 9.3.7.

However, development of these processes—all now deemed ready for commercial use—was also paralleled by development of technologies that seek to exploit the relationship between tar yields and heating rates more fully by much faster heat-up to the maximum temperature. This approach, more correctly termed "flash pyrolysis" than the Lurgi–Ruhrgas, Toscoal, and COED processes, is well illustrated by Occidental Petroleums's carbonization technology (Sass, 1972; Adam *et al.*, 1974), which can, when required, deliver a hydrocarbon-rich high-Btu gas in lieu of tars. Fast cracking is here effected by contacting finely comminuted (<75-μm) coal with a turbulent stream of hot char and thereby raising the coal to the desired reaction temperature at an estimated 275°C/s (Fig. 9.3.5). Maximum yields of oil accrue at ~580°C, and maximum generation of a ~23.3 MJ/m^3 fuel gas occurs at ~870°C. Table 9.3.8 shows some typical product distributions.

Of the same genre, although still experimental, is the use of flash heating in *hydropyrolysis*. SRT-Rocketdyne technology (Rockwell International, 1977) has generated up to 0.45 m^3 hydrocarbon liquids per tonne (2.8 bbl/t) by heating dry, H-entrained <150-μm coal to 530°C in 50–100 ms, and the Coalcon process (Smith and Wailes, 1975), from which only slightly lower conversions (0.35–0.40 m^3/tonne) were recorded when reacting fluidized coal with H_2 at up to 560°C/7 MPa, suggests that conversion levels as high as Rockwell's could quite conceivably be reached in commercial facilities [30].

HT Carbonization

Carbonization at ~900°C is designed to manufacture blast-furnace cokes [3], which can only be produced from so-called "metallurgical" (or "caking") coals. The "retorts" used for this purpose are wall-heated ovens—typically 15–17 m long, 6–6.5 m high, and 30–55 cm wide [32]—that are operated in batteries in which 20 or more such ovens alternate with heating flues through which hot (~1000°C) combustion gas is circulated. The carbonization cycle depends

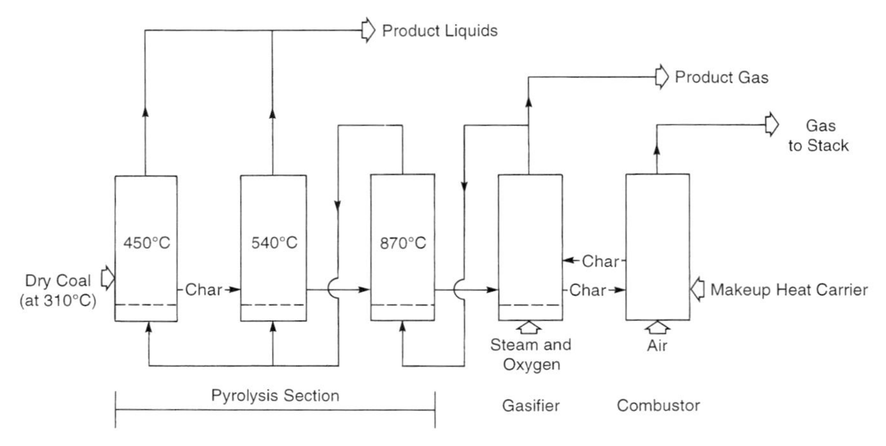

FIGURE 9.3.4 A schematic of the COED/COGas process.

TABLE 9.3.7 COGas Product Yields from Three Coals (wt%)

	hvb IL	hvb UT	Subbit. WY
Char	60.7	59.0	50.0
"Oils"	18.7	21.9	11.2
Aq. liquor	5.8}		11.6
Gas[a]	14.6}	21.5	27.2

[a] Excludes gasifier products.

TABLE 9.3.8 Product Distributions from Occidental Petroleum's Flash Pyrolysis

At ~580°C:	Dry feed (wt %)
Char	56.7
Oils[a]	35.0
Gas	6.6
Aq. liquor	1.7
At ~870°C:	**Dry gas[b] (%)**
H_2	26.8
CO	30.0
CO_2	8.5
CH_4	22.4
C_2+	12.3

[a] Includes ~5% preasphaltenes + 20–30% asphaltenes; approx. 50% fairly low-mol.-wt. pentane solubles.
[b] Free of N_2 and H_2S.

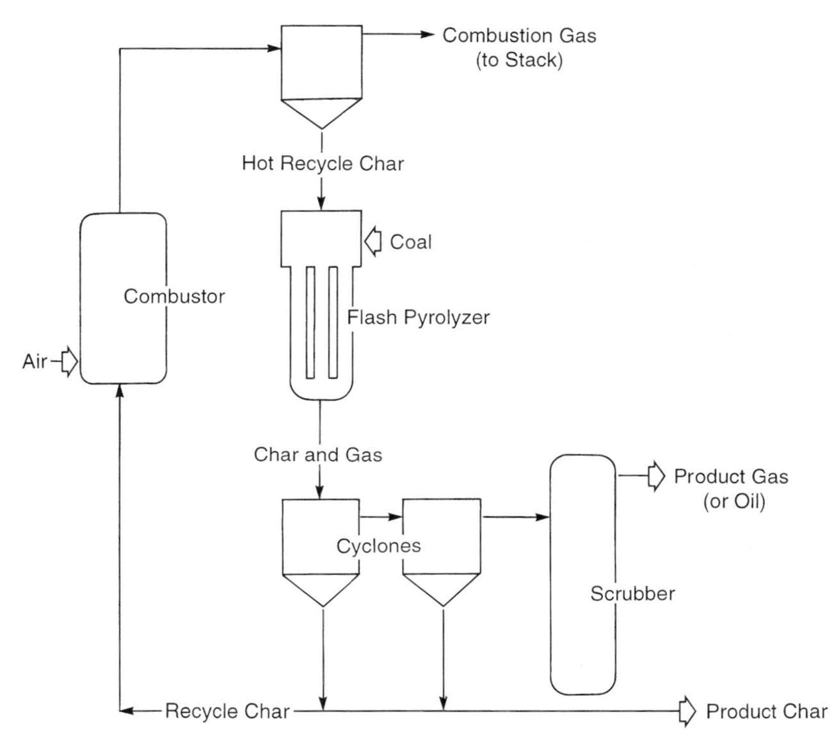

FIGURE 9.3.5 A schematic of Occidental Petroleum's flash pyrolysis process.

on oven width (≡ ~35 min/cm); and because the by-product tar is in fairly long contact with hot carbon before it leaves through external collector pipes, it suffers extensive dealkylation and aromatization. HT tars are therefore much more homogeneous than their LT counterparts, hold a broad spectrum of recoverable industrial chemicals, and are for that reason more elaborately processed.

Much as in fractionation of crude oils, HT tars are separated into light (<220°C), middle (220–375°C), and heavy (375–450°C) oils [33] by heating them in continuous pipestills from which they are released into a fractionating tower [34]. Batch stills with 11.5–30 m^3 capacities are obsolescent.

The light oils consist mostly of benzene (45–72%), toluene (11–19%), xylenes (3–8%), styrene (1–1.5%), and indene (1–1.5%) and are either blended into motor gasolines and aviation fuels or fractionated to furnish industrial solvents or feedstocks. Objectionable sulfur compounds, nitrogen bases, and unwanted unsaturates are, when necessary, removed by

1. acid-washing, neutralizing, and recovering the clean oils, and steam distilling or centrifuging them to remove ill-defined polymeric matter; or,
2. as is sometimes preferred, hydrotreating them at 300–400°C/2–3 MPa over a Co/Mo or Ni/W catalyst

Before separating the purified oils into their components, residual saturated hydrocarbons—in particular, cyclohexane, methylhexanes, *n*-heptane, and thiophenes—are removed by azeotropic or extractive distillation, selective solvent extraction, and/or crystallization.

From the middle oils, tar acids, tar bases, and naphthalene are recovered, and the residual oils are processed to meet specifications for diesel fuels, kerosine, or creosote [35]. The tar acids, mainly phenol, cresols, and xylenols, are isolated by adding dilute aq. NaOH to the crude fraction, separating the resultant "carbolate" layer, and passing stream through it to remove any remaining hydrocarbons. The acids are then freed by treating the carbolate with CO_2 or dilute H_2SO_4 and distilled *in vacuo*. Tar bases—pyridine, picolines, lutidines, aniline, quinoline, methylquinolines, and isoquinoline [36]—are similarly recovered by treating the acid-free oils with dilute H_2SO_4, regenerating the bases with excess NaOH or aqueous lime, and fractionating the mix.

Naphthalene is recovered from the resultant neutral middle oils by fractional distillation, and purified by crystallization or azeotropic distillation with cresol. A very pure naphthalene can be produced by hydrogenating the crude compound at 500–550°C over Al_2O_3-supported platinum oxide before fractionating and recrystallizing it.

Distillation of heavy oils is taken to temperatures that depend on what kind of residual pitch is to be produced. But whatever that temperature, the distillates are important sources of two-, three- and four-ring aromatics—in particular, anthracene, phenanthrene, carbazole, acenaphthene, fluorene, and chrysene—which are isolated by secondary fractionation and purified by azotropic distillation and recrystallization. Distillation with monohydric alcohols thus separates anthracene and phenanthrene from carbazole (Andrews, 1954), whereas distillation with diethylene glycol separates anthracene and phenanthrene from carbazole and fluorene (Feldman and Orchin, 1952).

The remaining heavy oils are, like some petroleum residua, marketed as fuel oils or blended with pitches to meet specifications for road tars and/or so-called "refined" tars required for manufacture of tar-based paints and tar-saturated insulating felts.

Residual pitches *per se* are complex mixtures of polycondensed aromatics that are endowed with unique resistance to water and weathering, and are therefore widely used as preservatives. Other important uses are as binders in the manufacture of briquettes, carbon electrodes, carbon catalyst supports, etc.

Hydrogenation

Contemporary hydrogenation techniques, which increase the H/C ratio of a feed and thereby furnish higher yields of hydrocarbon liquids than could be obtained by cracking or hydrocracking, are outgrowths of Bergius' classic studies of high-pressure interactions between coal and elemental H_2 at 450–500°C. By 1927 these studies [37] had defined an industrial technology that offered practical means for converting coal into gasolines, aviation fuels, and heating oils, and played a vital role in meeting the liquid fuel needs of several countries during the Second World War [38].

To avoid erosion of high-pressure equipment by injection of pulverized coal, the hydrogenation feed was a 1 : 1 or 2 : 3 w/w coal/oil slurry, and depending on whether a low- or high-rank coal was used, the first step entailed liquid-phase hydrogenation at 475–485°C/25–30 MPa or 485°C/35–70 MPa. The slurrying oil was process-derived heavy oil; the catalysts were Bayer masses [39] or tin oxalate promoted by NH_4Cl; and hydrogenation was accomplished by sending the slurried coal through a set of reactors, from each of which hydrocarbon vapors were recovered. 97–99% of the organic coal substance was in this manner transformed into

1. middle oils (b.p. 180–325°C), which represented 50–60% of the product stream and afforded the feedstock for a second hydrogenation step, and
2. heavy oils (b.p. > 325°C), of which some were recycled for slurrying fresh coal, and the remainder was used as diluent for HOLD material [40] that was then centrifuged for recovery of hydrocarbon liquids from unreacted solids.

Typical hydrogen consumptions ranged from ~160 m^3/100 liters of middle oils from lignites and subbituminous coals to ~195 m^3/100 liters of such oils from bituminous coals.

In a second processing step, the middle oils were hydrogenated in the vapor phase over fixed catalysts at 450–460°C/30 MPa, and reaction products were cooled to 30°C by heat exchange against incoming feed, depressurized to yield separate hydrocarbon-rich and -lean gas streams, and fractionated. Middle oils accruing from fractionation were recycled. Yields of light oil depended on the catalyst, but the best—a tungsten sulfide catalyst prepared from ammonium sulfotungstate or WS_2 supported on HF-activated Fuller's earth—converted >50% middle oils to gasoline and aviation fuel in a single 2- to 4-min pass through the reactor.

Because accompanying laboratory work was centered on improving plant performance and, in any event, lacked the technical means needed for detailed characterization of the hydrogenated products, some fundamental aspects of

Bergius hydrogenation are still uncertain. However, it is possible to infer its principal features from plant operations; these sources suggest:

1. That pressures higher than the 25–30 MPa deemed optimal for liquid-phase hydrogenation would have enhanced conversion and reduced H_2 expenditures by inhibiting the formation of hydrocarbon gases
2. That yields of hydrocarbon liquids produced under similar processing conditions—especially yields of middle oils—fell slowly with increasing coal rank
3. That yields from coals of similar rank depended on the H/C ratios of the coals
4. That the proportion of asphaltenes in middle oils, as well as generation of hydrocarbon gases, increased with rank
5. That the nature of the catalyst affected oil yields and compositions more than its quantity: whereas tin oxalate proved superior to molybdena for hydrogenating bituminous coal, the reverse was true for lignite hydrogenation; and when tin oxalate, usually in amounts of 0.02–0.06 wt%, was used in the latter case, it had to be promoted with 1–1.2% NH_4Cl

Like LT carbonization and tar processing, Bergius hydrogenation technology was effectively relegated to history by the abundant supplies of acceptably priced oil that became available in the late 1940s. But as indicated in Chapter 10, better understanding of hydrogenation chemistry gained in later studies has defined H-transfer reactions that offer important technical and economic advantages over the earlier processing.

NOTES

[1] As a rule, chemical means are only employed when useful materials can be recovered from otherwise rejected matter. An example is the extraction of elemental sulfur from H_2S separated from raw natural gas.

[2] Separate development of the coal and petroleum industries made each generate its own technical jargon, and virtually identical technologies are therefore referred to and described in different terms. Controlled thermal decomposition is thus termed *carbonization* by coal technologists and *thermal cracking* by oil processors.

[3] An example is decarboxylation. In the case of coal, LT decomposition is also indicated by the formation (and evolution) of HCOOH, CH_3COOH, H_2S, etc.

[4] Although this designation seems to be reserved for hydroprocessing of the heavy fractions of a crude oil, virtually identical processing, but then usually referred to as *hydropyrolysis,* has been conducted with coal.

[5] Unlike other, often better hydrogenation catalysts (such as Ni, Pd, or Pt), these formulations are less easily poisoned by S- and N-bearing reaction by-products, and therefore offer longer useful catalyst life.

[6] All liquids boiling above ~220°C at atmospheric pressure are now considered suitable feeds for catalytic cracking (Berger and Anderson, 1981). The defining point is the upper temperature to which a gasoline fraction is distilled for use in automobile engines.

[7] This was later adapted for processing hydrocarbon liquids obtained from liquefaction of coal (see Chapter 10).

[8] See in this connection H-transfer in coal liquefaction processes (Chapter 10).

[9] As a rule, the temperatures at which oil *forms* and the total oil *yields* are inverse and direct functions, respectively, of the amount of planktonic material bacterially reworked during deposition of the kerogen.

[10] Hydropyrolysis of Eastern U.S. shales under pressure furnished up to 250% of the Fischer assay yield, but did not significantly improve oil yields from Western (Green River) shales.

[11] The first recorded retorting of oil shales dates to 1694 (when English Patent #330 was issued to a Martin Hale, Esq.), but commercial operations began only in the mid-1800s. After the Second World War, availability of crude oils forced the gradual closure of these plants, most recently in Australia (1952) and Spain (1966). However, as late as 1972, shale oil production still ran to some 10 million tonnes in the Republic of China and 50,000 tonnes in Brazil. Because the insolubility of kerogen, in part ascribed to impermeability of the shale matrix, precludes solvent extraction and at present also rules out direct hydrogenation, retorting remains the only potentially viable processing option.

[12] In support of this statement it was noted that tests in a zone that contained an estimated 7800 bbl (1240 m^3) of (potential) shale oil in place and released ~1000 bbl (159 m^3) of that potential, recovered only ~60 bbl (9.5 m^3). However, because the hot combustion gas was injected into the zone at 705°C, it is conceivable that poor oil recovery was due to excessive thermal cracking: regrettably, *gas* production was apparently not recorded.

[13] Lurgi–Ruhrgas plants were commercially operated in Argentina, Britain, Germany, Japan, and the former Yugoslavia. In Japan the facility is reportedly used to crack petroleum fractions to olefins.

[14] The Petrosix retort closely resembles, and may indeed have been adapted from, the Lurgi–Spulgas retort, which was widely used in Germany in the 1930s and 1940s for LT carbonization of coal (Barritt and Kennaway, 1954; Wilson and Clendenin, 1963).

[15] In this connection it is, however, important that heat requirements for retorting demand a minimum kerogen content. For *zero* net energy gain, the minimum is ~2.5% (Burger, 1973). But a more realistic minimum concentration would be ~5% (corresponding to an oil yield of 25 liter/tonne) or, having regard for prevailing crude prices, ~10% (≡ 50 liters/tonne). Significantly, Lee (1991) considers ~42 liters/tonne to be the minimum for economic viability.

[16] As already noted in Chapter 5, hydropyrolysis of Eastern U.S. shales under pressure furnished up to 250% of the Fischer assay yield, but did not significantly improve oil yields from Western (Green R.) shales.

[17] This process came on-stream in 1962 in a 400-m^3/d facility at Cities Services' Charles Lake (Louisiana) refinery, and after 1976 was operated there in two units with a combined capacity of ~12,700 m^3/d. It was subsequently adapted for upgrading primary liquids from coal liquefaction (see Chapter 10).

[18] In the terminology applicable to coal, such retorting is referred to as *carbonization.*

[19] This designation is intended to emphasize the superiority of the char over "smoky" coal, which is often still burned in open domestic fireplaces.

[20] Metallurgical coals possess more or less well-developed thermoplastic properties that manifest themselves at temperatures between ~400 and 500°C and can be assessed in the laboratory by standardized screening tests. If such preliminary evaluation is satisfactory, test programs are carried out in coke ovens to determine actual behavior and coke quality. For a detailed

discussion of this topic, the reader is referred to Elliott (1981). For a briefer review, see Berkowitz (1985, 1994).

[21] A summary of classic and contemporary LT coal carbonization technologies has been presented by Berkowitz (1994).

[22] In several cases, procedures for processing *petroleum* hydrocarbons were actually adapted from methods developed for upgrading *coal tars*.

[23] A detailed review of German wartime practices using this approach was published by Holroyd (1946).

[24] Such demand was expected to develop from the needs of large steam-raising facilities, notably central power plants.

[25] Large-scale test firing of chars from contemporary LT carbonizers in pulverized-fuel and fluidized-bed combustion systems confirms that LT chars could prove as efficient a steam-raising fuel as their precursor coals; that substitution of LT chars for coal would generally need little more than burner-tip adjustments; and that such substitution would also modestly reduce material handling requirements because of the higher net heat value of the char. The *tars* generated by LT carbonization would be as amenable to hydroprocessing as heavy fractions from crude oil (see, in this connection, properties of coal liquids, Chapter 10).

[26] These measurements reflect experiences of "classic" carbonization, in which heating rates were (and are) <5°C/min. The accepted definition of "volatile matter" is therefore an arbitrary one.

[27] This effect may reflect more extensive fragmentation of the coal macromolecule, but is probably better attributed to faster explusion, and therefore to briefer contact of primary volatile matter with hot carbon. Thermal cracking of hydrocarbons on carbon is well known.

[28] This was also noted as effective means for retorting oil sands and processing oil shales.

[29] COED: an acronym for "char-oil-energy development."

[30] Energy economics since the late 1970s have severely restricted further development of such coal processing, and hydroflash pyrolysis has thus far only been tested in very small units, the biggest a 225 kg/day facility. The technology is therefore still in its infancy.

[31] A brief account of the manufacture and properties of metallurgical coke has been presented by Berkowitz (1994). More complete information can be found in Eisenhut (1981) and Perch (1981).

[32] Because heating rates affect the type and intensity of thermoplasticity (which determines coke quality), and the rates at which heat can suffuse a coal charge depend on oven width, there is a close connection between that width and coal behavior. This tends to make coke-oven batteries captive to a coal source, and is one of the reasons why coke makers are reluctant to consider alternative coals for their facilities. At the very least, a fairly lengthy in-plant test program is needed to determine the suitability of a possible substitute.

[33] If a hard rather than soft pitch is desired, distillation is taken to 550°C.

[34] Although the primary fractions possess greater value as sources of specific chemicals, an appropriate choice of cut points can also deliver distillate oils that are, in part, directly interchangeable with the corresponding petroleum fractions.

[35] Specifications for diesel fuel (b.p. < 390°C) focus on viscosity and ignition characteristics, but are otherwise flexible. Kerosine (b.p. ≈ 180–320°C), also referred to as solvent naphtha, is used as a cleaning fluid, heating oil, and illuminant, as well as a jet fuel (*if* refined to consist mostly of C_{10}–C_{16} paraffins and aromatics). Creosote, an important wood preservative, is used as low- or high-residue creosote, depending on whether it contains little or substantial proportions of material boiling at >355°C. Processing of neutral middle oils after recovery of naphthalene therefore requires little more than blending different cuts from fractional distillation.

[36] These can also be obtained, although in much smaller amounts, from some light oil cuts.

[37] Details of this work, which brought Friedrich Bergius the 1931 Nobel Prize in Chemistry, are reported at length in German literature of the 1918–1925 period and were later described in several English-language post–World War II technical reports that focused on German hydrogenation plants active between 1927 and 1945. An excellent historical essay (Stranges, 1987) discusses some of the problems that Bergius faced—as well as his successes.

[38] The operating procedures of German coal hydrogenation plants (which, by 1943, collectively produced some 3.5×10^6 tonnes of gasoline, aviation fuels, and other liquid hydrocarbons per year) were published shortly after the end of the Second World War in a series of U.S. and British technical reports (see Oriel *et al.*, 1945; Cockram, 1945; Ellis, 1945; Holroyd, 1947; Sherwood, 1947).

[39] Termed red mud, luxmasse, or Bayermasse, this was a by-product of bauxite processing and sufficiently cheap to be massively used as a throwaway once-through catalyst.

[40] This *Heavy Oil Let-Down* material was the bottoms product collected in a hot catchpot, and consisted of some very heavy oil mixed with mineral matter and small amounts of unreacted coal.

REFERENCES

Adam, D. E., S. Sacks and A. Sass. *Chem. Eng. Progr.* **70**(6), 74 (1974).

Al-Soufi, H. H., Z. F. Savaya, H. K. Mohammed, and A. Al-Azawi. *Fuel* **67**, 1714 (1988).

Andrews, J. W. U.S. Pat. No. 2,675,345 (1954).

Bahmüller, H. *Braunkohle* **9**, 486 (1957).

Barritt, D. T., and T. Kennaway. *J. Inst. Fuel (London)* **27**, 229 (1954).

Batchelder, H. R., R. B. Filbert, Jr., and W. H. Mink. *Preprints, 135th ACS Natl. Mtg., Boston* (1959).

Berger, B. D., and K. E. Anderson. "Modern Petroleum" 1981. PennWell Publ. Co.: Tulsa, Oklahoma.

Berkowitz, N. *The Chemistry of Coal,* 1985. Amsterdam: Elsevier.

Berkowitz, N. *An Introduction to Coal Technology,* 2nd ed., 1994. New York: Academic Press.

Bishop, W. *Proc. Symp. Heavy Oil: Upgrading and Refining, Can. Soc. Chem. Eng., Calgary, AB* (1990).

Burger, J. *Rev. Inst. Fr. Petr.* **28**, 315 (1973).

Cane, R. F. The mechanism of pyrolysis of torbanite, oil shale and cannel coal. *Inst. Petrol. London* **2**, 592 (1951).

Carlson, C. S., A. W. Langer, J. Steward, and R. M. Hill. *Ind. Eng. Chem.* **53**, 1067 (1958).

Carlson, F. B., L. H. Yrdumian, and M. T. Atwood. Proc. *2nd. Symp. Clean Fuels from Coal, Inst. Gas Technol., Chicago, IL* (1975).

Chilton, C. H. *Chem. Eng.* **65**(2), 53 (1958).

Cockram, V. *CIOS XXX-102 (Scholven Plant), London* (1945); *CIOS XXX-104 (Welheim Plant), London* (1945); *CIOS XXX-105 (Gelsenberg Plant), London* (1945).

Cook, E. W. *Fuel* **53**, 16 (1976).

Das, K. Solvent and supercritical fluid extraction of oil shale: A literature review. *DOE/METC-89/4092, U.S. Dept. Energy* (1989).

Decroocq, D. *Catalytic Cracking of Heavy Petroleum Fractions,* 1984. Paris: Editions Technip.

Dell, M. B. *Ind. Eng. Chem.* **51**, 1297 (1959).

Eddinger, R. T., J. F. Jones, J. F. Start, and L. Seglin. *Proc. 7th Internatl. Coal Sci. Conf., Prague* (1968).

Eisenhut, W. *Chemistry of Coal Utilization,* 2nd Suppl. Vol. (M. A. Elliott, ed.), Chapter 14, 1981. New York: Wiley.

Elliott, M. A. (ed.). *The Chemistry of Coal Utilization,* 2nd Suppl. Vol., 1981. New York: Wiley-Interscience.

Ellis, J. F. *CIOS XXXii-92 (Bohlen Plant), London* (1945).
Feldman, J., and M. Orchin. U.S. Pat. No. 2,590,096 (1952).
Fitzgerald, D., and D. W. van Krevelen. *Fuel* **38**, 17 (1959).
FMC Corporation. *NTIS PB-169 562/AS and 563/AS* (1966); *NTIS PB-173 916/AS and 917/AS* (1967); *Rept. No. FE/1212/F US~ERDA, Washington, D.C.* (1975).
Fookes, C. J. R., and C. K. Walters. *Fuel* **69**, 1105 (1990).
Fookes, C. J. R., G. J. Duffy, P. Udaja, and M. D. Chensee. *Fuel* **69**, 1142 (1990).
Franks, A. J., and B. D. Goodier. *Colo. School of Mines Quarterly* **17**(4), Suppl. A (1922).
Gray, M. R. *Upgrading of Heavy Oils and Residua,* 1994. New York: Dekker.
Holroyd, R. *CIOS CCCII-107 (Leuna Plant), London* (1945); *U.S. Bur. Mines Inf. Circ. No. 7370* (1946); *BIOS Overall Rept. No. 1, London* (1947).
Johnson, W. F., D. K. Walton, H. H. Keller, and E. J. Couch. *Colo. School of Mines Quarterly* **70**(3), 237 (1975).
Lee, S. *Oil Shale Technology,* 1991. Boca Raton, FL: CRC Press.
Maier, C. G., and S. R. Zimmerly. *Utah Univ. Res. Invest. Bull.* **14**, 62 (1924).
Matar, S. *Synfuels—Hydrocarbons of the Future,* 1982. Tulsa, OK: Pennwell Publ. Co.
Matthews, R. D., and H. Feldkirchner. *ACS Symp. Ser.* **230**, 139 (1983).
Nowacki, P. (ed.). *Oil Shale Technical Handbook,* p. 180, 1981. New Jersey: Noyes Data Corporation.
Ogunsola, O. M., and N. Berkowitz. *J. Fuel Process. Technol.* **45**, 95 (1995).
Oriel, J. A., I. H. Jones, and H. H. Weir. *CIOS XXVIII-40 (Wesseling Plant), London* (1945).
Paige, W. A. *Proc. 5th Synth. Fuels from Coal Conf., Oklahoma State University, Stillwater, OK* (1975).
Perch, M. In *Chemistry of Coal Utilization, 2nd Suppl. Vol.* (M. A. Elliott, ed.), Chapter 15, 1981. New York: Wiley.
Peters, W. *Chem. Ing. Technol.* **32**(3), 178 (1960).
Pruden, B. B., J. M. Denis, and G. Muir. *Paper #80, 4th UNITAR/UNDP Conf., Edmonton, AB* (1989).
Pursglove, J., Jr. *Coal Age* **62**, 70 (1957).
Rammler, R. W. *Can. J. Chem. Eng.* **48**, 552 (1970); see also Rammler, P. Schmalfeld, and H. J. Weiss. U.S. Pat. No. 4,098,674 (1978); Weiss, H. J., and P. Schmalfeld. *Erdöl Kohle* **46**(2), 235 (1989).
Roberts, M. J., D. M. Rue, and F. S. Lau. *Fuel* **71**, 1433 (1992).
Robinson, E. T. *Proc. 12th Oil Shale Symp., Golden, CO 1979,* p. 195.
Rockwell International (Rocketdyne Division). *Ann. Rept. No. FE-2044-11, Dept. Energy, Washington, D.C.* (1977).
Sass, A. *Proc. 65th Ann. Mtg., AlChemE, New York* (1972).
Schobert, H. H. The Chemistry of Hydrocarbon Fuels, 1990. London: Butterworths.
Selvig, W. A., and W. H. Ode. *U.S. Bur. Mines Bull. No. 571* (1957).
Sherwood, P. W. *FIAT Final Rept. No. 952* (1947).
Smith, I. W., and P. C. Wailes. *Proc. Symp. Energy & Liquid Fuels, Inst. Chem. Eng., Adelaide, Australia* (1975).
Sommers, H. *VGB Kraftwerkstechnik* **5**, 306 (1974).
Squires, A. M. U.S. Pat. No. 3,855,070 (1975).
Strangeby, P. C. and P. L. Sears. In *Oil Sand & Oil Shale Chemistry* (O. P. Strausz and E. M. Lown, eds.), p. 101, 1978. New York: Verlag Chemie.
Stranges, A. N. *Fuel Process. Technol.* **16**, 205 (1987).
Taciuk, W. U.S. Pat. No. 4,180,455 (1977).
Taciuk, W., L. R. Turner and B. C. Wright. *Proc. 4th UNITAR-UNDP Conf.* **5**, 439 (1989).
Terres, E. *Natl. Pet. News* **38**(6), R84 (1946).
U.S. Office of Technology Assessment. *An Assessment of Oil Shale Technologies, OTA-M-118,* 1980. Washington, D.C.: U.S. Congress.

van Driesen, R. P., V. A. Strangio, A. Rhoe, and J. J. Kolstad. *Energy Processing/Canada* **13**, July/Aug. (1987).

Wenzel, F. W. In *Proc. Internatl. Symp. Hvy. Oil & Resid. Upgrading* (C. Han and C. Hsi, eds.), 1992. Beijing: Internatl. Acad.

Whitcomb, J. A., and R. G. Vawter. In *Science and Technology of Oil Shale* (T. F. Yen, ed.), 1976. Ann Arbor, MI: Ann Arbor Sci. Publ.

Wilson, P. J., and J. D. Clendenin. In *Chemistry of Coal Utilization* (H. H. Lowry, ed.), Suppl. Volume, Chapter 10, 1963. New York: Wiley.

Yen, T. F., and G. Chilingarian. *Oil Shale,* 1976. Amsterdam: Elsevier.

CHAPTER 10

Conversion

In a technical sense, "conversion" implies a major change of status—a change in physical state and/or an extensive alteration of composition—and *any* chemical reaction could therefore be deemed to cause conversion. But applied to fossil hydrocarbons, the term is generally used in a more restricted sense to denote (i) *gasification,* which transforms a solid or liquid substance into gaseous fuels, or (ii) *liquefaction,* which converts a solid or semisolid heavy hydrocarbon into a lighter "synthetic" oil by controlled molecular degradation.

Both technologies attest to the inherent versatility and potential interchangeability of fossil hydrocarbons.

1. GASIFICATION

The Fundamentals

Unlike pyrolysis, which thermally cracks a heavy hydrocarbon into an H-rich liquid and a correspondingly C-enriched solid fraction, and thereby yields a gas as a by-product, gasification converts the entire organic feed into a useful gas, and is therefore deployed for generation of a syngas or of fuel gases with varying heat contents.

Formally expressed, the net chemical reactions that bring this change about are relatively simple. If the feed is a volatile hydrocarbon that completely vaporizes while being raised to the gasification temperature (usually 700–900°C), the primary reaction can be written as

$$C_nH_m + nH_2O \rightarrow nCO + [n + (m/2)]H_2,$$

and accompanying processes will involve hydrogenation, gas shifting and methanation, as in

$$\begin{aligned} &C_nH_m + [2n - (m/2)]H_2 \rightarrow nCH_4 \\ &CO + H_2O \rightarrow CO_2 + H_2 \\ &CO + 3H_2 \rightarrow CH_4 + H_2O. \end{aligned}$$

This sequence is initiated by chemisorption of the hydrocarbon on solid surface sites (such as might be provided by a catalyst) and involves fragmentation of the sorbate molecules, i.e.,

$$C_nH_m \rightarrow CH_x,$$

which then allows

$$CH_x + H_2O \rightarrow CO_2 + H_2$$
$$CH_x + H_2 \rightarrow CH_4.$$

However, if the feedstock is a heavy hydrocarbon—a heavy gas oil, residuum, bituminous substance, or coal, all of which break down during the approach to reaction temperature—gasification is usually effected by interaction with oxygen and steam, and two sequences then come into play. The H-rich fraction will vaporize and react as in

$$C_nH_m + [n + (m/4)]O_2 \rightarrow nCO_2 + (m/2)H_2$$
$$C_nH_m + nH_2O \rightarrow nCO + [(m/2) + n)]H_2$$
$$C_nH_m + nCO_2 \rightarrow 2nCO + (m/2)H_2,$$

and these processes will be accompanied by

1. partial combustion of the carbon-rich solid fraction, as in

$$C + O_2 \rightarrow CO_2$$

2. CO_2 reduction via the *Boudouard* reaction,

$$C + CO_2 \rightarrow 2CO$$

3. the *carbon–steam* reaction,

$$C + H_2O \rightarrow CO + H_2$$

4. the *shift* reaction,

$$CO + H_2O \rightarrow CO_2 + H_2$$

In some circumstances, direct carbon hydrogenation as in

$$C + 2H_2 \rightarrow CH_4$$

and CH_4 "reforming" via

$$CH_4 + H_2O \rightarrow CO + 3H_2$$
$$CH_4 + CO_2 \rightarrow 2CO + 2H_2$$

will also proceed.

Accompanying secondary reactions involve organically bound sulfur and nitrogen, which always tend to concentrate in heavy residues when hydrocarbons are cracked. These reactions are exemplified by

$$S + O_2 \rightarrow SO_2$$
$$SO_2 + 3H_2 \rightarrow H_2S + 2H_2O$$
$$C + 2S \rightarrow CS_2$$
$$CO + S \rightarrow COS$$
$$N_2 + 3H_2 \rightarrow 2NH_3$$
$$N_2 + xO_2 \rightarrow 2NO_x$$

and generate environmentally objectionable pollutants that also tend to poison catalysts used in downstream processing of the raw gas.

However, formal net reaction equations obscure the mechanistic complexities of the processes they are purported to represent. Combustion thus entails chemisorption of oxygen at active carbon sites (C^*) and subsequent detachment of CO, which then interacts with O_2, as in

$$C^* + O_2 \rightarrow C(O) + O$$
$$C^* + O \rightarrow C(O)$$
$$C(O) \rightarrow CO + C^* \text{ (or C)}$$
$$CO + O_2 \rightarrow CO_2 + O$$
$$CO + O \rightarrow CO_2.$$

The carbon–steam reaction follows a similar course by proceeding via

$$C^* + H_2O \rightarrow C(O) + H_2$$
$$C(O) \rightarrow CO + C^* \text{ (or C)}$$
$$C(O) + CO \rightarrow CO_2 + C^*$$

and consequently contributes to gas shifting. And carbon hydrogenation is believed to proceed by successive addition of H atoms on carbon atoms at exposed crystallite edges.

The thermodynamics and kinetics of these processes impose important constraints on gasification rates [1], govern the compositions of the product gas streams, and consequently make it necessary to establish optimum processing parameters by experimentation. Formal reaction equations do, however, identify gasification options and define some useful efficiency criteria.

The simplest option for gasification of a solid or semisolid hydrocarbon is combustion followed by the Boudouard reaction. Together, put as

$$2C + O_2 \rightarrow 2CO,$$

this defines *incomplete combustion* due to insufficient oxygen; and if combustion is sustained with air, restricted air flow into a deep fuel bed yields a nitrogen-rich, so-called "producer gas" that approaches the theoretical

$$C + 0.5\ (O_2 + 4N_2) \rightarrow CO + 2N_2$$

with ~33% CO and a heat value of ~5.6 MJ/m^3 (150 Btu/ft^3).

If such hydrocarbon is gasified with *air in the presence of steam,* the carbon–steam reaction comes into play with generation of CO and H_2, and the outcome is a *water gas* with correspondingly greater heat value. However, unless steam rates are carefully controlled, the endothermic carbon–steam reaction will lower fuel-bed temperatures and promote gas shifting—and that will raise the CO_2 content of the product gas at the expense of its CO concentration.

If combustion is sustained with *oxygen* instead of air, the same reactions will occur, but then generate a N_2-free *synthesis gas* [2] that is almost entirely composed of CO, CO_2, and H_2 and, after removal of CO_2, offers a unique feedstock for a wide spectrum of chemical syntheses (see Section 2).

In all three cases, a gross measure of process efficiency is afforded by carbon conversion—i.e., by the fraction of input carbon appearing in C-bearing gases—and this can be determined from the volumes of CO, CO_2, and hydrocarbon gases leaving the reactor. However, in an autothermal process [3], in which all CO_2 except the amount formed by gas shifting must be attributed to combustion, it is more realistic to consider only [C] in useful gases. If residual unreacted carbon and [C] in syngas, gaseous hydrocarbons, and hydrocarbon liquids [4] are denoted by C_R, C_S, C_H, and C_L, the concentration of "available" carbon is given by

$$[C_A] = [C_R + C_S + C_H + C_L],$$

and a more appropriate indicator of process efficiency is then given by the fraction of carbon converted to useful forms, i.e.,

$$f_C = [C_S + C_H]/[C_R + C_S + C_H + C_L].$$

Gasification Technology

Manufacture of useable fuel gases by incomplete combustion of coal began in the mid-1800s with generation of producer gas in reactors designed by Sir William Siemens (Fig. 10.1.1). These units were replaced in the late 1800s by systems that furnished a *water gas* with ~11 MJ/m^3 (300 Btu/ft^3) by interaction of hot coke with steam. And by the 1920s, water-gas generators had, in turn, come to be superseded by pressure producers into which air and steam could be continuously injected through grates that permitted uninterrupted discharge of ash as molten slag. Exemplified by the renowned Leuna reactor (Fig. 10.1.2),

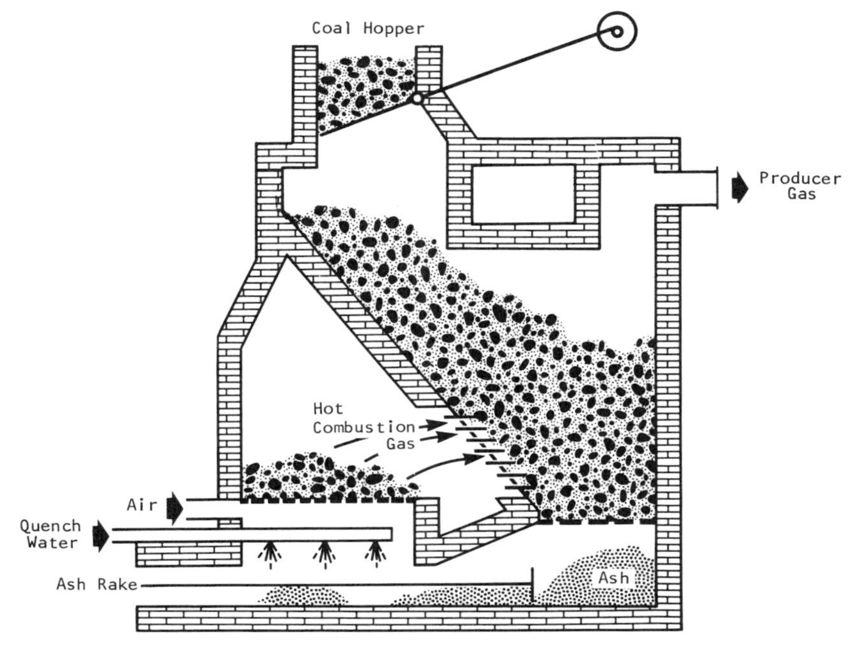

FIGURE 10.1.1 An early (ca. 1860) producer gas generator.

these furnished ~350 m^3 of a 6.33 MJ/m^3 gas and 75–85 liters tar/tonne of (dry) coal, and are the direct antecedents of modern fixed-bed gasifiers, which began to be introduced in the 1930s and which are, with minor modifications, still the mainstays for gasification of solid and some semisolid fossil hydrocarbons.

Oil Gasification

Methods for gasifying heavy oils and bituminous materials, which came into being in the 1930s and 1940s by making use of experiences gained from coal processing, quickly developed a more distinctive character by embodying elements of petroleum processing that reflected the growing understanding of hydrocarbon pyrolysis in different regimes. The outcome was a set of gasification procedures that, in effect, associated the minutiae of gasification with the H/C ratio or specific gravity of the feedstock. The transformation of light oils, naphthas, and heavy oil fractions such as HGOs and residua into "high-Btu" fuel gases by steam reforming, hydrogasification, and/or hydrocracking are

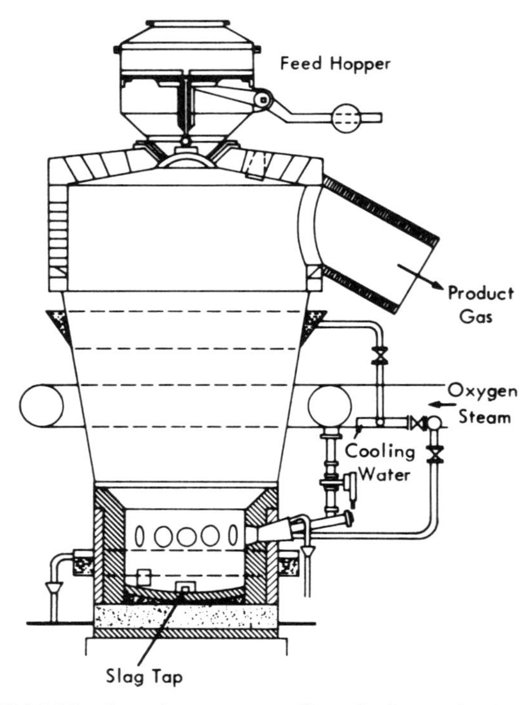

FIGURE 10.1.2 An early pressure gasifier—the Leuna slagging gasifier.

cases in point. All three are designed to generate a gas with a heat content comparable to that of natural gas and therefore designated as *SNG* [5].

Steam Reforming

Conversion of light hydrocarbon streams (e.g., CH_4, C_3H_8, C_4–C_7 LPGs, or naphthas) into a high-Btu fuel gas by steam reforming proceeds by vapor-phase interaction with H_2O, as in

$$C_nH_m + nH_2O \rightarrow nCO + (n + m/2)H_2,$$

which, as already noted, is accompanied by hydrogenation, shifting, and methanation. Figure 10.1.3 is a schematic of that process.

Reaction rates are affected by feed compositions and tend to fall in the order n-paraffins > iso- and cyclic paraffins > aromatic hydrocarbons; and among n-paraffins, the reactivity increases with the number of carbon atoms. Equilibrium gas compositions are likewise influenced by feed compositions

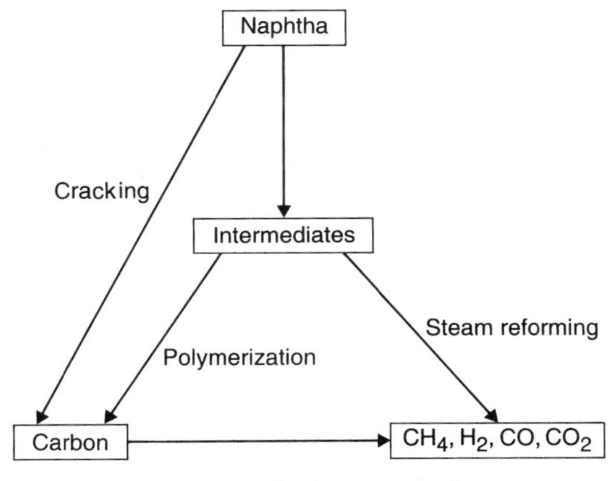

FIGURE 10.1.3 A schematic of reforming.

and, as shown in Figs. 10.1.4–10.1.6, are largely governed by the operating temperature, pressure, and steam-to-feed ratio.

The preferred feeds for steam reforming are LPGs and light oils boiling below 200°C; operating temperatures range from ~700 to 900°C when hydrogen is to be generated, or from 400 to 500°C if a useable fuel gas is to be produced [6], and the process as a whole is normally conducted in three stages. In the first, the feed is catalytically hydrotreated at 350–400°C to remove sulfur compounds that might poison downstream catalysts, and H_2S is fixed by sorption on ZnO or FeO. In the second, it is reformed by gasification; and in the third, the product gas stream is stripped of residual CO_2 and H_2O. The catalyst used in the second (steam-reforming) step is as a rule a Ni/SiO_2 or $Ni/\gamma\text{-}Al_2O_3$ formulation that is occasionally promoted by a refractory oxide such as MgO, TiO_2, or ZrO_2, and catalyst activity depends on the Ni content as well as on the accessible surface area of the catalyst. However, very high Ni concentrations can be counterproductive: Specific reaction rates have been observed to increase rapidly up to 10–20% Ni, but thereafter to remain constant and often to fall beyond 40–50% (Rostrup-Nielsen, 1973). Catalyst life is influenced by operating temperatures that, if excessive, cause complete decomposition of the hydrocarbon and consequently deactivate the catalyst by carbon deposition.

Modern steam reforming is exemplified by the British Gas Corporation's Catalytic Rich Gas (CRG) process (Davies *et al.*, 1967; Pelofsky, 1977), which converts light naphthas at 450–500°C/1–3 MPa into a gas composed of 61% CH_4, 16% H_2, 22% CO_2, 1% CO and therefore amenable to direct methanation at 300–400°C under pressures up to 2.8 MPa. If hydrogasification is inserted between reforming and methanation, the reformate can be made to consist of

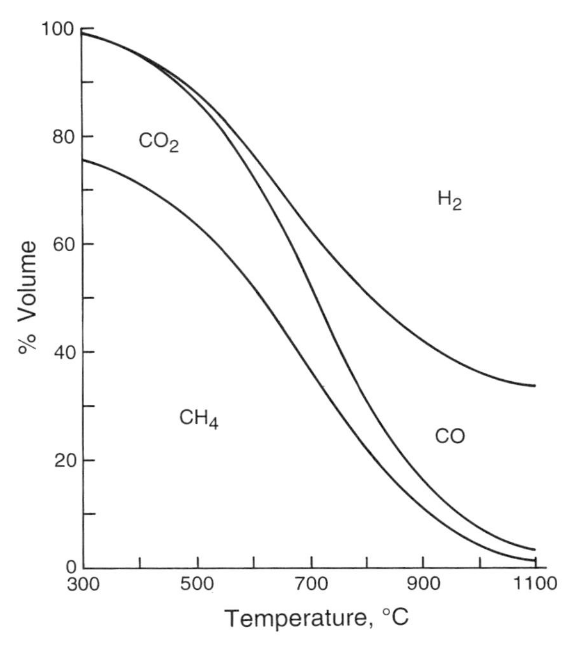

FIGURE 10.1.4 The equilibrium gas composition as a function of temperature (Davies *et al.*, 1967).

90% CH_4, 1% H_2, 4% CO_2, and 5% butane, and then needs almost no shifting before methanation.

Schemes that use one- or two-step reforming sequences followed by one or more methanation stages are illustrated by the Japanese MRG (methane-rich gas) process (Ishiguro, 1968; Thornton *et al.*, 1972) and the Gasynthan

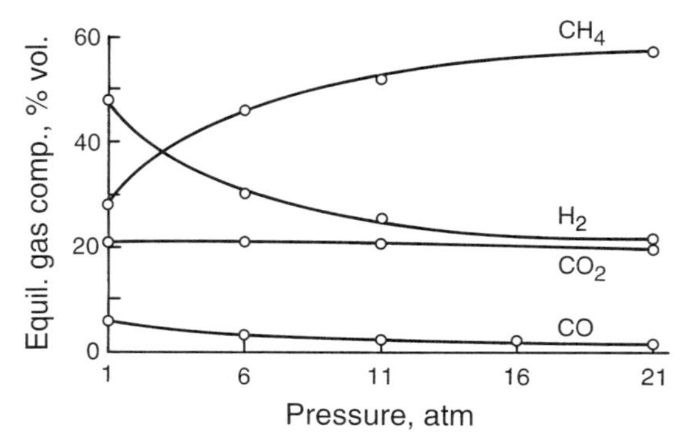

FIGURE 10.1.5 The equilibrium gas composition as a function of pressure (Voodg and Tielrooy, 1967).

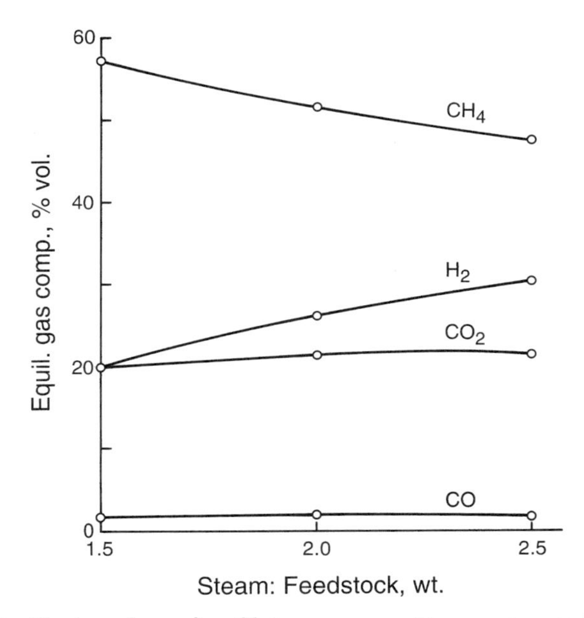

FIGURE 10.1.6 The dependence of equilibrium gas compositions on steam-to-feed ratios (Ishiguro, 1968).

process (Jockel and Triebskorn 1973). The former is designed to generate SNG from light naphthas, which it does by catalytically steam reforming the desulfurized feed into a CH_4-rich gas at 400–500°C/3–3.5 MPa. Subsequent conversion to a SNG with 98% CH_4 is accomplished in a second reforming section at 290–350°C/2–3 MPa. The Gasynthan scheme, which was developed by Lurgi and BASF for production of rich fuel gases from light hydrocarbon feeds, uses similar processing parameters and also yields ~98% CH_4, 1% H_2, and 1% CO + CO_2.

Hydrogasification

An alternative technique, which could also gasify light hydrocarbons but is more appropriate for converting heavy oil fractions into CH_4 and C_2H_6, is hydrogasification—a technique tantamount to severe hydrocracking (or hydropyrolysis). This reacts the feed in an H_2 atmosphere at 700–750°C under pressures up to 7 MPa, and thereby fragments aliphatic components into methylene radicals that rapidly react, as in

$$(\text{—}CH_2\text{—}) + H_2 \rightarrow CH_4$$
$$(\text{—}CH_2\text{—})_2 + H_2 \rightarrow C_2H_6.$$

Concurrently it eliminates heteroatoms as NH_3, H_2O, and H_2S. However, unless it is conducted at or near 7 MPa, aromatics will tend to survive without substantial conversion to naphthenes (and subsequent ring opening), and the proportions of benzene and toluene then generated will depend on the feed composition and the initial H_2 : feed ratio.

In modern oil refineries, hydrogasification is conducted in conjunction with other technologies and, as exemplified by the schemes operated by UOP (1973) and Fluor (1973; Pelofsky, 1977), is usually integrated into a sequence such as

1. initial atmospheric-pressure distillation of crude oil;
2. vacuum distillation to separate the >370°C stream into a heavy gas oil and >540°C bottoms;
3. converting the gas oil into naphtha and lighter products by hydrotreating;
4. stripping the naphtha of nitrogen, sulfur, and other objectionable matter by hydrogenation;
5. gasifying naphtha and lighter fractions with H_2 to produce CH_4, CO, CO_2, and H_2O; and
6. finally purifying the resultant gas by absorbing CO_2 and H_2S into mono- or diethanolamine.

In such operations, hydrogasification can consequently be used to process crude oils and crude oil fractions, as well as heavy bitumen- or asphalt-derived oils with unacceptably high S-contents, without prior coking.

Hydrocracking

This technique has been discussed as a processing method in Chapter 9, but is briefly noted here because it frequently precedes gasification of heavy oils that would produce unacceptable amounts of coke if directly gasified, and would therefore furnish substantially less useable gas. Hydrocracking is resorted to in such cases in order to convert a heavy feed into lighter, more easily processed ones, and allows the by-product coke to be separately gasified (see later discussion).

Fluor's hydrogasification system (Fig. 10.1.7), as well as a Kellogg process (Finneran, 1972) and two sequences proposed by the Institute of Gas Technology (Anon, 1973a, 1973b) illustrate this use of hydrocracking.

Partial Oxidation

Gasification of hydrocarbon liquids with oxygen and steam is mainly used to convert heavy oils and residua into a crude syngas, and closely resembles the gasification of coal and other carbonaceous solids. The initial stage entails

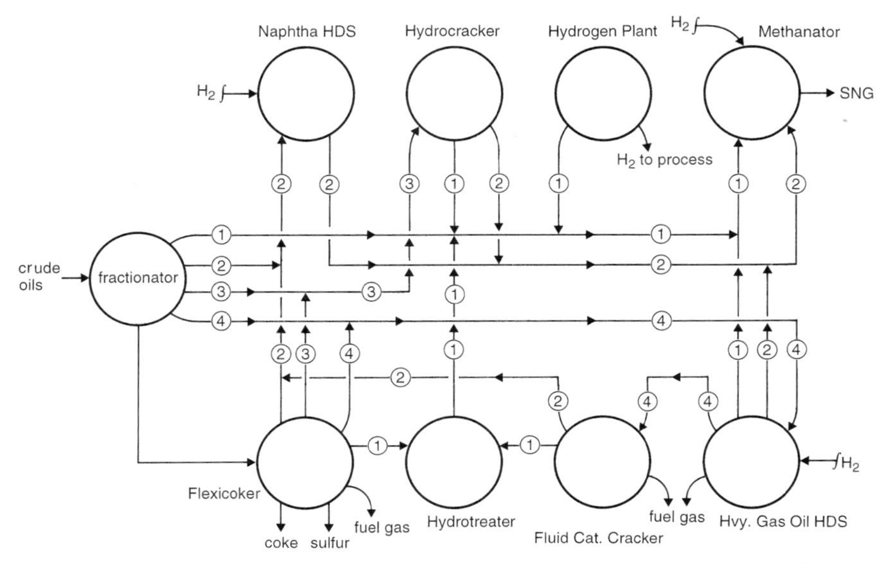

FIGURE 10.1.7 Schematic of Fluor hydrogasification. Streams: 1, $<C_5$; 2, naphtha; 3, light gas oil; 4, heavy gas oil.

partial cracking of the feedstock into carbon, H_2, CH_4, and hydrocarbon radicals; this is followed by radical interactions with O_2, H_2O, and CO_2, as in

$$
\begin{aligned}
&{}^*C_nH_m + [n + (m/4)]O_2 \rightarrow nCO_2 + (m/2)H_2 \\
&{}^*C_nH_m + nH_2O \rightarrow nCO + [n(m/2)]H_2 \\
&{}^*C_nH_m + nCO_2 \rightarrow 2nCO + (m/2)]H_2.
\end{aligned}
$$

The combustion reaction provides heat for endothermic interaction of hydrocarbon fragments with H_2O and CO_2 to form $CO + H_2$, and this is accompanied by steam gasification of carbon,

$$C + H_2O \rightarrow CO + H_2,$$

as well as by the Boudouard reaction and CH_4 "reforming," as in

$$
\begin{aligned}
&CH_4 + H_2O \rightarrow CO + 3H_2 \\
&CH_4 + CO_2 \rightarrow 2CO + 2H_2.
\end{aligned}
$$

Product gas compositions are therefore determined by

$$CO + H_2O \rightarrow CO_2 + H_2,$$

and the net overall process can be written as

$$C_nH_m + (n/2)O_2 \rightarrow nCO + (m/2)H_2.$$

Partial oxidation is intended to generate a syngas that can be methanated or used for production of NH_3, CH_3OH, and/or hydrocarbon liquids (see Section 2), and is not limited by the aromaticity of the feed. But if it is followed by *methanation* of the syngas, major stoichiometric differences between it and generation of SNG by hydrogenation—for which H_2 is produced by partial oxidation—become evident. Thus, whereas the net stoichiometry of partial oxidation and catalytic methanation of a $C_{10}H_{22}$ hydrocarbon is

$$1.9\ C_{10}H_{22} + 9.45\ O_2 \rightarrow 10.0\ CH_4 + 9.0\ CO_2 + 0.9\ H_2O,$$

the net stoichiometry for direct hydrogenation is

$$1.43\ C_{10}H_{22} + 2.14\ O_2 + 4.29\ H_2O \rightarrow 10\ CH_4 + 4.29\ CO_2.$$

Because the latter equation includes the net $C_{10}H_{22}$, O_2, and H_2O needed for the production of H_2, feedstock and O_2 requirements for partial oxidation plus methanation are manifestly greater than those for hydrogenation. Nevertheless, partial oxidation is seen as particularly attractive for converting heavy oils and residua to a syngas destined for uses other than synthesis of CH_4.

Commercial deployment of partial oxidation is illustrated by the Shell and Texaco processes, which differ little except in details of equipment design and secondary processing.

For the Shell process (Kuhre and Sykes, 1973), the feed—on occasion a naphtha, but usually a crude oil, heavy fuel oil, residuum or asphalt—is injected with O_2 and steam into a refractory-lined steel reactor in which it is vaporized and partly cracked while being brought to reaction temperature (~1320°C). The resultant mix of carbon, CH_4, H_2, and hydrocarbon radicals is then combusted with limited oxygen to generate H_2 and CO_2, as in

$$C_nH_m + (n + 0.25m)\ O_2 \rightarrow nCO_2 + 0.5m\ H_2.$$

which is followed by endothermic gasification with CO_2 and steam to furnish $CO + H_2$, as in

$$C_nH_m + nCO_2 \rightarrow 2nCO + 0.5mH_2$$
$$C_nH_m + nH_2O \rightarrow nCO + (0.5m + n)H_2.$$

Careful mixing of the reactants ensures that the gasification reactions absorb most of the heat generated by combustion, and because this limits overall gas temperatures to 1300–1500°C, the net make of CO_2 and H_2O is very small.

TABLE 10.1.1 Shell Partial Oxidation: Syngas Compositions (vol %)[a]

	Naphtha	Heavy resid	Fuel oil
H_2	51.7	46.2	45.4
CO	41.8	46.9	47.0
CO_2	4.8	4.3	5.2
CH_4	0.3	0.3	0.6
N_2 + Ar	1.4	1.4	1.4
H_2S + COS	70 ppm	0.9	0.4

[a] Pelofsky (1977).

Product gas stream compositions (Table 10.1.1) are mainly determined by the equilibrium conditions of the shift reaction at 1250–1350°C and depend only to a slight extent on the feed.

The Texaco process (Schlinger, 1967) generates a syngas with >95–98% H_2 + CO by partial oxidation of residual fuel oils at 1100–1500°C, quenches the product gas with hot water in the bottom section of the reactor (Fig. 10.1.8), and then passes it directly to a shift converter, where the necessary

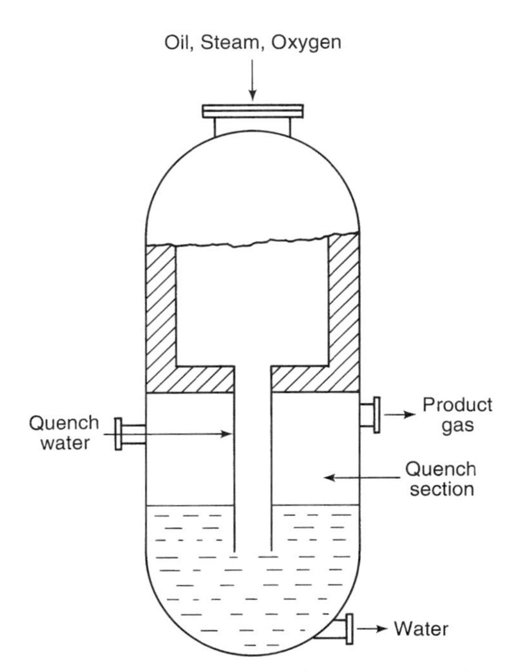

FIGURE 10.1.8 Schematic of the Texaco gasifier.

shifting proceeds at 60–80°C over a cobalt–molybdenum catalyst claimed to be unaffected by H_2S in the raw syngas [7].

Because the feeds are mixed with O_2 and steam before injection into the gasifier, Texaco partial oxidation allows rapid switching from one type of feed to another and usually only requires no more than changing the feed gun in the burner. That the Texaco gasifier is also at the heart of a very similar *coal* gasification process underscores its versatility.

Coal Gasification

The "Established" Systems

Unlike liquid and some semisolid hydrocarbons, solid carbonaceous matter is invariably gasified via partial oxidation. Depending on whether the objective is a fuel gas or a syngas destined for methanation or synthesis of other hydrocarbons, gasification is accomplished with air/steam or O_2/steam at atmospheric or elevated pressures.

The direct descent of current fixed-bed gasifiers from early pressure producers (Fig. 10.1.2) is clearly seen in the Lurgi (Ricketts, 1961; Lurgi Express, 1975), Woodall–Duckham (Odell, 1947), and Wellman (Hamilton, 1963; Howard-Smith and Werner, 1976) reactors.

The Lurgi reactor (Fig. 10.1.9) is, like its immediate antecedents, a water-cooled pressure shell that is intermittently charged with coal through an upper lock hopper. Steam and air (or O_2) are injected through a rotary grate, which concurrently crushes and withdraws clinkered ash into a lock hopper at the bottom; gasification proceeds at 925–1035°C/3–3.5 MPa; and because these conditions favor carbon hydrogenation, a cleaned CO_2-free syngas from gasification with O_2 consists of ~50% H_2, 35% CO, and 15% CH_4.

Since its introduction in 1936, the Lurgi reactor has undergone several design changes that enable it to process all but strongly caking coals. Current standard (~4-m i.d.) reactors can gasify up to 700 tonnes/day at gas makes of ~5000 $m^3\ m^{-2}\ h^{-1}$, but require prior removal of fine (<6-mm) coal [8].

Closely resembling the Lurgi, but operating at atmospheric pressure, are the Woodall–Duckham (Odell, 1947) and the virtually identical Wellman reactors (Hamilton, 1963; Fig. 2.1.10). Both are water-cooled vessels that are charged from a lock hopper and distributor through one or more pipes while air/steam (or O_2/steam) is injected through tuyeres in a revolving basal grate that withdraws ash into a bottom dump. Because both, like the Lurgi reactor, devolatilize the coal with consequent formation of tar before gasifying the residual char, heat values of the gas depend on whether or not tar vapors are separated from it. If gasification is conducted with air and steam [9], typical values range from ~6.3 MJ/m^3 for a cold gas from which tar has been extracted,

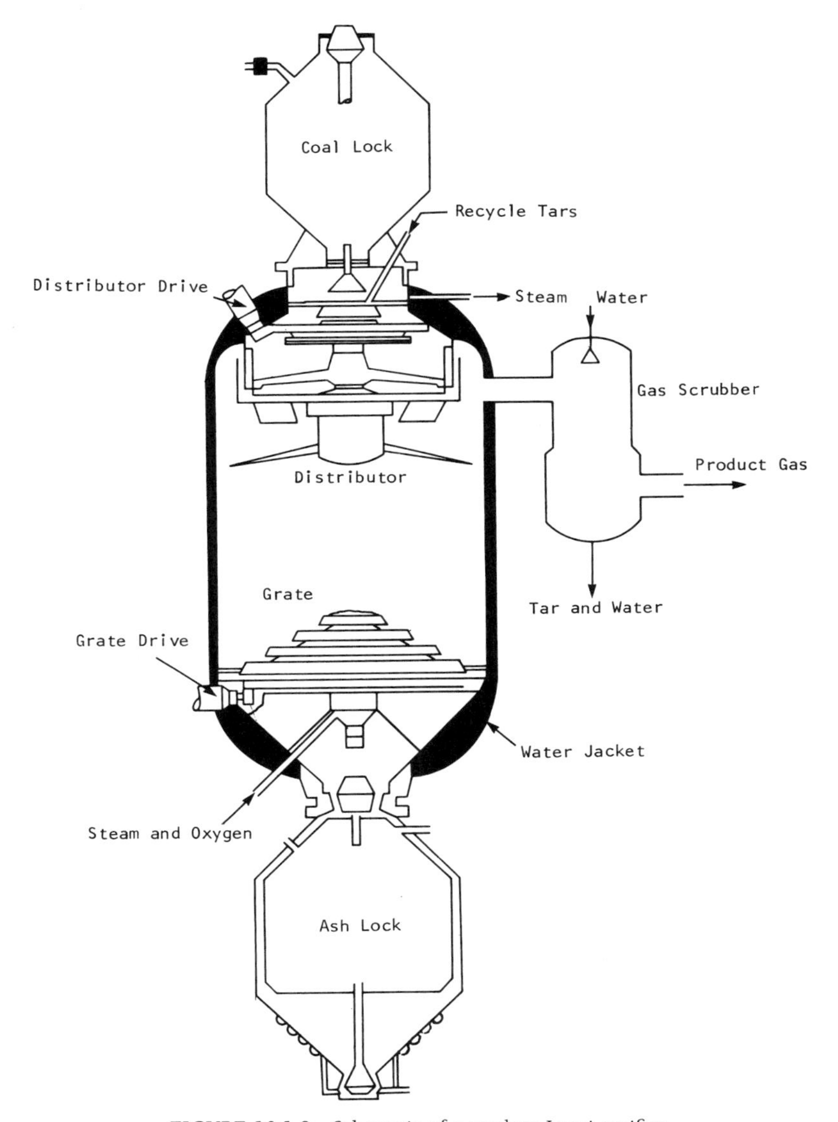

FIGURE 10.1.9 Schematic of a modern Lurgi gasifier.

to 6.9 MJ/m^3 for a hot detarred gas and 7.45 MJ/m^3 for a hot raw gas. The heat value of a hot raw gas produced by gasification with O_2 and steam would run to ~10.8 MJ/m^3.

Woodall–Duckham units can process up to 100 tonnes/day and furnish up to 0.32×10^6 m^3/day, while the slightly smaller two-stage Wellman can gasify

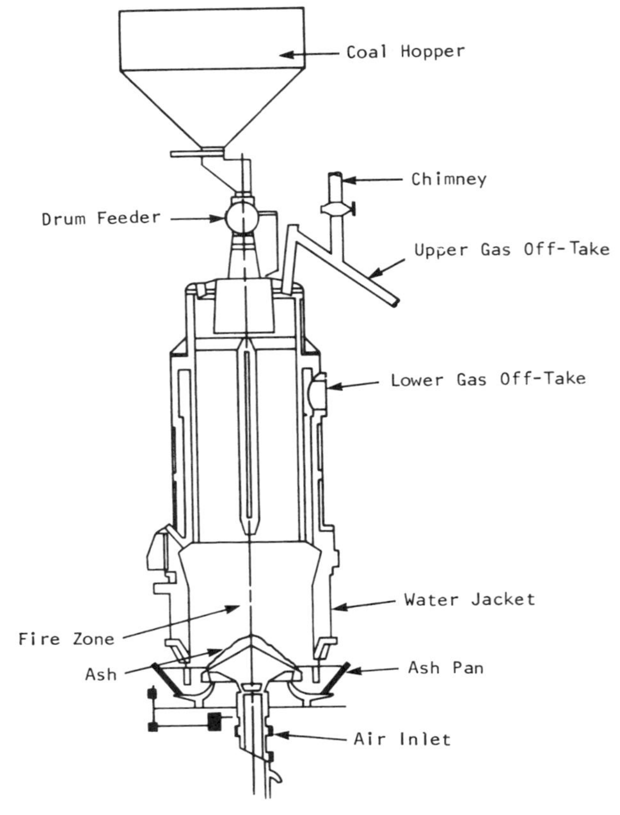

FIGURE 10.1.10 Schematic of the Wellman gasifier.

up to 70 tonnes/day with production of $\sim 0.23 \times 10^6$ m^3/day. Although neither can accept <6-mm coal or coals with high (>1200°C) ash fusion temperatures, both are advantaged by being operable between full and one-third rated capacity without appreciable loss of efficiency.

But in addition to fixed-bed gasifiers, two other reactor systems—the Koppers–Totzek generator, which processes oxygen-*entrained* coal (Farnsworth *et al.*, 1973), and the Winkler generator, which gasifies *fluidized* coal (Odell, 1947; Anon., 1974)—have also firmly established themselves since the mid-1930s.

Koppers–Totzek generators (Fig. 10.1.11), which are widely used for manufacture of syngas, are squat refractory-lined vessels into which O_2-entrained coal is injected through opposed burner heads [10]. Steam, introduced around these heads, shrouds the reaction zone and protects vessel walls from ex-

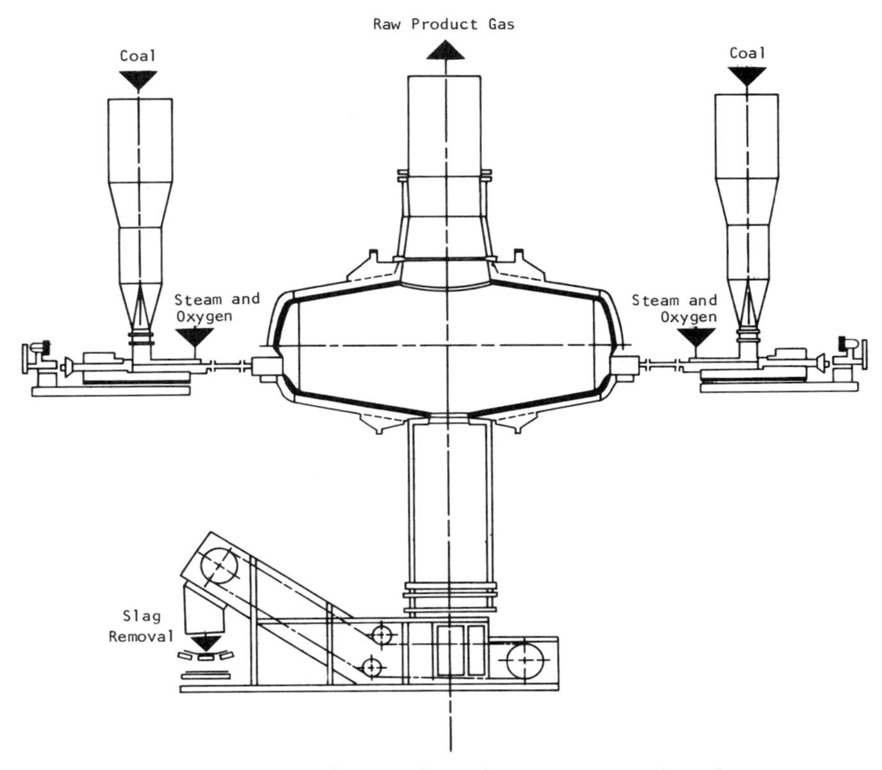

FIGURE 10.1.11 Schematic of a modern Koppers–Totzek gasifier.

cessive heat; product gas leaves through a collector pipe, which surmounts a waste heat boiler; and ash is withdrawn through the bottom as a molten slag.

Koppers–Totzek units operate at atmospheric pressure, make no tar, and yield a substantially CH_4-free gas with ~55% CO, 30–32% H_2, and 12% CO_2. The relatively low carbon inventory (and consequent high O_2 consumption) is offset by an ability to accept all types of coal, by flexible removal of ash (which can be tapped off as a slag or discharged as fly ash), and by high operating temperatures (1800–1925°C) that permit very fast gasification. A unit with four burner heads can gasify up to 770 tonnes/day and produce 1.2×10^6 m^3 syngas per day.

Contemporary Winkler reactors (Fig. 10.1.12) are shaft units 20–25 m high, which can process <10-mm coal at rates exceeding 1000 tonnes/day. Feed coal, entering from a pressurized hopper, is fluidized by a primary blast

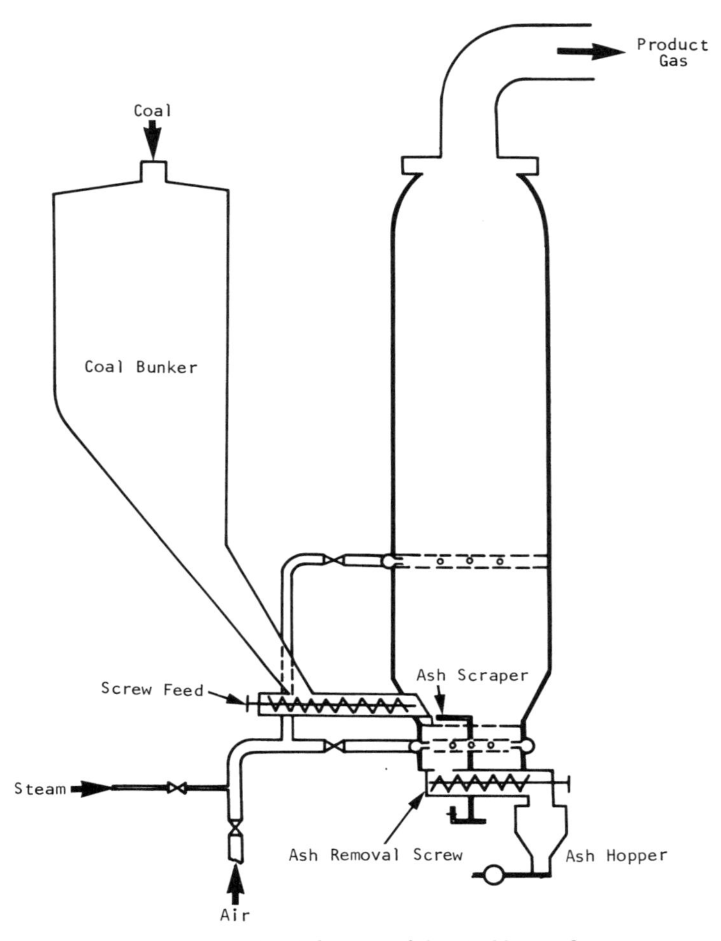

FIGURE 10.1.12 Schematic of the Winkler gasifier.

immediately above the bottom grate; a secondary blast above the fully fluidized bed serves to gasify any still unreacted char; ash is extracted from the gas stream by an external cyclone; and bottom ash is withdrawn by a basal rotating scraper. Depending on whether a producer gas or syngas is to be made, either an air/steam or an O_2/steam blast is used. Outputs can be varied over a wide range without appreciable loss of efficiency, but the construction of the reactor limits operating temperatures to <980°C, and that restricts gasification to lignites and subbituminous coals [11]. A typical syngas contains 36–37% H_2, 45% CO, and 16% CO_2.

Of more recent origin and usually viewed as a "second-generation" system (see below), but well established because it was adapted from a proven technology, is the Texaco reactor (see Fig. 10.1.8, earlier), which has long been used to gasify heavy oil fractions. This unit accepts an aqueous slurry of <75-μm coal, which is injected with preheated O_2 at ~550°C/1.5 MPa, and gasifies the coal in slurry-derived steam at 1100–1375°C/2–8.5 MPa. Raw syngas is taken off above a quench section, and the molten ash, which forms under these conditions and flows into the quench section, is withdrawn from there as a granular suspension in water. Modern 4.5-m-high Texaco reactors can reportedly gasify up to ~1700 tonnes/day with production of 2.8×10^6 m^3 syngas.

The "Second-Generation" Systems

Despite the proven performance and adequacy of commerically available gasifiers [12], the range of gasification *concepts* was greatly widened in the 1960s and 1970s by R&D responses to projections that anticipated serious shortages of acceptably priced gas well before the end of the century. These responses centered on development of "second-generation" systems that could gasify coal at rates of $5–10 \times 10^3$ t/d and maximize CH_4 formation by operating at ~7 MPa [13]—and did indeed outline several innovative reactor configurations [14].

Illustrative of the novel directions taken in fixed-bed gasification are the Bi-Gas reactor (Glenn and Grace, 1978) and the CO_2-Acceptor process (McCoy *et al.*, 1976).

The prototype Bi-Gas reactor (Fig. 10.1.13) was a water-cooled 16.5-m-high slagging gasifier operating at ~7.6 MPa. Fresh steam-entrained pulverized (<75-μm) coal was partly gasified in a dense-phase upper section at ~925°C by syngas from the lower section; unreacted char leaving with the product-gas stream was separated in a cyclone and then reinjected into the lower section, where it was almost completely gasified with O_2 and steam at ~1480°C; and ash was withdrawn as molten slag into a water reservoir from which it was periodically removed through a lock hopper. Crude syngas taken off near the top of the reactor typically contained 19–20% CH_4.

In the CO_2-Acceptor process (Fig. 10.1.14), a more innovative chemistry used calcined dolomite as a heat-transfer medium as well as to remove CO_2 by (exothermic) formation of a carbonate. Crushed coal and $MgO \cdot CaO$ from the regenerator section were first sent to a devolatilizer, where the mix was fluidized with steam and gas from the gasifier at ~815°C/2–3 MPa. The net reactions taking place in this unit were

$$2C + H_2 + H_2O \rightarrow CH_4 + CO$$
$$CO + H_2O \rightarrow CO_2 + H_2$$
$$MgO \cdot CaO + CO_2 \rightarrow MgO \cdot CaCO_3,$$

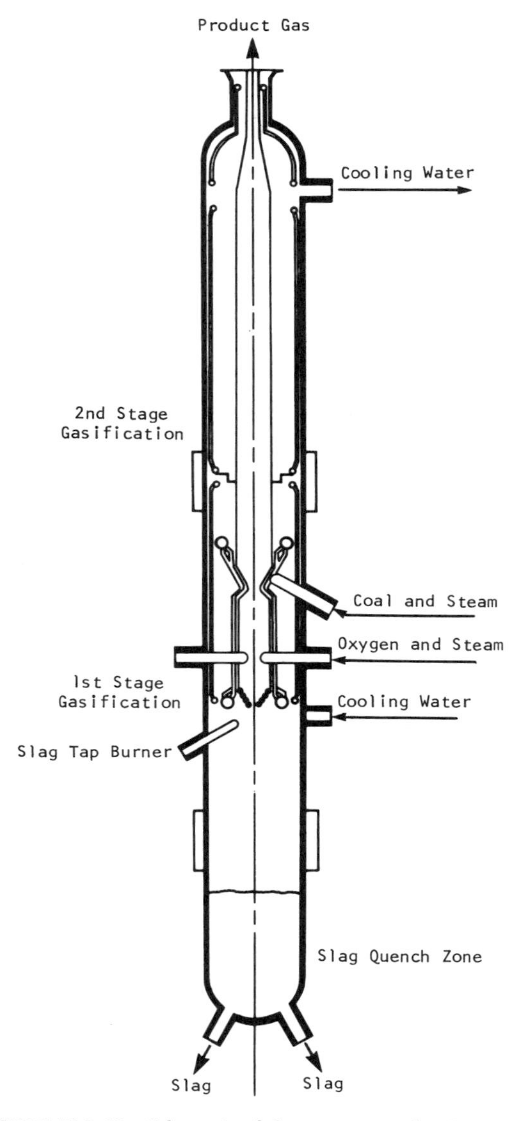

FIGURE 10.1.13 Schematic of the experimental BiGas reactor.

and as these reactions proceeded, much of the spent dolomite, being heavier than the partly gasified char, moved down toward the bottom of the fluidized bed. From the upper part of the bed, unconsumed char was then transferred to the gasifier, where it reacted with steam and calcined dolomite as in

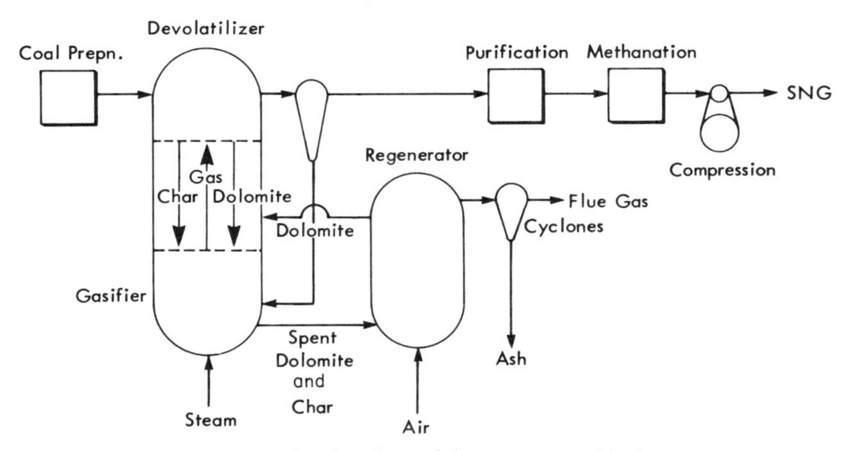

FIGURE 10.1.14 Simplified flowsheet of the experiment CO_2-Acceptor process.

$$C + H_2O \rightarrow CO + H_2$$
$$CO + H_2O \rightarrow CO_2 + H_2$$
$$MgO \cdot CaO + CO_2 \rightarrow MgO \cdot CaCO_3,$$

and the product gas was returned to the devolatilizer, from which it finally left as raw syngas. Unreacted char from the gasifier and spent dolomite from the devolatilizer and gasifier were returned to the regenerator, where the dolomite was calcined by burning the char in air at ~1075°C.

The CO_2-Acceptor process gains significant advantages from requiring no off-site O_2 or H_2 and enhancing gas shifting by heat from carbonate formation (and consequently promoting generation of a syngas with $H_2 : CO \approx 3$). There is therefore no need for downstream gas shifting before methanation. However, the use of dolomite restricts operation of the gasifier and devolatilizer to <870°C; and as in Winkler gasification, this limits feedstock choices to lignites and subbituminous coals.

Another approach to CH_4-rich syngas relied on *hydrogasification* not unlike the process used to transform heavy oils into such gas (see earlier discussion) and is illustrated by the HYGAS reactor (Fig. 10.1.15). By directly hydrogenating the coal with H_2 and steam at ~8–10.5 MPa, this generated a raw gas with ~37.5% CH_4, 45% H_2, and 13% CO that could be methanated without any prior shifting (Schora *et al.*, 1973; Lee, 1975).

The feed was a preheated (~425°C) coal/oil slurry that was pumped into the dryer section of the gasifier at ~7.5–10.5 MPa; and as the oil vaporized and left with raw product gas, the coal descended through countercurrent H_2 and steam, reacting in a first stage at 650°C and in a second at 925–980°C. Unconsumed char, representing ~50% of the coal charge, was withdrawn at

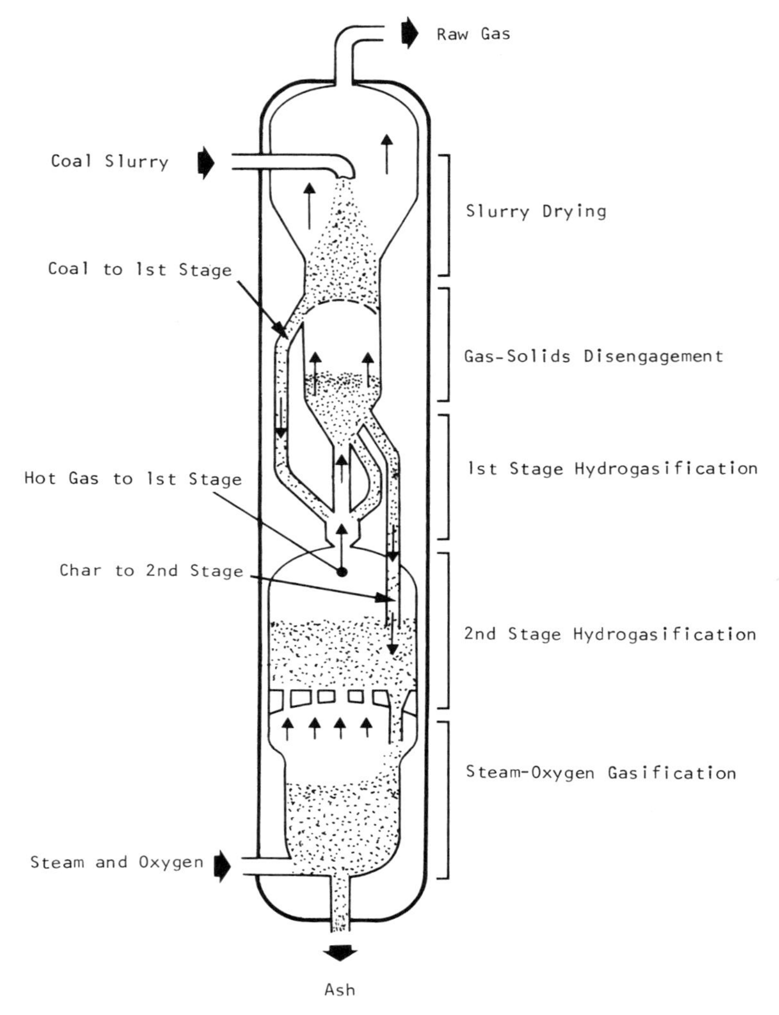

FIGURE 10.1.15 Schematic of the experimental HYGAS gasifier.

the bottom of the gasifier and used to produce a crude H_2 stream—either by the steam/iron process [15] or by fluidizing it with O_2 and steam to generate a syngas, shifting this completely to H_2, and stripping it of CO_2.

Systems that seek to generate a syngas by *catalyzed* gasification are illustrated by the Kellogg Molten Salt process (Cover *et al.*, 1973), which gasified the coal with steam in a recirculating Na_2CO_3 melt, and thereby utilized the ability

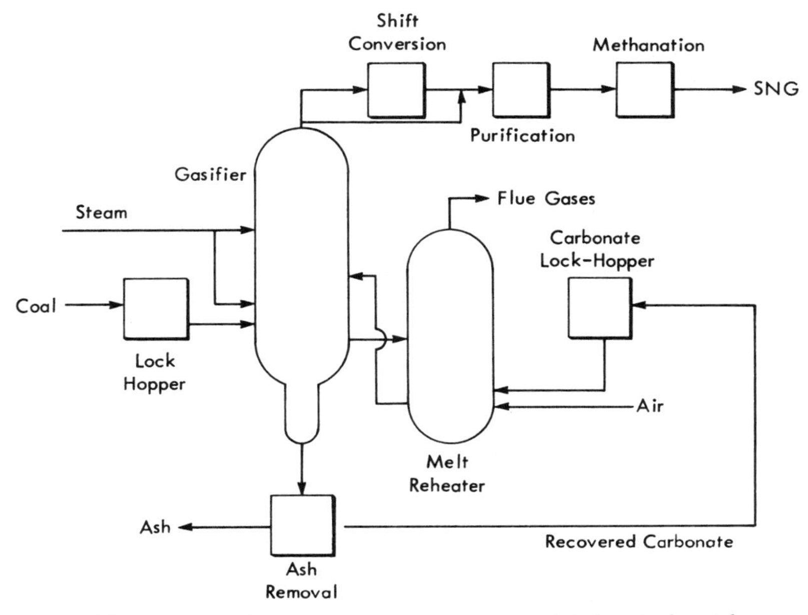

FIGURE 10.1.16 A simplified flowsheet for the experimental Kellogg Molten Salt process.

of this salt to catalyze the carbon–steam reaction. Figure 10.1.16 shows a schematic flowsheet of this scheme [16].

Crude syngas was generated in the gasifier section by injecting coal and steam into the Na_2CO_3 melt at ~1000°C/3–8.5 MPa, and continuously moving the melt with unreacted char and ash into a reheat section [17] where it was renewed by burning the residual fuel in air at 1200°C. Ash was extracted by simultaneously withdrawing a slipstream with ~8% ash from the reheated melt, cooling it to 230°C with warm (~38°C) aqueous Na_2CO_3, flashing the resultant slurry to atmospheric pressure until all Na_2CO_3 had dissolved, and filtering it. The clear solution was then reacted with CO_2 in order to precipitate solid $NaHCO_3$, which was recycled to the gasifier.

Despite its complexity, the Kellogg process was widely thought to be technically attractive, but its development was slowed by containment problems posed by the highly corrosive melt. Whether these could have been satisfactorily resolved if R&D on second-generation systems had not been discontinued in the late 1970s is still uncertain [18].

In retrospect, greater interest may therefore attach to the ATGas scheme (Karnavos *et al.*, 1974), which envisaged processing pulverized coal and lime-

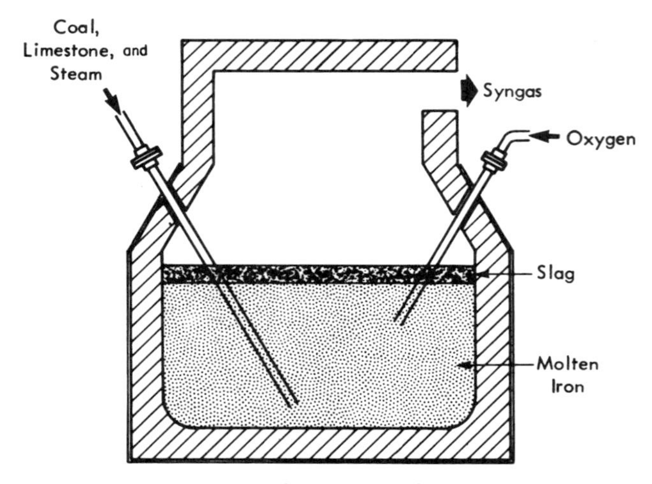

FIGURE 10.1.17 The ATGas gasification concept.

stone with O_2/steam in molten iron at 1375–1425°C (Fig. 10.1.17) and thereby generating a syngas almost entirely composed of CO and H_2 by thermal cracking of volatile matter and concurrent steam gasification of the char. Sulfur in the feed would be captured by Fe, then transferred to a lime slag that would also flux the ash and could be skimmed off the top of the bath. Any carbon dissolved in Fe would quickly oxidize to CO and contribute to the output gas. Significantly, most of the technology needed for the ATGas concept is well established by separate unit processes in the iron and steel industry, but whether these components could be appropriately combined in a single operation remains to be demonstrated [19].

The ingenuity of some second-generated gasification techniques makes it from some points of view regrettable that easing of gas supplies and pricing in the late 1970s effectively ended their further development. But as matters stand, it is doubtful whether any of them are likely to be brought to commercial status in the foreseeable future: that will probably be determined by the success of established reactor systems that are now being redesigned with larger capacities and capabilities for operating as "environment-friendly" slagging gasifiers [20].

Catalysis of Gasification

In principle, greater capacity can more easily accrue from *accelerated* gasification than from larger gasifiers. Considerable attention has therefore been focused on catalyzing gasification reactions with a view to improving selectivity and allowing acceptable gasification at lower temperatures, thereby reducing the energy expenditures associated with gasification [21].

That these objectives might be attainable is suggested by wide-ranging studies of catalytic effects of alkali metal and alkaline earth metal salts (specifically KOH, K_2CO_3, Na_2CO_3, $Ca(OH)_2$, CaO, $CaCO_3$) and of transition metals (notably Fe and Ni) on gasification of carbon-rich solids. These materials were physically mixed with the carbon source, impregnated into it from solution, or introduced by cation exchange, and then evaluated by carbon gasification with steam, H_2 and CO_2 at 500–1000°C and pressures up to 10 MPa. Catalyst effects on reaction rates were assessed thermogravimetrically as well as by measurements on fixed and fluidized beds (Johnson 1981).

Comparisons of the results of these investigations are difficult; the use of dissimilar carbon sources as well as of different catalyst preparations and reaction conditions does indeed make some of the findings sometimes appear inconsistent. Nevertheless, it is possible to draw at least two important general inferences from them.

1. Most effectively promoting steam gasification of carbon seem to be metals of the Pt group, oxides of transition metals, oxides of Cu, Pb, and V, and carbonates, chlorides, and oxides of alkali and alkaline earth metals. In some instances, these were found to as much as triple specific gasification rates (Kröger and Melhorn, 1938; Haynes *et al.*, 1974a; Wilson *et al.*, 1974).
2. Hydrogasification rates can be doubled by K_2CO_3, $KHCO_3$, and $ZnCl_2$ (Gardner *et al.*, 1974) and significantly improved by metallic iron (which is, however, quickly poisoned by sulfur).

Concurrently, studies of gasification catalysis have also provided useful information about reaction mechanisms. For example, promotion of carbon gasification by O_2 and/or steam is thought to entail participation of the catalyst in oxidation/reduction cycles at carbon surfaces (Spiro *et al.*, 1983; Kim *et al.*, 1989; Miura *et al.*, 1989; Salinas-Martinez de Lecea *et al.*, 1990; Moreno-Castilla, 1990) and hence to involve reactions of the form

$$2\,M_xO_y + O_2 \rightarrow 2\,M_xO_{y+1}$$
$$M_xO_{y+1} + C \rightarrow M_xO_y + CO.$$

Catalysis by an alkali carbonate might therefore be envisaged to proceed as in

$$M_2CO_3 + C + O_2 \rightarrow M_2O + 2\,CO_2$$
$$M_2O + n/2\,O_2 \rightarrow M_2O_{1+n}$$
$$M_2O_{1+n} + n\,C \rightarrow M_2O + n\,CO;$$

and this allows some specific reaction routes to be outlined. Thus, if an intermediate intercalation compound (C_nM) is postulated, steam gasification would probably follow a path such as

$$M_2CO_3 + 2\,C \rightarrow 2\,M + 3\,CO$$
$$2\,M + 2n\,C \rightarrow 2\,C_nM$$
$$2\,C_nM + 2\,H_2O \rightarrow 2n\,C + 2\,MOH + H_2$$
$$2\,MOH + CO \rightarrow M_2CO_3 + H_2.$$

Or, without postulating intercalation compounds, the Boudouard reaction might be thought to proceed as in

$$M_2CO_3 + 2\,C \rightarrow 2\,M + 3\,CO$$
$$2\,M + 2\,H_2O \rightarrow 2\,MOH + H_2$$
$$2\,MOH + CO \rightarrow M_2CO_3 + H_2.$$

Industry trials of carbon gasification with alkali metal salts, which generally disperse well on carbon surfaces and remain molten at reaction temperatures, have confirmed that heavy (10–20 wt%) dosages of alkali hydroxides and carbonates can lower optimum temperatures and pressures for steam gasification of bituminous coals from 975–1050°C/7 MPa to 650–750°C/3.5 MPa, and also demonstrated that 5–10 wt% $Ca(OH)_2$ will reduce optimum hydrogasification temperatures by as much as 150–200°C. However, effective catalysis demanded fairly elaborate preparation of the feedstock, and considerable difficulties were encountered in catalyst recovery and regeneration, which the massive catalyst quantities necessitate on economic grounds. It is therefore by no means certain that catalyzed gasification, as now understood, could in practice offer appreciable advantages over uncatalyzed operations.

Gas Shifting

Transformation of syngas into a specific hydrocarbon or hydrocarbon mix requires adjustment of its $CO:H_2$ ratio to the appropriate stoichiometric value. Thus, for synthesis of ammonia by

$$N_2 + 3\,H_2 \rightarrow 2\,NH_3,$$

for which the syngas is only provides hydrogen, *all* CO must be abstracted, and for synthesis of methanol by

$$CO + 2\,H_2 \rightarrow CH_3OH$$

or of methane by

$$CO + 3\ H_2 \rightarrow CH_4 + H_2O,$$

the required $CO:H_2$ ratios are obviously 1:2 and 1:3, respectively.

In principle, these compositional adjustments could be accomplished by membrane-separation technology. But for large volumes, this is costly as well as inconvenient, and reliance is therefore placed on "shifting"—that is, reacting the necessary volume of syngas with steam and removing all CO_2 formed by

$$CO + H_2O \rightarrow CO_2 + H_2.$$

As a rule, this operation is carried out in uncooled reactors over a fixed catalyst, and the reaction heat—approximately 32 MJ/kg mol CO → CO_2—is used to maintain catalyst activity as well as to enhance reaction rates. However, because increasing temperatures lower the equilibrium conversion of CO, it is often preferable to accelerate the reaction rate by raising the system pressure and ensure maximum CO oxidation by appropriate adjustment of the H_2O: feed gas ratio.

Conventional *high-temperature* catalysts for gas shifting at 350–550°C are chromia-promoted iron oxide formulations composed of 56–58% Fe, 5–6% Cr, 2–4% C, and 0.07–0.2% S; but also frequently used are

1. *low-temperature* preparations made up of 0–40% Al_2O_3, 15–30% CuO, and 32–62% ZnO, which allow operation in the range 200–250°C (Allen, 1973), or
2. *medium-temperature* catalysts composed of 3–4% CoO and 13–15% MoO_3 on Al_2O_3, which promote shifting at 260–450°C, but must be used in a sulfided state and require at least 20 ppm sulfur in the feed gas (Auer *et al.*, 1971).

These alternative catalysts afford considerable freedom for shifting and have led to the common practice of conducting *extensive* shifting, where that is necessary, in two stages—the first accomplishing the bulk of the shift over a high-temperature catalyst at 370–425°C, and the second proceeding over a low-temperature catalyst at 190–230°C.

Gas Cleaning

If a syngas is destined for use in hydrocarbon synthesis, its preparation entails a sequence of cleaning operations quite different from what is needed to prepare natural gas or synthetic high-Btu gas for domestic consumption as a fuel (see Chapter 8). Specifically, it then becomes necessary to

1. cool the gas to near-ambient temperatures to extract particulates and tarry substances (if such were generated, e.g., by coal gasification in a Lurgi or Wellman reactor);
2. remove residual light hydrocarbons by scrubbing with a wash oil or passage through metal-impregnated active carbon to extract thiophenes, mercaptans and CS_2; and
3. eliminate traces of NH_3 generated from N compounds in the feedstock by scrubbing with fresh or slightly acidified water.

Removal of acid gases, notably H_2S and CO_2 is thereafter accomplished by reversible absorption in an amine or alkali carbonate (see Chapter 8).

2. CARBON MONOXIDE HYDROGENATION

The importance of gasification technology lies not only in its ability to transform heavy fossil hydrocarbons into more desirable or useful fuel gases with heat values between ~6 and 35 MJ/m^3 [22]. Even more important is its role in generating a syngas that, after appropriate shifting, offers a unique building block for synthesizing an extraordinarily wide range of hydrocarbons and hydrocarbon derivatives, some by stereospecific reactions that deliver a single product and others yielding a spectrum of related substances.

Stereospecific Syntheses

Ammonia

The synthesis of ammonia from $N_2 + 3H_2$ is, in a sense, a special case because it does not require CO. It does, however, almost always proceed from a syngas from which CO is removed by gas shifting or absorption of CO as, for example, by the COsorb process, and is then accomplished by passing a 3 : 1 mix of purified H_2 and N_2 through a converter (see Fig. 10.2.1) in which NH_3 forms at 500–600°C/20–100 MPa over an iron-promoted catalyst such as Al_2O_3-supported K_2O. From the exiting gas, anhydrous NH_3 is extracted by refrigeration, and the stripped stream is recycled.

Methanol

The synthesis of methanol from $CO + H_2$ dates from 1913, when a German patent for it was issued to the Badische Anilin & Soda Fabrik (BASF), and was for many years conducted at 100–150 MPa over cerium, cobalt, chromium

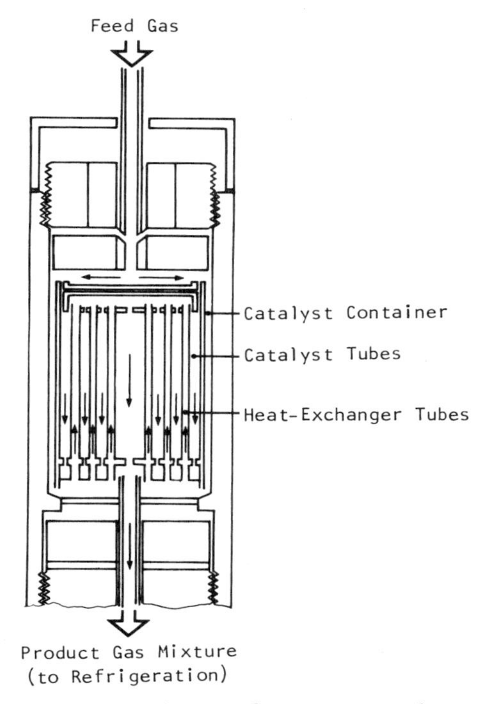

FIGURE 10.2.1 Schematic of an ammonia synthesis unit.

molybdenum, or manganese catalysts. In modern plants, which mostly use a medium-pressure process developed by Imperial Chemical Industries (ICI), it is achieved by catalyzed reaction at 250–400°C/10–60 MPa.

If the feed is prepared by reforming natural gas, some CO_2 must be added during reforming in order to generate the stoichiometrically required 1 : 2 $CO : H_2$ ratio via

$$CH_4 + H_2O \rightarrow CO + 3\,H_2$$
$$CH_4 + 2\,H_2O \rightarrow CO + 4\,H_2$$
$$\text{and} \quad CO_2 + H_2 \rightarrow CO + H_2O.$$

However, if a coal-based syngas is used, this ratio is directly adjusted by shifting. In either case, plant operations other than the final synthesis step are thus not unlike those involved in manufacture of NH_3.

Noteworthy in this connection is that methanol, aside from its other industrial uses, is now also commanding attention as a feedstock in a Mobil Oil technology that converts it to *gasoline* (Meisel *et al.*, 1976; Jüntgen *et al.* 1981).

TABLE 10.2.1 Mobil Oil's $CH_3OH \rightarrow$ Gasoline Process: Typical Processing Conditions[a,b]

Temperature (°C)	
Inlet	360
Outlet	415
Pressure (bar)	21
Yields (wt % of charge):	
Hydrocarbons	36.2
Water	63.2
CO, CO_2	0.3
Coke, other	0.3
Hydrocarbon products (wt %)	
Light gas	1.4
Propane	5.5
Propene	0.2
n-Butane	3.3
Isobutane	8.6
Butenes	1.1
C_{5+} gasoline	79.9
Gasoline, incl. alkylate (wt %)	85.0
LPGs (wt %)	13.6
Fuel gas (wt %)	1.4

[a] Meisel *et al.* (1976).
[b] Feedstock: crude methanol with ~17% H_2O.

Chemically, this begins with dehydration of methanol to the dimethyl ether as in

$$2\ CH_3OH \rightarrow CH_3OCH_3 + H_2O$$

and proceeds to formation of aromatic hydrocarbons over shape-selective zeolite catalysts, specifically ZSM-5 [23], whose pore openings (~0.6 nm) and channel structure tend to limit outputs to C_5–C_{10} [24]. A typical product stream thus consists of aromatics with C_6 ~2%, C_7 ~16%, C_8 ~39%, C_9 ~28%, and C_{10} ~13% and is characterized by research octane numbers between 90 and 100. The conditions under which this transformation is effected are shown in Table 10.2.1.

Methanation → "Synthetic Natural Gas"

The transformation of a syngas into a synthetic natural gas (or SNG) with >35 MJ/m^3, as in

$$CO + 3\,H_2 \rightarrow CH_4 + H_2O,$$

is, in effect, reverse steam-reforming of methane, but is accompanied by

$$2\,CO + 2\,H_2 \rightarrow CH_4 + CO_2$$
$$CO_2 + 4\,H_2 \rightarrow CH_4 + 2\,H_2O.$$

All three reactions are heavily favored by high pressures and by temperatures below 1000°C, and are therefore usually conducted at 260–370°C/2–7 MPa over freshly precipitated Raney nickel or a Ni–kieselguhr catalyst. However, the process is highly exothermic and releases ~2.4 MJ/m^3 syngas converted to CH_4, and catalyst deterioration by sintering and/or carbon deposition can only be avoided by close temperature control. Technical studies [25] of methanation have therefore given much attention to reactor designs that included fixed-catalyst bed systems cooled by heat-exchange surfaces (Dirksen and Linden, 1963) or gas recycling (Forney *et al.*, 1965); fluidized catalyst beds cooled by heat-exchange surfaces (Schlesinger *et al.*, 1956; Dirksen and Linden, 1960); reactors in which the catalysts are supported on and cooled by heat-exchanger tubes (Haynes *et al.*, 1974b); and a liquid-phase methanation reactor in which the catalyst is fluidized by inert gas that also serves as a heat sink (Sherwin *et al.*, 1973; Frank *et al.*, 1976). In suitably cooled methanators, catalysts will remain active for as long as five years and accomplish almost 100% conversion.

The status of current methanation technology is attested by satisfactory performance in full-scale trials at the British Gas Board's Westfield plant, where a 13.8 MJ/m^3 Lurgi-generated syngas was converted to 35.4 MJ/m^3 SNG at 74 000 m^3/d by "staged" methanation (Elgin and Perks, 1974), as well as by Lurgi methanation tests in SASOL's facilities at Sasolberg (Moeller *et al.*, 1974; Eisenlohr *et al.*, 1975). In *staged methanation,* the syngas is passed through a series of reactors in which the catalyst beds are cooled by internal coils, gas is cooled between stages, and part of the product stream is recycled. In *Lurgi methanation,* the syngas is sent through ZnO to remove the last traces of sulfur compounds, reacted in a first stage at 450°C (which accomplishes ~80% methanation), and completed in a second stage. Cooling in this stage is achieved by recycling a portion of the product gas.

The Fischer–Tropsch Syntheses

Conversion of a syngas to liquid hydrocarbons and related oxygenated compounds over variously promoted Group VIII catalysts has its origins in the classic researches of Franz Fischer and Heinz Tropsch in the 1920s, and was brought to commercial status in the 1930s and early 1940s. The major products of these conversions are straight-chain alkanes, alkenes, ketones, acids, and minor amounts of alcohols, all of which result from reactions that are formally represented by

$$(2n + 1)\ H_2 + n\ CO \rightarrow C_nH_{2n+2} + n\ H_2O$$
$$(n + 1)\ H_2 + 2n\ CO \rightarrow C_nH_{2n+2} + n\ CO_2$$
$$2n\ H_2 + n\ CO \rightarrow C_nH_{2n} + n\ H_2O$$
$$\rightarrow C_nH_{2n+1}OH + (n - 1)\ H_2O$$
$$(n + 1)\ H_2 + (2n - 1)\ CO \rightarrow C_nH_{2n+1}OH + (n - 1)\ CO_2$$
$$n\ H_2 + 2n\ CO \rightarrow (C_nH_{2n+1}—)_x + n\ CO_2.$$

These processes are, however, always accompanied by shifting and a reverse Boudouard reaction [26], and processing conditions that make for high CO conversion will therefore also allow some CO_2 hydrogenation by reverse shifting, as in

$$(3n + 1)\ H_2 + n\ CO_2 \rightarrow C_nH_{2n+2}\ 2n\ H_2O.$$

As well, straight-chain hydrocarbons and their oxygenated homologs will sometimes undergo partial secondary isomerization and/or cyclization to branched-chain and aromatic compounds.

This multiplicity of possible reactions requires narrowing the spectrum of potential products by careful selection of, and close control over, process variables, and in that context, the significance of temperature–pressure–catalyst combinations makes it important to differentiate between five quite different CO hydrogenation procedures.

Medium-Pressure Synthesis

Maximum yields of gasoline, diesel, and aviation fuels are generated by reacting an appropriately shifted syngas at 220–340°C/0.5–5 MPa over a *fused* or *nitrided fused iron* catalyst promoted by K_2O. (The nitrided catalyst favors formation of low-molecular-weight hydrocarbons and, because of its greater resistance to attrition, is also preferred for fluid-bed operations). Higher $H_2 : CO$ ratios increase the gasoline fraction of the product stream and also improve gasoline quality, but as shown by data from South Africa's SASOL plants (Tables 10.2.2 and 10.2.3), the overall compositions of the product slates are also markedly influenced by whether synthesis proceeds over a fixed or fluidized catalyst.

High-Pressure Synthesis

Reaction of syngas at 100–150°C/5–100 MPa over a *ruthenium* catalyst generates mainly straight-chain paraffin waxes with melting points to 132–134°C

TABLE 10.2.2 **Distribution of Hydrocarbons from Medium-Pressure FT Synthesis (wt %)**[a]

	ARGE reactor[b]		SYNTHOL reactor[c]	
	% of total	% in olefins	% of total	% in olefins
C_1	7.8	—	13.1	—
C_2	3.2	23	10.2	43
C_3	6.1	64	16.2	79
C_4	4.9	51	13.2	76
C_5–C_{11}	24.8	50	33.4	70
C_{12}–C_{20}	14.7	40	5.1	60
C_{20+}	36.2	15		
Alcohols	2.3		7.8	
Acids			1.0	

[a] Frohning and Cornils (1978).
[b] Fixed catalyst; $H_2 : CO = 1.7$; $T = 220$–$240°C$; gasoline yield ~32%.
[c] Fluid catalyst; $H_2 : CO = 3.5$; $T = 320$–$340°C$; gasoline yield ~70%.

and molecular weights up to 10^5 (Pichler, 1938; Pichler and Firnhaber, 1963). However, formation of lower-molecular-weight hydrocarbons can be encouraged by increasing the $H_2 : CO$ ratio of the feed gas, raising the reaction temperature, and lowering the pressure. How far such product modification can be driven is indicated by reports that reaction of a 4 : 1 $H_2 : CO$ syngas at 225°C over 0.5% Ru on Al_2O_3 yielded ~96% CH_4 at ~82% CO conversion (Schultz *et al.*, 1967).

TABLE 10.2.3 **Influence of Operating Parameters on Composition of Product Slates from Medium-Pressure FT Synthesis**[a]

	Average mol. wt. of products	Oxygenated products	Olefins
Increasing			
Pressure	↑	↑	↓
Temperature	↓	↓	↑
CO conversion	↑	↓	↓
Flow rate	↓	↑	↑
Gas recycle ratio	↓	↑	↑
$H_2 : CO$ ratio	↓	↓	↓

[a] ↑ increasing; ↓ decreasing.

Iso Synthesis

Branched-chain hydrocarbons—mainly consisting of C_4 and C_5 isoparaffins—are conveniently obtained by reacting a syngas at 400–500°C/10–100 MPa over *thoria* or a K_2CO_3-promoted *thoria–alumina* catalyst (Pichler and Ziesecke, 1949). Under favorable conditions, 1000 m^3 feed gas can thus deliver as much as 85 kg isobutane and 25 kg higher branched-chain hydrocarbons.

Synthol Synthesis

At temperatures substantially above 400°C, iso synthesis promotes coproduction of aromatics, and below that temperature it tends to encourage formation of oxygenated compounds. The latter are, however, more readily obtained from synthol synthesis (Fischer and Tropsch, 1923), in which the feed gas is reacted at 400–450°C/~14 MPa over *alkalized iron*, or from either of two modifications of that method.

In one, termed *synol* synthesis (Wensel, 1948), the syngas is reacted at 180–200°C/0.5–5 MPa over a highly reduced ammonia catalyst: if conversion per pass through the reactor is suitably restricted, product slates with up to 45% oxygenated compounds (mostly straight-chain alcohols) can be generated.

In the other modification, known as *oxyl* synthesis (Roelen *et al.*, 1943), the reaction proceeds at 180–200°C/2–5 MPa over *precipitated Fe;* and if synthesis conducted in two stages, up to 95% CO is converted to a product mix with >30% oxygenated compounds, also mostly straight-chain alcohols.

Oxy Synthesis

More correctly termed *hydroformylation,* this reaction involves interaction of the syngas with an olefin at 100–200°C/10–50 MPa over a cobalt carbonyl catalyst. It requires an approximately equimolar CO/H_2 mixture and a syngas : olefin ratio of ~2, and generates aldehydes by adding CO and H_2 across the olefinic bond, as in

$$R{-}CH{=}CH_2 + CO + H_2 \rightarrow R{-}\underset{\displaystyle H}{\underset{|}{CH}}{-}\underset{\displaystyle CHO}{\underset{|}{CH_2}}.$$

Since the early 1950s, oxo synthesis has become the most common method for producing C_3–C_{16} aldehydes, which are then reduced to the corresponding alcohols and used in the manufacture of detergents. Other products made

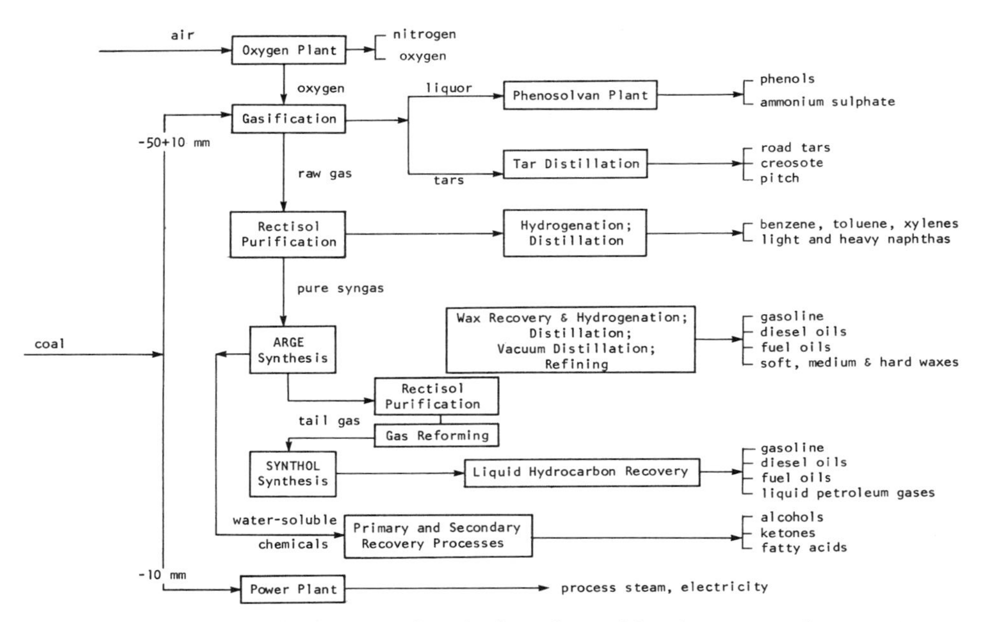

FIGURE 10.2.2 Simplified flowsheet of South Africa's first coal-based SASOL complex, commissioned in 1952. Two similar, but larger, plants began operations in 1980 and 1984.

from the alcohols are esters—e.g., phthalates required for plasticizing resins such as vinyl—but considerable quantities are also marketed as solvents.

With European and Japanese coal-based Fischer–Tropsch plants gradually phased out after World War II, the only commercial facilities currently attesting to the versatility of medium-pressure FT syntheses are three complexes operated by SASOL (South African Coal, Oil & Gas Corporation) [27].

The first of these facilities was brought on stream in 1955 at Sasolburg, and now transforms some 2×10^6 tonnes subbituminous coal/year into motor fuels, heating oils, industrial solvents, and by-products that include a ~24 MJ/m^3 fuel gas mainly used as a domestic fuel. The gasifier section is made up of 25 continuously operated Lurgi reactors that deliver raw syngas at $\sim 10.5 \times 10^6$ m^3/d, and syntheses are effected in fixed-catalyst-bed ARGE reactors over alkaline precipitated Fe, as well as in SYNTHOL reactors that employ entrained catalysts formulated from highly reduced alkaline magnetite.

Figure 10.2.2 shows a simplified flowsheet of the Sasolburg complex; syngas processing conditions are set out in Tables 10.2.4 and 10.2.5; and typical outputs were shown earlier in Table 10.2.2.

The success of the Sasolburg plant has encouraged the construction of two substantially similar, but three times larger facilities (SASOL II and III), which

TABLE 10.2.4 Operating Conditions for FT Syntheses in SASOL's ARGE Reactors

Catalyst charge	32–36 tonnes					
Temperature	220–225°C					
Pressure	2.5 MPa					
Fresh feed gas rate	20–22 × 10^3 m^3/h					
Recycle ratio[a]	2.2–2.5					
Syngas conversion	60–66%					
	Gas composition (%)					
	H_2	CO	CO_2	CH_4	C_mH_n	N_2
Feed gas	53.7	31.6	1.0	13.2	—	0.5
Tail gas	40.5	23.5	7.1	27.0	0.9	1.0

[a] Tail gas : fresh gas ratio.

TABLE 10.2.5 Operating Conditions for FT Syntheses in SASOL's SYNTHOL Reactors

Catalyst charge	90–125 tonnes[a]					
Temperature	320–330°C					
Pressure	2.3 MPa					
Fresh feed gas rate	80–100 × 10^3 m^3/h					
Recycle ratio[b]	2					
Syngas conversion	70–75%					
	Gas composition (%)					
	H_2	CO	CO_2	CH_4	C_mH_n	N_2
Feed gas	61.6	22.2	7.2	5.0	—	4.0
Tail gas	29.7	1.9	17.5	26.2	11.0	13.7

[a] Circulating at an hourly rate of 4.5–7.5 × 103 tonnes.
[b] Tail gas : fresh gas ratio.

commenced operations in, respectively, 1980 and 1984. Each employs 70 Lurgi reactors plus five on standby, and gasifies >6 × 10^6 tonnes bituminous coal/year.

Hydrocarbon Synthesis from CO + Steam

A potentially important, but almost totally neglected variant of CO hydrogenation, which differs from Fischer–Tropsch processing in requiring *no external*

TABLE 10.2.6 Unsaturated Hydrocarbons from the CO/Steam Reaction[a]

Boiling range (°C)	Yield (wt %)	s.g. (g/cm^3)	Unsaturates (wt %)
70–200	33.7–34.6	0.760–0.771	36.8–50.4
200–250	20.8–22.2	0.784–0.795	35.0–56.0
250–300	15.2–29.0	0.800–0.820	30.0–49.0

[a] Kravtsov *et al.* (1974).

hydrogen, is hydrocarbon synthesis from CO and *steam* (Kölbel and co-workers, 1952, 1957, 1958; Maekawa *et al.*, 1977). Summarily written as

$$3\ CO + H_2O \rightarrow (-CH_2-) + 2\ CO_2,$$

this proceeds, presumably after antecedent internal gas shifting, at 190–250°C over cobalt, nickel, or ruthenium catalysts, or at 250–300°C over iron. Optimum pressures lie between 800 and 1000 kPa for pure CO, but increase somewhat with falling CO concentrations in an impure feed gas, and under suitable conditions, a 75% olefinic product slate with 45–60% C_8–C_{10} hydrocarbons can be obtained (Kravtsov *et al.*, 1974). Some representative data from laboratory studies are summarized in Tables 10.2.6 and 10.2.7.

An investigation of the role of iron catalysts in what has come to be known as Kölbel–Engelhardt synthesis (Kotanigawa *et al.*, 1981) suggests that formation of hydrocarbons depends on a dynamic equilibrium between transient oxidation states of Fe and on the development of surface-active "free" carbon species in the catalyst.

TABLE 10.2.7 C_5–C_8 Hydrocarbons from CO/Steam Reaction[a]

	Yield (wt %) over		
	Fe/Cu/ kieselguhr	Fused iron	Fe/Cu/ kaolin
Paraffins	25.6	25.6	50.7
Olefins	74.3	74.2	49.3

[a] After Kravtsov *et al.* (1974).

3. UNDERGROUND (U/G) COAL GASIFICATION

The possibility of gasifying coal *in situ*—a topic that is attracting renewed attention as means for extracting useful energy from deeply buried or otherwise uneconomic coal, tar sands, and certain oil shales—was envisaged as early as 1868 by Sir William Siemens and independently discussed by Mendeleev (1888) in a series of papers published in St. Petersburg. But even though such a process was described in an early British patent (Betts, 1909), which dealt with injection of air and steam into an ignited coal seam through boreholes, almost nothing was done [28] until 1931, when the Soviet government of the time, influenced by Lenin's enthusiasm for the idea (Lenin, 1913), initiated development of workable methods for u/g gasification [29].

Intensive subsequent work in the Soviet Union led to operation of several large u/g facilities that generated a 4–4.5 MJ/m^3 fuel gas for power stations and various industrial consumers, and in the decade following the end of the Second World War, u/g gasification field tests were also conducted in the Western world [30]. However, most of the USSR installations were abandoned after 1955, when abundant supplies of natural gas removed incentives for further development work; field work in the West was likewise suspended; and since then, only limited activity, resumed in the early 1970s, has been reported from the United States (Schrider *et al.*, 1974; Brandenburg *et al.*, 1975; Stephens *et al.*, 1976) and Canada (Berkowitz and Brown, 1977).

The Principles of u/g Gasification

The technical basis of u/g gasification is almost deceptively simple, and is illustrated in Fig. 10.3.1, which shows two boreholes (I, II) sufficiently well linked to allow substantially unhindered flow of gas between them. If coal near the bottom of I is now ignited, and combustion is sustained by injecting air, three distinctive zones quickly establish themselves. In the first, combustion produces CO_2, which is progressively pushed toward II by incoming air that also drives the combustion front ahead. In the second, CO_2 reacts with the partly devolatilized coal to form CO by the Boudouard reaction, while H_2O in the coal or injected with combustion air generates CO + H_2 by steam gasification of carbon. And finally, in the third zone, pyrolysis of coal immediately ahead of the reduction zone releases volatile matter that cracks and thereby contributes CO, H_2, CH_4, and C_{2+} hydrocarbons to the gas stream.

This process continues until all coal within its reach is consumed, and what eventually surfaces through the production hole (II) is a fuel gas with heat values up to $\sim$6.5 MJ/m^3.

Design parameters for a practical u/g scheme are, however, always site-

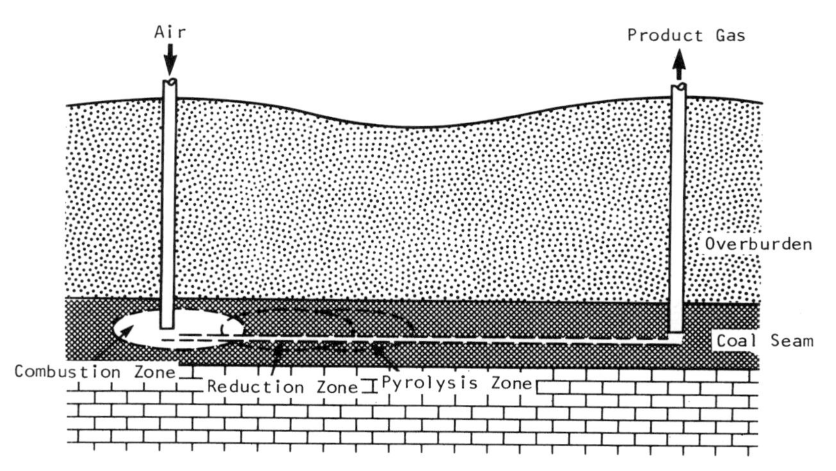

FIGURE 10.3.1 Principles of underground coal gasification. The same approach could be used to gasify bitumens and other heavy hydrocarbons.

specific, and operable gas producers require considerable preparation in the field. Even when the scheme is oriented to take advantage of the natural fracture system—that is, oriented so that forward combustion can advance in the direction of the major cleat [31]—it is almost always mandatory to increase the permeability of the coal by some form of seam fracturing, and several alternatives are, in fact, available for this. A single channel can be created by directional drilling or reverse combustion—i.e., in the latter case sustaining initial combustion in hole I by drawning air to it through II. Or a large number of smaller, more tortuous fractures can be created by hydrofracking or electrolinking [32, 33].

As well, there is need to prevent influx of formation waters, which would lower temperatures in the combustion and reduction zones, correspondingly shift the CO/CO_2 equilibrium, and thereby lower the heat value of the product gas [34]. And during operations, water and tar vapors must be prevented from forcing the combustion front to override the coal—i.e., to advance along the *top* of the seam rather than within it, and so drastically curtail gas production.

Nonetheless, provided the seam is thicker than 1.5 m, the efficiency of energy extraction by current, still fairly rudimentary u/g gasification methods exceeds the average efficiency of mining. With adequate control over operation of the generator, substantially constant gas quality can be maintained, and over 75% of the in-place coal can be gasified with an effective energy recovery of 50–75% (Gibb, 1964).

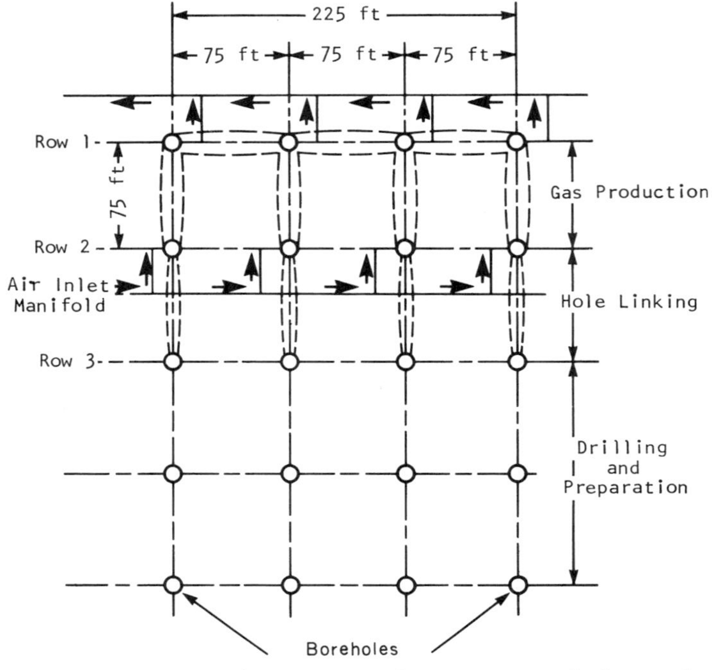

FIGURE 10.3.2 Schematic of a gas generator for operations in a flat-lying coal seam.

Generator Systems

How u/g gasification is accomplished depends on whether it is conducted in a flat-lying or pitching seam. The former case is exemplified in Fig. 10.3.2 by a generator that fueled a power station near Moscow: in this scheme, the boreholes were spaced at intervals of ~25 m and very high energy recoveries were recorded by periodically changing the gasification *direction*. Gasification of steeply inclined seams is illustrated in Fig. 10.3.3 by a facility that began to be operated near Lisichansk in 1955, and is of particular interest because linkage paths connecting the first-stage injection holes were established by directional drilling.

Generator Outputs

The USSR experience and North American field work conducted in the 1970s (de Crombrugghe, 1959; Elder, 1963; Capp *et al.*, 1963; Kreinin and Revva, 1966; Nadkami *et al.*, 1974; Gregg and Olness, 1976) have defined the major

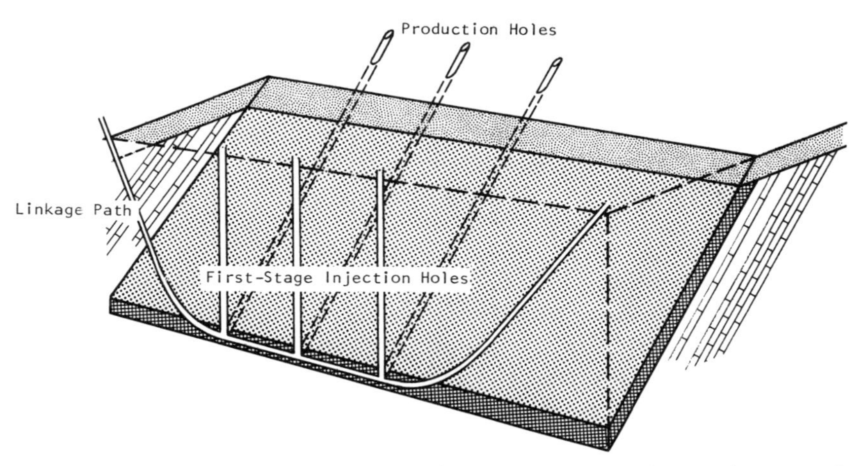

FIGURE 10.3.3 Schematic of a gas generator for operations in a pitching coal seam. In this illustration, the linkages were established by directional drilling.

operating variables that affect gas yields, gas compositions, and conversion efficiencies, and outlined operating procedures that promise to yield a richer gas, enable better control over water influx, and eliminate unacceptable gas leakages through overburden fractures. As well, computer modeling has led to a better understanding of the physical and chemical changes that accompany u/g gasification. Improved monitoring equipment that enables close surveillance of the process has, in fact, already made possible generation of gases with 5.2–5.8 MJ/m^3; and there now also exist [35] some data bases for assessing any potentially adverse environmental impacts of u/g gasification.

4. COAL LIQUEFACTION

The Fundamentals

Laboratory studies of hydrogen-transfer reactions (Fieldner and Ambrose, 1947) such as

$$+ 2 \overset{CH_3}{} \rightarrow 3 \quad + 2CH_4$$

in which the donor tetrahydronaphthalene molecule can provide up to four

H atoms, have led to the development of conversion technologies that offer significant technical and economic advantages over Bergius high-pressure hydrogenation [36].

With highly aromatized feeds such as oil residua, bitumen, or coal, uncatalyzed H transfer proceeds only at temperatures between 350 and 500°C—i.e., at temperatures at which such feeds thermally crack—and transfer processes must therefore be understood to depend, like other forms of hydrogenation, on formation of molecular fragments that are stabilized by abstraction of H atoms from the donor. However, unlike Bergius hydrogenation, these processes do not necessarily need a catalyst or very high pressures; and perhaps even more important, they allow close control over H traffic in the reaction system. It can therefore be a relatively simple matter to decouple successive levels of H insertion into a host molecule, and thereby gain much operational freedom as well as an ability to minimize H expenditures.

"Solvent Refining"

One example of such decoupling—tantamount to *prematurely arrested* hydrogenation—presents itself in solvent refining, a technology designed to produce environmentally clean solid fuels by transforming coal into low-melting, soluble solids.

Early empirical procedures for such refining can be seen in the Pott-Broche process (Kröger, 1956)—which was operated between 1938 and 1944 in a small (125 t/d) plant at Welheim (Germany) for producing the pure feedstocks needed for manufacture of carbon electrodes, etc.—and in a modified version of that process due to Uhde and Pfirrmann (Uhde, 1936; Pfirrmann, 1939).

In the former, a 1 : 2 w/w paste of finely pulverized coal and middle oils [37] was heated for 1 h at 415–430°C/10–15 MPa, cooled to 150°C, and filtered through ceramic cartridges that removed residual solids. Spent donor oils were recovered from the filter cake by vacuum distillation at 360°C, rehydrogenated, and recycled. The soluble product represented 65–70% of the organic coal material charged to the process, softened between 200 and 240°C, and contained less than 0.06% ash.

The Uhde–Pfirrmann version generated an equivalent, but appreciably improved, extract by reacting the coal–oil slurry under gaseous H_2 at 410°C/30 MPa. The higher pressure shortened the reaction period to ~30 min, converted 70–80% of the coal into soluble material, and yielded an extract that contained substantially more H (and correspondingly less N, O, and S), melted in a lower temperature range (60–120°C), and possessed combustion characteristics that reportedly allowed its use as a fuel in stationary diesel engines.

Contemporary solvent-refining techniques are direct descendants of these

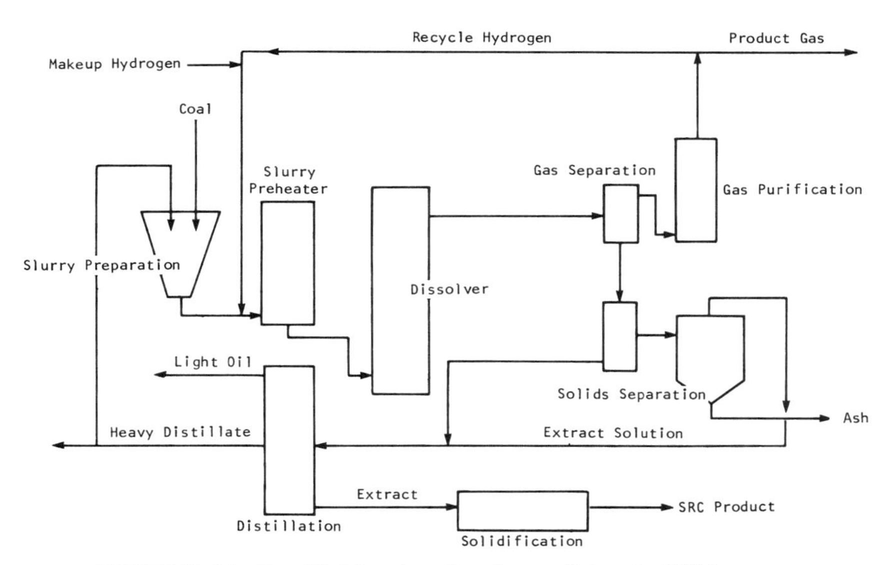

FIGURE 10.4.1 Simplified flowsheet for solvent refining: the SRC-I process.

processes, but better grounded in basic experimental data, and therefore enable product slate compositions to be varied within wide limits. Thus, whereas an addition of ~2% hydrogen can suffice to solubilize ~85–90% of the coal and generate ~80% solvent-refined coal (SRC), a slightly greater H transfer (~3%) can raise total conversion to ~93%, increase the yield of hydrocarbon *liquids and gases* at the expense of SRC, and endow the "refined" solid with a much lower melting range. Operating conditions for solvent refining are therefore generally determined by interest in maximizing yields of ash- and sulfur-free material for minimum hydrogen expenditures.

Information from two pilot facilities [38] shows how these objectives can be reached. In the preferred operational mode, pulverized (90% <75-μm) coal with $<3\%$ H_2O was slurred with 3–4 times its weight of recycle solvent, pressurized to ~12 MPa with H_2, and heated for ~40 min at 440–455°C. The slurry was then cooled to 310–315°C, stripped of hydrocarbon gases, unreacted H_2, H_2S, and H_2O, and depressurized to ~0.8 MPa before being passed through a rotary drum or leaf filter. Spent solvent in the filtrate was recovered by vacuum distillation at 315°C, rehydrogenated and recycled, and the extract, which, depending on its H content, melted between 150° and 200°C, was solidified by cooling to room temperature.

Figure 10.4.1 shows a flowsheet for such processing, Table 10.4.1 summarizes processing conditions, and Table 10.4.2 details the product slates obtained under these (or similar) conditions from three different hvb coals.

TABLE 10.4.1 SRC Processing Conditions, Wilsonville Plant

Coal feed rate (kg/h)	227
Coal concentration in slurry (wt %)	25–33
Slurry feed rate (liter/min)	13.6
Dissolver temperature, in/out (°C)	435/440
Dissolver pressure (MPa)	11.5
Filtration rate (liter/h m^2 of filter)	450
H_2 consumption (wt % daf)	2–2.2

TABLE 10.4.2 Typical Product Slates from Solvent Refining of Coal[a] (Wilsonville plant, wt % daf Coal)

	#1	#2	#3
Solvent-refined extract	63	71	63
Unreacted residue	7	7	7
C_1–C_4 gases	6	7	7
C_5–175°C distillates	6	4	4
175–400° distillates	10	6	12
H_2S	2	2	2
$CO + CO_2$	6	3	2
H_2O	6	3	3
[H_2 consumption	2.4	2.0–2.2	2.5]
Total conversion	93	92	92

[a] Coal #1: Illinois; #2: Kentucky; #3: Pennsylvania; all are of hvb rank.

Liquefaction

If H-transfer is not prematurely terminated, as it is in solvent refining, it can almost completely transform the organic carbon of coal into hydrocarbon liquids and small amounts of C_1–C_5 gases. This transformation has been conceptually put in the form

$$\text{coal} \rightarrow \text{preasphaltenes} \rightarrow \text{asphaltenes} \rightarrow \text{oils},$$

in which "oils" are variously identified with pentane-soluble maltenes or hexane-soluble carbenes/carboids (see Table 10.4.3), and are termed "primary" liquids that, because of their chemical complexity, are most conveniently defined by their solubilities. However, because such sequential changes must of necessity encompass regressive as well as progressive reactions, a more realistic formulation is written as in Fig. 10.4.2. This reflects potential revers-

TABLE 10.4.3 "Primary" Liquids from H Transfer to Coal

	Soluble in	Insoluble in
Preasphaltenes	tetrahydrofuran	benzene, toluene
Asphaltenes	benzene, toluene	*n*-hexane
Carboids	*n*-hexane	carbon disulfide
Carbenes	carbon disulfide	*n*-pentane

ibility at different stages of liquefaction due to concurrent thermal and hydrocracking reactions, and properly identifies liquefaction as a two-stage process:

1. *Solubilization,* controlled by the type and intensity of pyrolytic and H-transfer reactions in the reactant system, followed by
2. *Secondary hydrogenation,* which is governed by reaction conditions and drives the process toward structurally simpler liquids with lower molecular weights

Specific information about the chemistry and kinetics of liquefaction is still fragmentary and at times contradictory. Thus, disgreeing with one study, from which it was concluded that yields of asphaltenes plus lighter hydrocarbons fell systematically with increasing coal rank (Neavel, 1976), another (Gorin, 1981) found grossly divergent rank dependencies of liquefaction yields, apparently due to different regional metamorphic histories and consequent disparate rank/reactivity relationships [39]. Observations such as these indicate a much

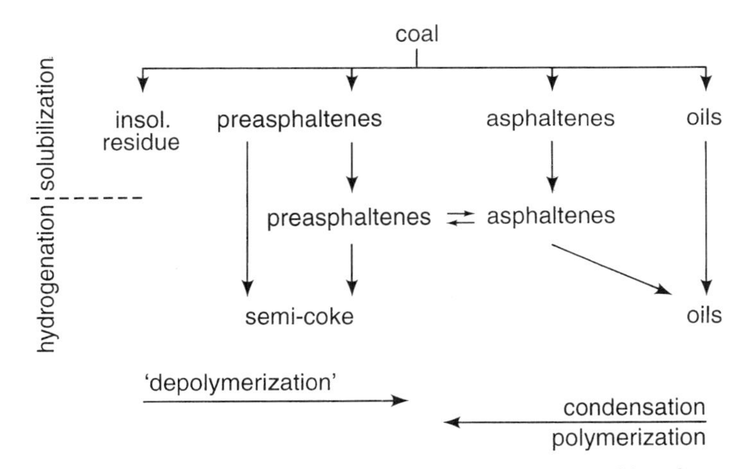

FIGURE 10.4.2 Conceptual reaction sequences in "complete" coal liquefaction.

TABLE 10.4.4 Coal Desiderata for Liquefaction

Property	Influences	Desired level
Rank	Liquids yield	Medium
Ash content	Operations and handling	Low
Moisture content	Thermal efficiency	Low
Hydrogen content	Liquids yield & H consumption	High
Oxygen content	Gas make & H consumption	Low
Extractability[a]	Liquids yield & quality	High
Aliphatic character	Liquids yield & quality	High
Reactive macerals[b]	Liquids yield	High
Particle size	Operations	Pulverized

[a] In effect, "solubility" in potent solvents.
[b] Principally vitrinites and exinites.

greater diversity of coal chemistries than is usually recognized, and necessitate careful selection of candidate coals and processing conditions in light of screening tests and some empirically established desiderata (Table 10.4.4).

The dichotomy between coal solubilization and hydrogenation shown in Fig. 10.4.2 makes it possible to decouple the two stages and either

1. limit H transfer from the donor to generation of heavy fuel oils, and implement the second stage only if lighter (transportation) fuels are required; or
2. drive conversion directly to predominantly light- and medium-gravity oils by H transfer in reaction regimes very similar to those used in catalytic hydrogenation or hydrocracking

In either case, product yields and yield structures [40] will be governed by choices of feedstock and process engineering, as well as by operating parameters [41].

Conversion techniques that limit H-transfer to generation of relatively heavy fuel oils and assume that upgrading to lighter oils could, when required, be accomplished by conventional hydrotreating are chemically equivalent to taking solvent-refining beyond the production of low-melting solids, and technically unremarkable. What they can achieve is exemplified in Table 10.4.5 by some input/output data for the Lummus Corporation's Clean Fuels From Coal (CFFC) process (Sze and Snell, 1974, 1976), which seeks to convert bituminous coals with high ash and sulfur contents into fuel oils with $<0.1\%$ ash and $<0.5\%$ sulfur [42].

However, in the context of this review, greater interest attaches to techniques that drive conversion to an *all-distillate* product slate.

TABLE 10.4.5 Compositions of a Bituminous Coal and Corresponding CFFC Product Slate[a]

	Coal	CFFC product
Elemental composition (wt %)		
Carbon	66.3	88.8
Hydrogen	4.8	6.9
Nitrogen	1.2	1.2
Sulfur	3.9	0.34
Oxygen	11.1	—
Ash	9.7	0.05
Specific gravity (15°C)		1.100
Pour point (°C)		18.3
Flash point (°C)		102
Heating value (MJ/kg)		39.31
Distillation (°C):	IBP	172.2
	10%	242.8
	30%	281.1
	50%	421.1
	67%	482.1

[a] Simone (1976).

These are illustrated by HRI's H-Coal process (U.S. Dept. of Commerce, 1974), which was adapted from an oil hydrogenation method (see Chapter 9) and employs an ebullated-bed reactor in which a coal–oil slurry encounters H-rich gas while moving upward through a partially fluidized bed of pelleted cobalt molybdate/Al_2O_3 (Fig. 10.4.3). The reaction system is maintained by continuous injection of fresh slurry, and by deploying coal and catalyst pellet sizes (<150 μm and 1.5–6 mm, respectively) that make it possible for coal liquids, unreacted coal, and ash to leave without carrying catalyst with them. The height of the catalyst bed is controlled by the quantity of catalyst and the coal concentration in the slurry, and acceptable catalyst activity is maintained by replacing spent catalyst at ~0.5 kg/t coal passing through the reactor. Temperatures are maintained by the circulating catalyst and by recycling much of the heavier oil, which is used to thin incoming slurry to the desired consistency.

In one of the two modes in which the H-Coal process has been operated, hydrogenation is conducted in a single reactor at ~450°C/17.5 MPa. With coal fed at up to 1600 kg m^{-3} expanded catalyst h^{-1}, liquids yields ran to 0.53–0.70 m^3/t daf coal, and overall conversion to oil plus C_1–C_3 hydrocarbon gases exceeded 90%.

In the other mode, the CTSL (Catalytic Two-Stage Liquefaction) process,

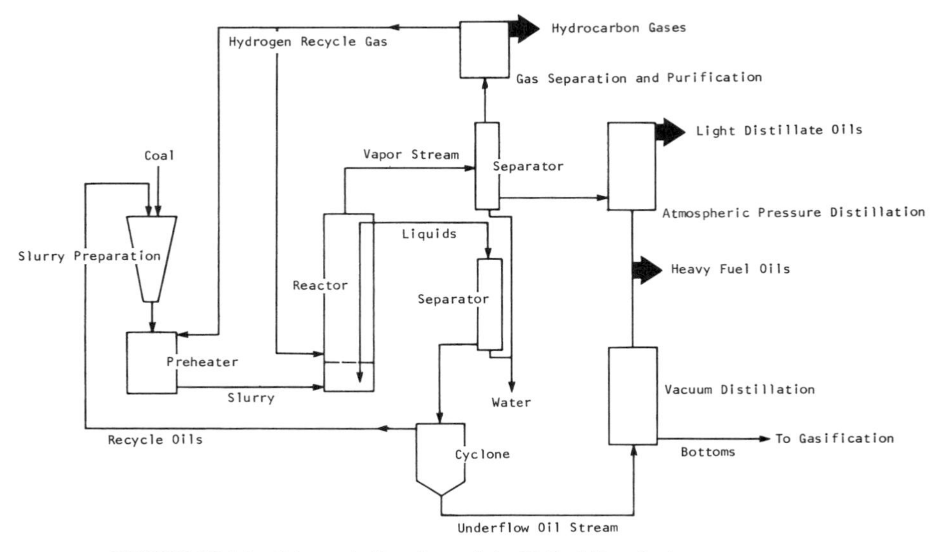

FIGURE 10.4.3 Schematic flowsheet of the H-Coal liquefaction process.

two closely coupled reactors, operated at 400 and 420–425°C, respectively, made it possible to convert all preasphaltenes, asphaltenes, and residual reactive coal material into distillate liquids. This slightly increased total conversion to oils plus C_1–C_3 hydrocarbons and enhanced the "hydrogen efficiency," defined as wt% C_4–525°C distillate/wt% H_2 consumed.

Under active development since 1968, and demonstrated in a 550 t/d pilot plant, which provided data for scale-up, both versions of the H-Coal process were pronounced ready for commercial use in the early 1980s. Conventional refining procedures would be used for further processing of the synthetic crude oils they generate.

Similar paths, only differing in procedural details, are followed in Gulf Oil's SRC-II and Catalytic Coal Liquids (CCL) processes.

SRC-II is an offspring of SRC-I solvent refining technology, uses the substantially more severe hydrogenation conditions needed for generating hydrocarbon liquids (Schmid and Jackson, 1976), and recycles heavy product fractions to the reactor. Product filtration as in SRC-I is replaced by fractionation of the oil stream, and process hydrogen is generated by gasifying distillate bottoms. Table 10.4.6 details a typical product slate from SRC-II processing of bituminous coal, and Table 10.4.7 lists properties of the SRC-II oils.

CCL processing technology (Chung, 1974) resides primarily in reacting the coal–oil slurry with H_2 over a proprietary catalyst that is claimed to possess

TABLE 10.4.6 SRC-II Product Yields from Bituminous Coal (wt % Dry Coal)

C_1–C_4 hydrocarbon gases	16.6
Naphtha (b.p. < 195°C)	11.4
Middle distillates (195–250°C)	9.5
Heavy distillates (250–450°C)	22.8
Total liquids (to 450°C)	43.7
Resid. oils (bp > 450°C)	20.2
Carbonaceous solids	3.7
Water	7.2
Ash	9.9
CO, CO_2, H_2S	3.4
	104.7

exceptionally long life and to be virtually unaffected by ash components that slowly poison other catalysts.

A more innovative approach has, however, been taken in Exxon's Donor Solvent (EDS) process, which accomplishes liquefaction by noncatalytic H transfer from a process-derived recycle oil and hydrotreats the spent donor before recycling it (Furlong *et al.*, 1976). The operating sequence (see Fig. 10.4.4) begins with reacting coal at 425–465°C/10.5–14 MPa with gaseous H_2 and a process-derived (205–455°C) distillate oil, then passes the product slurry through heat exchangers, flashes it to atmospheric pressure to recover unreacted H_2 and C_1–C_3 hydrocarbons, hydrotreats it, and finally fractionates it into naphtha and distillate oils, of which a portion is returned to the primary reactor. Vacuum bottoms are transferred to a flexicoker [43] from which additional heavy oil is recovered, and depending on whether H_2 or a fuel gas for in-plant use is required, the residual coke is gasified with O_2 or air.

TABLE 10.4.7 Properties of SRC-II Oils

	Light distillates	Fuel oils
Boiling range (°C)	~30–200	~200–400
API gravity	39	5
Flash point (°C)		75.5
Elementary composition (wt %)		
Carbon	84.0	87.0
Hydrogen	11.5	7.9
Nitrogen	0.4	0.9
Sulfur	0.2	0.3
Oxygen	3.9	3.9

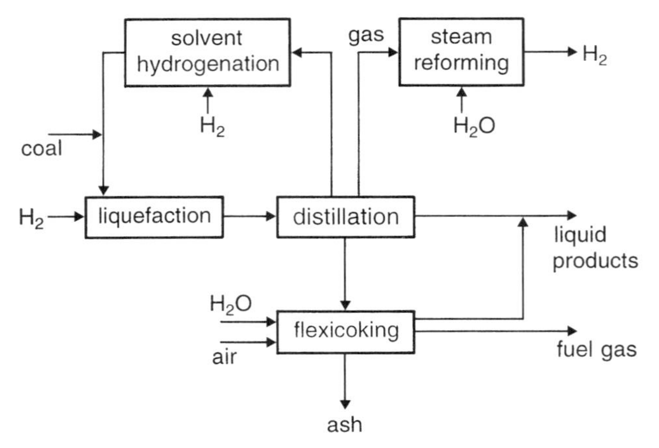

FIGURE 10.4.4 Schematic flowsheet of the EDS coal liquefaction process.

Overall yields from the EDS scheme average ~0.6 m^3 of low-S <565°C distillate oils per tonne dry coal (see Table 10.4.8),, but compositions of the oils depend on processing details. Total conversion and the yields of hydrocarbon liquids increase with liquefaction pressure and solvent quality, and yields of oils in the C_4–naphtha range can be improved at the expense of heavier oils by increasing hydrogen expenditures to >6% on feed coal.

A similar two-stage approach to all-distillate oils characterizes the Consolidation Coal Co.'s Synthetic Fuels (CSF) process (Gorin *et al.*, 1971; Phinney, 1974), in which the first stage is limited to interaction with a light process-derived distillate oil at 400–425°C/1–3 MPa. Additional H_2 is only used at this stage if substantial (>80%) solubilization of the feed requires more than ~0.6% H transfer [44]. Further hydrogenation of the extract in a second step, and fractionation of product oils, is envisaged being accomplished by a hydrocracking scheme (see Fig. 10.4.5).

Coprocessing

In common with second-generation gasifier systems, further development of coal liquefaction technology was shelved in the late 1970s, and its resumption will presumably have to await a more favorable economic climate [45]. Some attention has, however, meanwhile been focused on possible gains from replacing process-derived recycled H-donor oils with *petroleum residua*—and, in effect, *coprocessing* such residua and coal into distillable oils. Although seemingly no more than a simple modification of liquefaction technology, this could

TABLE 10.4.8 Composition of a Bituminous Coal and Corresponding EDS Product Slate[a]

	Coal[b]	Heavy naphtha		Fuel Oil (>200°C)	
		a[c]	b[d]	a[c]	b[d]
Elementary composition (wt % cry coal)					
Carbon	69.7	85.6	86.8	89.4	90.8
Hydrogen	5.1	10.9	12.9	7.7	8.6
Nitrogen	1.8	0.2	0.05	0.6	0.25
Sulfur	4.2	0.5	0.005	0.5	0.05
Oxygen	9.5	2.8	0.2	1.8	0.3
Ash	9.7				
Density (g/cm^3)		0.87	0.80	1.08	1.01
Paraffins (wt %)			10		3
Naphthenes + aromatics (wt %)			90		97
Boiling range (°C)		70–200	70–200	200–540	200–540
Heat value (MJ/kg)		42.6	44.9	39.8	42.1

[a] Alpert and Wolk (1981).
[b] hvb coal (Illinois #6 seam).
[c] Raw.
[d] After EDS hydrotreating.

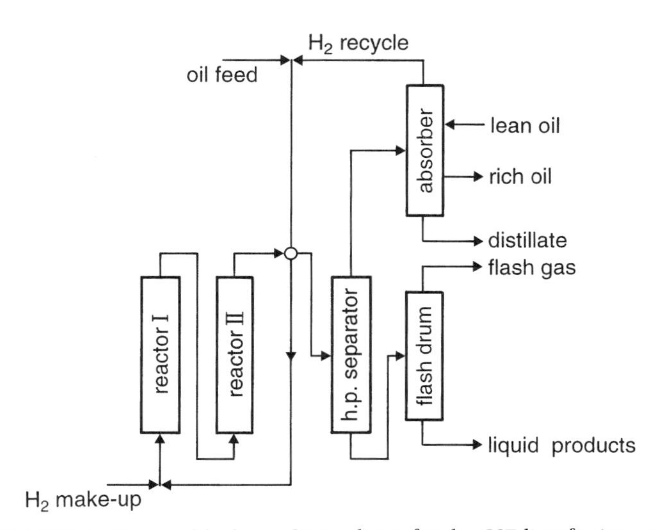

FIGURE 10.4.5 Proposed hydrocracking scheme for the CSF liquefaction process.

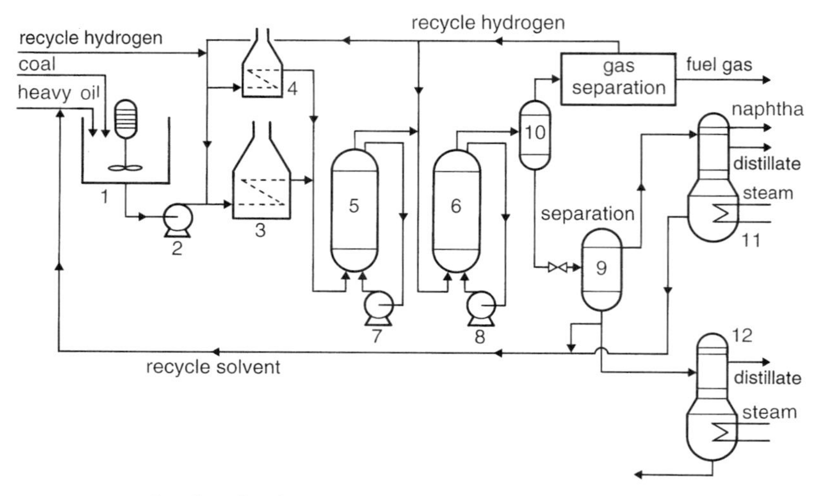

FIGURE 10.4.6 Flowsheet for the HRI two-reactor coprocessing system. 1, slurry preparation; 2, slurry pump; 3, slurry preheater; 4, hydrogen heater; 5, 6, first- and second-stage reactors; 7, 8, ebullating pumps; 9, 10, atmospheric and high-pressure separators; 11, atmospheric-pressure fractionation; 12, vacuum distillation.

make better use of capital equipment and offer benefits from potential synergy that would enhance oil quality and/or increase oil yields by as much as 5% over yields from separate processing of the two components (McRae, 1988).

How coprocessing, which would rely more on hydrocracking than on H-transfer from a donor, might be conducted is exemplified by the operation of a modified H-coal two-reactor scheme (U.S. Dept. of Energy, 1987), in which finely crushed coal and a petroleum residue are slurried, pressurized with gaseous H_2, and hydrocracked in a first reactor, where sulfur and nitrogen in the feed are also abstracted as H_2S and NH_3. The resultant heavy liquids are then hydrotreated in a second reactor, which delivers naphtha and distillate oils that require little more than filtration to remove residual solids. Figure 10.4.6 shows a flowsheet for this process.

Another coprocessing mode, reported to yield results as good as those from the H-coal scheme, has been used in the experimental Pyrosol process, which was designed in Germany for transforming bituminous coal and heavy petroleum residua into distillate oils (Boehm and St. Denis, 1987), and which has been successfully adapted for coprocessing Alberta subbituminous coal with vacuum bottoms from oil-sand bitumens (Boehm *et al.* 1989). The procedure involves mild noncatalytic hydrogenation of the feedstock and, in a close-coupled second stage, hydrocoking it—i.e., coking it under pressurized H_2.

This technique is claimed to generate a product stream with ~70% distillable oils for H_2 consumptions not significantly greater than ~1 wt% of the daf feed.

Catalysis of Liquefaction

Primary liquefaction—which results from insertion of donor-provided H into the coal macromolecule at points at which it is thermolytically severed, and from consequent dispersion of solubilized matter in the donor solvent—is assumed to be a noncatalytic process that ends when the thermal stability of H-capped molecular fragments equals that of the donor. The maximum amount of H_2 that can be inserted cannot therefore exceed ~2.5–3 wt% of the daf coal, and complete conversion to distillate oils can only be achieved by secondary processing that induces significant hydrocracking under more severe reaction conditions.

But there are indications that some forms of liquefaction are open to catalysis. Primary liquefaction under H_2, which is accompanied by *in-situ* hydrogenation of the depleted donor solvent, can thus be appreciably accelerated by FeS_x (Tarrer *et al.*, 1976), and more effectively speeded up by adding small amounts of elemental S to FeS_x in order to form ferromagnetic Fe sulfides with compositions in the range $FeS_{1.04}$–$FeS_{1.10}$ (Mukherjee and Choudhury, 1976). There have been reports of pronounced catalytic effects from Na (Given *et al.* 1980). And in a small (1 kg/h) bench unit, catalytic hydrocracking in $ZnCl_2$ has been observed to greatly accelerate the transformation of petroleum residua, bitumens, and coals into distillable oils with exceptionally high proportions of gasoline-type naphtha (Zielke *et al.*, 1976).

As well, studies of the effects of Lewis acid/Bronsted acid catalyst systems (such as HX–AlX_3–H_2) on conversion at temperatures as low as 190–210°C (Ross and Low, 1976) have shown

1. that aluminum halides tend to be overly active and to generate excessively light (mainly C_1–C_4) hydrocarbons;
2. that mercury and tin halides promote formation of heavy oils; but
3. that substantial gains could accrue from use of bismuth, gallium, antimony, titanium, and zinc halides, all of which strongly encourage formation of naphtha fractions in the gasoline boiling range.

LT primary conversion to pyridine- or THF-solubles in Lewis acid/Bronsted acid systems is believed to proceed by heterolytic rather than homolytic bond scission, and to involve a carbonium ion rather than free radical sequences.

TABLE 10.4.9 Fraction Compositions of a Solvent-Refined hvb Coal[a]

	C	H	O	N	S	mol. wt.[b]
Precursor coal[c]	80.1	5.8	10.3	1.4	2.4	
Whole extract	86.9	5.8	5.0	2.0	0.7	
Preasphaltenes (62.8%)	85.7	5.4	5.9	2.2	0.7	870
Asphaltenes (27.0%)	86.8	6.9	4.8	1.7	0.6	565
Oils (10.2%)	88.3	7.1	3.2	0.9	0.5	350

[a] Burk and Kutta (1976).
[b] As reported, but questionable (cf. Chapter 7).
[c] Illinois No. 6 (bituminous); analytical data expressed as wt.% daf coal.

THE "PRIMARY" COAL LIQUIDS

Composition and Properties

Even coals with very similar elemental and petrographic compositions can possess widely different structural chemistries [46], and depending as much on their precursor as on the hydrogenation depth and associated thermolytic process, the "primary liquids" accruing from conversion can range from solids and semisolids to heavy oils that are only fractionally distillable. But because H-supported thermolysis abstracts some aliphatics as C_1–C_3 hydrocarbons, and some H needed for the conversion can be supplied by naphthenic moieties in the coal, the solubilized material—composed of entities small enough in size to disperse and, in part, dissolve in the H donor—is always *more aromatic* than its precursor, and the more severe the reaction conditions in which it formed, the more polynuclear aromatics it contains [47].

TABLE 10.4.10 Composition of Liquids from EDS and H-Coal Processing (wt %; Precursor Coal as in Table 10.4.9)

	EDS[a]		
	<200°C	200–540°C	H-Coal[b]
Carbon	85.6	89.4	89.0
Hydrogen	10.9	7.7	7.9
Oxygen	2.8	1.8	2.1
Nitrogen	0.2	0.7	0.8
Sulfur	0.5	0.4	0.4

[a] After Furlong *et al.* (1976).
[b] After Callen *et al.* (1976).

TABLE 10.4.11 Comparison of an SRC and Coal Liquids from Two Liquefaction Processes with a Petroleum-Derived Oil[a]

	SRC-I	H-Coal	Synthoil	#6 Fuel oil
Elementary composition (wt %)				
Carbon	87.9	89.0	87.6	86.4
Hydrogen	5.7	7.9	8.0	11.2
Oxygen	3.5	2.1	2.1	0.3
Nitrogen	1.7	0.76	0.97	0.41
Sulfur	0.57	0.42	0.43	1.96
Ash	0.01	0.02	0.68	—
Simulated distillation fraction	Temperature (°C)			
IBP	—	250	222	175
15% vol	510	312	264	264
20% vol	>510	327	279	—
50% vol		404	379	478
70% vol		>517	>477	>532

[a] Bendoraitis *et al.* (1976).

Tables 10.4.9–10.4.11 illustrate these aspects. The first two concern a solvent-refined hvb coal and liquids generated from that SRC precursor by more severe H-Coal and EDS processing, and the third compares a petroleum-derived oil with coal liquids from two catalytic conversion processes and a noncatalytic procedure. The catalytic processes are here seen to deliver liquids very similar to the fuel oil in terms of boiling ranges, but to contain significantly higher proportions of asphaltenes. The SRC, generated by restricted H transfer, is better matched with a petroleum vacuum residue.

However, as elemental compositions tend to mask the chemical diversities of coal liquids, it is also pertinent to note more detailed information—in this case obtained by chromatographic fractionation of the liquids into sets of compounds with similar polarity and subsequent analysis of these fractions by NMR spectroscopy (see Table 10.4.12). A summary of such information by Crynes (1981) shows that

1. [0] typically runs to 4–5 wt% and presents few problems for further processing beyond an appreciable incremental hydrogen consumption in hydroprocessing (for O → H_2O);
2. [S] usually amounts to ~0.3–0.7 wt%, but ranges from a low of 0.05 to a high of 2.5 wt% and tends to be concentrated in high-boiling components (mostly in benzothiophene derivatives);
3. [N] normally amounts to 0.9–1.1 wt%, but is occasionally as low as 0.2 or as high as 2 wt%;

TABLE 10.4.12 Composition of Hydrotreated Liquefaction Products from a Kentucky hvb Coal[a] (Mobile Res.Dev.Co. 1976)

Fraction	wt %	Eluent	Major compounds
1[b]	0.4	Hexane	Saturates
2	15.0	Hexane/15% benzene	Aromatics
3[c]	30.0	Chloroform	Polar aromatics[e]
4	10.2	Chloroform/10% Et_2O	Simple phenols
5[d]	10.1	Et_2O/3% EtOH	Basic N-heterocycles
6	4.1	MeOH	Highly functional molecules[f]
7	6.4	$CHCl_3$/3% EtOH	Polyphenols
8	10.2	Tetrahydrofuran	Increasing [O], increasing N-basicity
9	8.5	Pyridine	??
10	5.1	Not eluted	??

[a] Mobil Res. Dev. Co. (1976).
[b] Fractions 1–3 designated as oils.
[c] Fractions 3–5 comprise material otherwise termed asphaltenes.
[d] Fractions 5–10 characterized by multifunctional moieties.
[e] Nonbasic N-, O-, and S-heterocyclics.
[f] Containg >10 wt % heterocyclics.

4. ash contents typically range from almost 0 to >3 wt%, depending on how efficiently the liquids have been separated from unreacted residues; particularly prominent components, all for the most part carryovers from catalysts, are Al, Fe, Si and Ti;
5. asphaltenes are mostly composed of acidic monofunctional moieties whose acidity stems from phenolic —OH and/or acidic N (as in pyrrole); molecular weights have been reported (Farcasiu *et al.*, 1976) to lie between 300 and 1000, but are questionable; and
6. hydrocarbons cover a wide range, but aromatic/hydroaromatic with 1–6 rings predominate; hydroaromatic species tend to be the most abundant moieties in liquids from catalytic processes with more substantial H-transfer, but alkyl groups are rarely longer than C_4.

Processing of Primary Coal Liquids

In preliminary trials that sought to assess their potential utility, coal liquids were feedstocks for various secondary petroleum processing techniques—in particular, hydrotreating (to remove hetero atoms), mild hydrogenation, hydrocracking, catalytic cracking, dealkylation, deasphalting, and reforming. These trials were also discontinued in the late 1970s, and results from them are therefore far from definitive. But they offer useful qualitative indications of probable behavior, and recent work, which implicitly tested the behavior of processed coal liquids against conventional crude oils and oil fractions (Tsonopoulos *et al.*, 1986), did, in fact, find substantial compliance.

Hydrotreating

Optimum temperatures and pressures for effective hydrotreating are 300–500°C and 10.5–17.5 MPa, respectively. However, since the gas make or the extent of hydrocracking (and consequent coke yields) increase steeply beyond

TABLE 10.4.13 Catalytic Cracking: Test Data[a,b]

	Midcontinent gas oil	As-distilled >345°C COED	Blend[c]	Hydrotreated >345°C COED
API gravity	26.1	11.2	22.4	15.3
Carbon	86.18	89.69		
Hydrogen	12.68	10.14	11.0	11.20
Sulfur	0.902	0.07	0.694	0.001
Oxygen	—	0.25	—	0.08
Nitrogen (ppm)	300	816	1050	181
Paraffins[d]	27.7	10.9	23.5	9.7
Monocycloparaffins	26.7	11.3	22.8	16.1
Polycycloparaffins	5.5	7.7	6.1	12.8
Monoaromatics	24.7	33.0	26.8	44.7
Diaromatics	8.7	13.7	10.0	9.5
Triaromatics	3.6	8.9	4.9	2.0
Tetraaromatics	1.7	9.6	3.6	3.5
Benzanthracene + 5 rings	1.4	4.9	2.3	1.7
Total polynuclear aromatics	15.4	37.1	20.8	16.7
ASTM distillation: Fraction	**Temperature (°C)**			
IBP	236	272		221
10%	319	374		346
30%	370	382		359
50%	416	403		373
70%	465	416		395
90%	552	466		444
95%		498		476

[a] Qader and Hill (1972).
[b] Analytical data in wt % unless otherwise stated.
[c] 25% as-distilled >345°C COED + 75% midcontinent gas oil.
[d] Paraffinic/cycloparaffinic parts of aromatic molecules included in aromatics.

TABLE 10.4.14 Reforming Tests: Properties of Feeds and Products[a]

	Feed #1	Product	Feed #2	Product
Temperature (°C)	513		520	
Pressure (MPa)	1.75		3.5	
H_2 generated (m^3/m^3)		206.5		174.5
C_1–C_3 (vol %)		7.5		6.2
C_4–C_5 (vol %)		5.2	0.4	5.8
C_6–C_9 (vol %)				
Paraffinic (vol %)	10.9	4.0	13.3	6.4
Naphthenic (vol %)	50.1	0.9	35.7	2.2
Aromatic (vol %)	28.7	73.4	33.4	70.2
C_{10+} (vol %)	10.3	6.0	17.2	6.6

[a] Peters (1976).

425°C, high temperatures must, if necessary, be neutralized by higher pressures. Favored catalysts are Co–Mo–AlO_3 and Ni–Mo–AlO_3 formulations—the latter preferred for removal of N; liquid loadings per kg catalyst can vary from 1.4 m^3 to as much as 3.9 m^3. Hydrotreating under these conditions has been reported to furnish good reformer feedstocks, gasoline blend cuts, jet fuels, and turbine fuels.

Hydrocracking

Because of their higher contents of heterocyclic and polynuclear aromatics, coal liquids are, as a rule, more difficult to hydrocrack than petroleum feedstocks. Typical conditions for reaction over hydrofining or dual-function catalysts are 350–500°C at 10.5–17.5 MPa, but processing SRC or similar heavy feedstocks requires a very porous catalyst support in order to accommodate ash and carbon accumulations that would otherwise quickly lower catalyst activity. Extant information indicates that mild hydrocracking of naphthas and gasolines from coal liquids provides good stock for further processing. Dealkylation can make them excellent sources of BTX and pure aromatics.

Catalytic Cracking

This processing mode is still relatively untested, but the few data at hand suggest that suitably prepared coal liquids can be cracked to a slate that compares well with a cracked midcontinent gas oil (see Table 10.4.13).

Reforming

For satisfactory reforming of coal-derived naphthas, preferred catalysts are Pt/AlO_3 formulations. Bimetallics, the more usual choice for reforming, have

TABLE 10.4.15 Steam Cracking Tests[a,b]

	H-Coal		COED
	Naphtha	Mid-distillate	Mid-distillate
Hydrogen	0.8	0.7	0.9
Methane	16.3	12.1	11.8
Ethylene	23.4	14.4	13.9
Propylene	8.5	5.9	6.5
Butadiene	3.0	1.5	1.3
Benzene	23.6	18.5	9.9
Fuel	24.4	46.9	55.8

[a] Peters (1975).
[b] Data in wt %.

very low tolerance for S and N, and would therefore demand severe prior hydrotreatment of the feed. Table 10.4.14 summarizes some data from tests with two coal-derived $<180°C$ naphthas. However, at 1.75 MPa catalysts aged more quickly than at 3.5 MPa—with aging evidenced by aromatics contents falling from 73 to ~55% over 100–150 h.

Steam Cracking

Extant information indicates that coal-derived oils would require hydrogenation before steam-cracking to petrochemical materials. Table 10.4.15, which compares an H-Coal naphtha, an H-Coal mid-distillate, and a heavier COED mid-distillate, illustrates this.

NOTES

[1] For example, since hydrocarbon reactivity increases with the number of C atoms in the molecule, light feeds such as CH_4 or C_3H_8 must be processed at substantially higher temperatures than naphthas; and because steam reforming becomes increasingly endothermic as temperatures rise, such feeds are not only penalized by a fivefold increase in energy expenditures between 650 and 850°C (Topsoe, 1966), but are also disadvantaged by diminishing heat values of their product gases, which fall by as much as 40% because higher temperatures favor gasification $\rightarrow CO + H_2$ over methanation.

[2] Now generally contracted to *syngas,* this designation reflects the unique versatility of $CO + H_2$ as a chemical building block.

[3] The other type of gasification process is *allothermal,* i.e., the necessary heat is supplied by *externally* generated heat carriers.

[4] In some gasifier systems, such as the Lurgi and Wellman reactors, gasification is preceded by thermal cracking. This generates tars that are separately recovered.

[5] Depending on personal preferences, this can be read as *synthetic* or *substitute* natural gas.
[6] The lower temperatures favor methanation over gasification.
[7] This catalyst was developed by BASF and can be used without prior removal of sulfur compounds from the raw syngas.
[8] A newer version of the Lurgi gasifier, likely to play a major role in future operations, is a slagging unit (Hebden *et al.*, 1964) in which the rotary grate is replaced by a swinging tap burner and slag tap. This allows molten ash to be intermittently withdrawn into a quench chamber before discharge into the bottom lock hopper. The higher temperatures needed for melting the ash are set by injecting a suitably proportioned O_2/steam mix through the tuyeres above the tap burner. The slagging unit is capable of 2–3 times greater output per unit cross-sectional area, and is also claimed to be more energy efficient.
[9] The Wellman reactor is also available as a single-stage (Wellman–Galusha) gasifier, which is widely used for generation of a producer gas.
[10] This positioning of the burner heads improves turbulence, which ensures more complete gasification. It also minimizes potential refractory problems because the highest temperatures are confined to the central core of the reactor.
[11] Coals of higher rank are generally not sufficiently reactive at these relatively low temperatures.
[12] South Africa's SASOL plants (see Section 2) show that large-scale coal gasification can be well served by current commercial reactors. Very much larger units are thought to require unacceptable output cuts whenever they need to be shut down for routine maintenance or repair.
[13] Such systems were thought to simplify plant operation and, by favoring CH_4 formation, to minimize gas shifting as well as obviate SNG compression prior to its injection into a pipeline.
[14] A survey at the beginning of the 1980s (Hebden and Stroud, 1981) identified more than 40 such second-generation systems—some little more than concepts, but others variously tested in process development units (PDUs) or larger pilot plants.
[15] This is a well-established process that calls for fluidizing char in a gas producer with air and steam at 1650°C/7–10.5 MPa and reacting the product gas with magnetite at 815°C/10.5 MPa. Spent gas is vented, and Fe_3O_4 is regenerated by contacting the reduced Fe/FeO with steam at 815°C/10.5 MPa.
[16] Since first reported in 1968, the flow diagrams and operating parameters of the process have been repeatedly modified, and various one- and two-vessel versions have been described. Formats proposed in the late 1970s have also contemplated high-pressure operation.
[17] This also transfers all sulfur by interaction of Na_2CO_3 with H_2S and other sulfur-bearing compounds.
[18] Similar problems seem to have beset the Rockgas process (Botts *et al.*, 1976), a variant of Kellogg's molten salt gasification.
[19] To date the ATGas process has only been demonstrated in a 60-cm i.d. bench reactor.
[20] An example of such redesign presents itself in reports of "environmentally friendly" commercial-scale coal gasification conducted by the British Gas Corporation with modified fixed-bed slagging Lurgi reactors (Cooke and Taylor, 1993). This showed that

1. sulfur in fuel gases destined for use in combined-cycle operations could be readily lowered by >99%, or in a syngas reduced by 99.95% to avoid downstream poisoning of methanation catalysts;
2. leachates from slags were well below acceptable limits in all but Fe, and trace element concentrations in the leachates were also far below legal thresholds; and
3. several viable alternatives are available for treatment of effluent waters: options tested included incineration, catalytic wet air oxidation, and biological treatment with reverse osmosis.

[21] Lower temperatures would also favor formation of methane, but that would only be beneficial if the syngas were destined for conversion into an SNG.

[22] In principle, any other abundant carbon source at hand—wood, paper, biomass—could also serve as a gasification feedstock. But without much prior preparation, these materials cannot provide acceptable carbon inventories in a gasifier, and would in most situations, be further disadvantaged by costs associated with accumulating adequate supplies at a commercial gasification facility. In contrast, the practicality of deploying coal-based gas is demonstrated by pre-1950 European practices as well as by technically more sophisticated and more recent operations. In South Africa, 6.5–7.5 MJ/m^3 producer gases from Wellman reactors are routinely used to fuel facilities for manufacture of glass, pottery, and steel. In Scotland, 6.7–9.3 MJ/m^3 gas from air/steam-blown Lurgi reactors has augmented other supplies of domestic gas. And in Germany, Lurgi-generated 5.2 MJ/m^3 producer gas was used to fuel a 170-MW combined-cycle prototype power plant, which was designed to provide scale-up data for future construction of a similar central generating plant.

[23] This designation identifies the companies (Socony and Mobil) who developed the catalyst. The technique by which methanol is transformed into gasoline is commonly referred to as the Mobil process.

[24] $>C_{10}$ aromatics are generally considered to be objectionable gasoline components because they possess relatively high melting points and consequently tend to cause carburetor icing in cold weather.

[25] It is sobering to recall that the first catalytic methanation of $CO + H_2$ over a nickel catalyst was achieved almost 100 years ago (Sabatier and Sonderens, 1902).

[26] Unless inhibited by appropriate choice of operating conditions and by catalysts that promote selective formation of higher alkanes, this reaction can have serious detrimental effects: it will quickly deactivate the catalyst by carbon deposition and plug the catalyst bed.

[27] In the late 1980s, these reportedly met well over 50% of the country's demand for liquid hydrocarbons.

[28] In 1913, Sir William Ramsay initiated preparatory work for a first u/g gasification test in England, but this was abandoned upon the outbreak of World War I and was never resumed—presumably because Ramsay's death in 1916 removed an active proponent of u/g gasification from the scene.

[29] Apparently unaware of Mendeleev's essays on the subject, Lenin had read of Sir William Siemens' proposals and enthusiastically supported their implementation—not because of potential economic and technical gains from u/g gasification, but because it would *"free human beings of the indignity of working underground"*!

[30] Between 1945 and 1955, such tests were conducted in the USA at Gorgas, AL (Elder *et al.*, 1951); in Britain at Newman Spinney, Derbyshire (Ministry of Fuel and Power, 1956); in Belgium at Bois-la-Dame (Institut Nationale de l'Industrie Charbonniere, 1952); in Italy at Banco-Casino (Loison and Venter, 1952); and under French auspices in Morocco at Djerada (Loison, 1952).

[31] Cleat is a highly regular fracture system, apparently caused by tectonic forces, in which smaller fissures (known as butt cleat) lie very nearly at 90° to the larger (major) fractures.

[32] Both are, of course, well-known techniques for enhancing oil and gas production from tight formations.

[33] All of these methods have been used—but all with varying degrees of success, and on the understanding that each has limitations that make choices between them dependent on site characteristics. Aside from its cost, directional drilling may thus be precluded by the radius of curvature which the drill bit requires for entry into the seam. Reverse combustion is slow and rarely advances at more than 1–3 m/d. Electrolinking cannot be used if the coal is not

sufficiently conducting. And hydrofracking can cause excessive vertical fracturing that would make it difficult to control gas leakages from the generator to the surface.

[34] For gasification at 800–950°C sustained by air injection, Soviet data indicate an optimum water influx rate of 1 m^3/t coal gasified. Heat values fall asymptotically from ~4.2 MJ/m^3 at that level to <2 MJ/m^3 when the water influx exceeds 5 m^3/t gasified.

[35] In the Canadian project (Berkowitz and Brown, 1977), excavation of the site 1 year after completion of the test showed complete restoration of the original water table, no measurable deterioration of groundwater quality, and no discernible deterioration of the biological quality of the agricultural soil at the surface—even though gasification was conducted at ~925°C in a shallow seam with less than 20-m overburden.

[36] H-transfer reactions began to be investigated in conjunction with detailed reappraisals of Germany's wartime hydrogenation practices. For the most part conducted by the U.S. Bureau of Mines, these suggested that the German hydrogenation specialists were unaware of the subtleties of H-transfer reactions—or were simply content with seeking to improve Bergius practices.

[37] As a rule, these were obtained from Bergius hydrogenation of coal and/or coal tars.

[38] One of these was a 50 t/d installation, operated by a subsidiary of Gulf Oil at Ft. Lewis, WA (Pastor *et al.*, 1976), and the other a 6 t/d pilot plant at Wilsonville, AL (Wolk *et al.*, 1976), operated for the Electric Power Research Institute (EPRI). Both installations came on stream in 1974 and used virtually identical process technology, but complemented each other by testing alternative ancillary techniques, in particular various options for filtration and SRC regeneration.

[39] This inference is underscored by a computer cluster analysis of 104 U.S. coals in terms of 15 coal characteristics (Yarzab *et al.*, 1980). The analysis discriminated among three populations—but found that useful predictive relationships between coal properties and percent conversion demanded a *different set of properties for each population.*

[40] Yield structure refers to the composition and physical properties of the product slate.

[41] However, because literature dealing with alternative liquefaction procedures often uses different criteria and/or terms when reporting liquefaction data, some caution is necessary when comparing seemingly similar processes. "Solubilization" is thus variously expressed as solubility in solvents as mild as benzene or toluene, or as aggressive as pyridine or cresol. What is designated as "primary coal liquids" ranges from high-molecular-weight semisolids, which would need extensive further hydrogenation or hydrocracking before being acceptable as refinery feedstocks, to light and medium oils, which could be refined without much additional upgrading. And conversion yields may refer exclusively to coal liquids or, in the extreme, include all by-product hydrocarbon gases, CO, CO_2, and water.

[42] The processing sequence of CFFC is very similar to several other procedures reported in the literature, but merits noting here because it replaces cumbersome filtration of the crude product stream with removal of suspended solid matter by an aliphatic *antisolvent*, in this case a 200–260°C kerosine cut. However, a number of U.S. patents (Hill, 1932; Pier, 1935; Schoenemann, 1936) make it evident that precipitation by antisolvents has been well known since the 1930s.

[43] As noted previously, this is proprietary Exxon technology that combines fluid coking with gasification of the residual coke.

[44] Additional hydrogen is only expected to be needed when processing subbituminous coals and lignites whose high oxygen contents consume H_2 in generation of H_2O.

[45] A detailed review of coal liquefaction (Alpert and Wolk, 1981) identified more than 12 major liquefaction systems in various stages of development in the United States, and a nearly equal number under study elsewhere (notably in Britain, Germany, and Japan). But although several were by the late 1970s deemed ready for industrial use, none has to date

actually come into use; and as the relevant literature citations (see Reference section) attest, very little *new* liquefaction-related material has been published since the end of the 1970s.

[46] Structural analyses of SRC fractions from two subbituminous coals (Whitehurst *et al.*, 1977) underscore the fact that seemingly very similar coals may possess dramatically different macromolecular structures (see Chapter 7)—and this suggests that optimal processing may demand closer chemical definition of liquefaction feedstocks than is now routinely called for.

[47] More severe conditions, which are usually associated with higher reaction temperatures and/or pressures, are sometimes necessitated by unacceptably high N and/or S contents of the feed coal.

REFERENCES

Allen, D. In *Fertilizer Science and Technology Series* (A. V. Slack and G. R. James, eds.), Vol. 2, 1973. New York: Dekker.

Alpert, S. B. and R. H. Wolk. *Chemistry of Coal Utilization,* 2nd Suppl. Vol. (M. A. Elliott, ed.), Chapter 28, 1981. New York: Wiley.

Anon. *Oil Gas J.* **71**(7), 36 (1973a).

Anon. *Oil and Gas J.* **15**, 32 (1973b).

Auer, W., E. Lorenz, and K. H. Grundler. *Proc. AlChemE 68th Natl. Mtg., Houston, TX* (1971).

Bendoraitis, R. B., A. V. Cabal, R. B. Callen, T. R. Stein, and S. E. Voltz. *EPRI Rept. No. 361-1, Phase 1* (1976).

Berkowitz, N., and R. A. S. Brown. *Bull. Can. Inst. Min. Metall.* **70**, 92 (1977).

Betts, A. G. Brit. Pat. No. 21, 674 (1909).

Boehm, F. G., and E. St. Denis. *Proc. AOSTRA Conf., Edmonton, Alberta* (1987).

Boehm, F. G., R. D. Caron, and D. K. Banerjee. *Energy Fuels* **3**, 116 (1989).

Botts, W. V., A. L. Kohl, and C. A. Trilling. *Proc. 11th Intersoc. Energy Conversion Eng. Conf., Lake Tahoe* (1976).

Brandenburg, C. F., R. P. Reed, R. M. Boys, D. A. Northrop and J. W. S. Jennings. *Proc. 50th Fall Mtg., AIME Soc. Pet. Eng., Dallas, TX* (1975).

Burk, E. H. Jr., and H. W. Kutta. *Proc. Coal Chem. Workshop, Session II, Stanford Research Inst.* (1976).

Capp, J. P., W. L. Rober, and D. W. Simon. *U.S. Bur. Mines Inf. Circ. 8193* (1963).

Chung, S. *Proc. Conf. Mat. Probl. & Res: Opportunities in Coal Conversion, Ohio State U., Columbus, OH* (1974).

Cooke, B. H., and M. R. Taylor. *Fuel* **72**, 305 (1993).

Cover, A. E., W. C. Schreiner, and G. I. Skapendas. *Chem. Eng. Progr.* **69**(3), 31 (1973).

Crynes, B. I. In *Chemistry of Coal Utilization,* 2nd Suppl. Vol. (M. A. Elliot, ed.), Chapter 29, 1981. New York: Wiley.

Davies, H. S., K. J. Humphries, D. Hebdden, and D. A. Percy. *J. Inst. Gas Engrs.* **7**, 707 (1967).

de Crombrugghe, O. *Ann. Mines Belges* **5**, 478 (1959).

Dirksen, H. R., and H. R. Linden. *Ind. Eng. Chem.* **52**, 584 (1960); *Inst. Gas Technol. Res. Bull. 31* (1963).

Eisenlohr, K. H., F. W. Moeller, and M. Dry. *ACS Adv. Chem. Ser.* **146** (1975).

Elder, J. L. In *Chemistry of Coal Utilization,* Suppl. Vol. (H. H. Lowry, ed.), p. 1023, 1963. New York: Wiley.

Elder, J. L., M. H. Fies, H. G. Graham, R. C. Montgomery, L. D. Schmidt and E. T. Wilkins. *U.S. Bur. Mines Rept. Invest. 4808* (1951).

Elgin, D. C. and H. R. Perks. *Proc. 6th Synth. Pipeline Gas Symp., Chicago, IL* (1974).

Farcasiu, M., T. O. Mitchell, and D. D. Whitehurst. *ACS Div. Fuel Chem., Prepr.* **21**(7), 11 (1976).
Farnsworth, J. F., H. F. Leonard, D. M. Mitsak, and R. Wintrell. *Kopper Co. Publ., Aug.* (1973).
Fieldner, A. C., and P. M. Ambrose. *U.S. Bur. Mines Info. Cir. 7417* (1947).
Finneran, J. A. *Oil Gas J.* **70**(29), 83 (1972).
Fischer, F., and H. Tropsch. *Chem. Ber.* **56**, 2428 (1923).
Fluor (Engineers and Contractors). SNG and low-S fuel oil from crude oils. *Company Report,* March (1973).
Forney, A. J., R. J. Denski, D. Bienstock, and J. H. Field. *U.S. Bur. Mines Rept. Invest. 6609* (1965).
Frank, M. E., M. B. Sherwin, D. B. Blum, and R. L. Mednick. *Proc. 8th Synth. Pipeline Gas Symp., Chicago, IL* (1976).
Frohning, C. D., and B. Cornils. *Hydrocarb. Process.,* Nov. (1978).
Furlong, L. E., E. Effron, L. W. Vernon, and E. L. Wilson. *Chem. Eng. Progr.* **72**(8), 69 (1976).
Gardner, N., E. Samuels, and K. Wilks. *ACS Adv. Chem. Ser.* **131**, 217 (1974).
Gibb, A. *Underground Gasification of Coal.* 1964. London: Pitman & Sons.
Given, P. H., W. Spackman, A. Davis, and R. J. Jenkins. *ACS Symp. Ser.* **139**, 3 (1980).
Glenn, R. A., and R. J. Grace. *Proc. 2nd. Synth. Pipeline Gas Symp., Pittsburgh, PA* (1978).
Gorin, E. In *Chemistry of Coal Utilization,* 2nd Suppl. Vol. (M. A. Elliott, ed.), p. 1845, 1981. New York: Wiley.
Gorin, E., H. E. Lebowitz, C. H. Rice, and R. J. Struck. *Proc. 8th World Petroleum Cong., Moscow* (1971).
Gregg, D. W., and D. U. Olness. *UCRL 52107, Lawrence Livermore Laboratory* (1976).
Hamilton, G. W. *Cost Eng.* **8**(7), 4 (1963).
Haynes, W. P., S. J. Gasior, and A. J. Forney. *ACS Adv. Chem. Ser.* **131**, 179 (1974).
Haynes, W. P., A. J. Forney, and H. W. Pennline. *ACS Div. Fuel Chem., Prepr.* **19**(3), 10 (1974b).
Hebden, D., A. G. Horsler, and J. A. Lacey. *Research Communication No. GC 112, Brit. Gas Council* (1964).
Hebden, D., and H. J. F. Stroud. In *Chemistry of Coal Utilization,* 2nd Suppl. Vol. (M. A. Elliott, ed.), Chapter 24, 1981. New York: Wiley.
Hill, W. H. U.S. Pat. No. 1,875,458 (1932).
Howard-Smith, I., and G. J. Werner. *Coal Conversion Technology,* 1976. New Jersey: Noyes Data Corp.
Institut National de l'Industrie Charbonniere. *U/g Gasification: Belgian Experience at Bois-la-Dame, Liège* (1952).
Ishiguro, T. *Hydrocarb. Process.* **47**, 87 (1968).
Jockel, H., and B. E. Triebskorn. *Hydrocarb. Process.* **52**, 93 (1973).
Jüntgen, H., J. Klein, K. Knoblauch, H.-J. Schroeter, and J. Schulze. In *Chemistry of Coal Utilization,* 2nd Suppl. Vol. (M. A. Elliott, ed.), Chapter 30, 1981. New York: Wiley.
Karnavos, J. A., P. J. LaRosa, and E. A. Pelczarski. *AlChemE Chem. Eng. Progr. Symp. Ser.* **70**, 245 (1974).
Kim, K. T., K. H. Kim, and J. O. Choi. *Fuel* **68**, 1343 (1989).
Kölbel, H., and K. Battacharyya. *Liebig's Ann. Chem.* **618**, 67 (1958).
Kölbel, H., and F. Engelhardt. *Brennst. Chem.* **33**, 13 (1952).
Kölbel, H., and E. Vorwerk. *Brennst. Chem.* **38**, 2 (1957).
Kotanigawa, T., S. K. Chakrabartty, and N. Berkowitz. *J. Fuel Process.* **5**, 179 (1981).
Kravtsov, A. V., V. Goncharov, S. I. Smol'yaninov, I. F. Bogdanov, and N. V. Lavrov. *Khim. Tverd. Topl.* **8**, 106 (1974).
Kreinin, E., and M. Revva. *UCRL Translation 1060, Lawrence Livermore Laboratory (1966).*
Kröger, C. *Erdöl Kohle* **9**, 441; 516; 620; 839 (1956).
Kröger, C., and G. Melhorn. *Brennst. Chem.* **19**, 257 (1938).
Kuhre, C. J., and J. A. Sykes, Jr. *IGT-SNG Symp. 1, Chicago, IL, March* (1973).

Lee, B. S. *Proc. 7th Synth. Pipeline Gas Symp., Chicago, IL* (1975).
Lenin, V. I. *Pravda* **19** (April 21, 1913).
Loison, R. *Proc. 1st Internat. Congr. u/g Gassification, Birmingham AL; CERCHAR, Paris* (1952).
Loison, R., and J. Venter. *CERCHAR Rept.* (Paris) and *INICHAR Rept.* (Liège) (1952).
Lurgi Express. *Info. Brochure No. 1018/10.75* (1975).
Maekawa, Y., S. K. Chakrabartty, and N. Berkowitz, *Proc. 5th Can. Symp. Catal., Calgary, Alberta* (1977).
McCoy, D. C., G. P. Curran, and J. D. Sudbury. *Proc. 8th Synth. Pipeline Gas Symp., Chicago, IL* (1976).
McRae, M. K. *Canad. Energy Res. Inst. Study #27* (1988).
Meisel, S. L., J. P. McCullough, C. H. Lechthaler, and P. B. Weisz. *Chem. Tech.* **6**(2), 86 (1976).
Mendeleev, D. I. *Severny Vestn. St. Petersburg* **8**(2), 27; **9**(2), 1; **10**(2), 1; **11**(2), 1; **12**(2), 1 (1888).
Ministry of Fuel and Power. *U/g Coal Gasification in the UK,* 1956. London: HMSO.
Miura, K., K. Hashimoto, and P. L. Silveston. *Fuel* **68**, 1461 (1989).
Mobil Res. Dev. Co. *Ann. Rept.* No. RP 410-1, *EPRI* A252, Feb. (1976).
Moeller, F. W., H. Roberts, and B. Britz. *Hydrocarb. Process.* **69**, April (1974).
Moreno-Castilla, C., F. Carrasco-Marin, and J. Rivera-Utrilla. *Fuel* **69**, 354 (1990).
Mukherjee, D. K., and P. D. Choudhury. *Fuel* **55**, 4 (1976).
Nadkami, R. M., C. Bliss, and W. J. Watson. *Chem. Technol.* **4**(4), 230 (1974).
Neavel, R. C. *Fuel* **55**, 237 (1976).
Odell, W. *U.S. Bur. Mines Info. Circ. 7415* (1947).
Pastor, G. R., D. J. Keetley, and J. D. Naylor. *Chem. Eng. Progr.* **72**(8), 27 (1976).
Pelofsky, A. H. (ed.). *Heavy Oil Gasification,* 1977. New York: Dekker.
Peters, B. C. *U.S. ERDA Tech. Progr. Rept. No. FE-1534-T-5* (1975).
Peters, B. C. *U.S. ERDA Tech. Progr. Rept. No. FE-1534-33* (1976).
Pfirrmann, T. W. U.S. Pat. No. 2,167,250 (1939).
Phinney, J. A. *Proc. Natl. Mtg. AlChemE, Pittsburgh, PA* (1974).
Pichler, H. *Brennst. Chem.* **19**, 226 (1938).
Pichler, H., and B. Firnhaber. *Brennst. Chem.* **44**, 33 (1963).
Pichler, H., and K. H. Ziesecke *Brennst. Chem.* **30**, 13, 60, 81 (1949).
Pier, M. U.S. Pat. No. 1,933,226 (1935).
Qader, S. A., and G. R. Hill. *ACS Div. Fuel Chem. Prepr.* **16**(2), 93 (1972).
Ricketts, T. S. *J. Inst. Fuel* **34**, 177 (1961).
Roelen, O, H. Heckel, and F. Martin. German Pat. No. 902,851 (1943).
Ross, D. S., and J. Y. Low. *SRI Repts. #1 and 2 under ERDA Contract E (49-18)-2202,* 1976. Washington, D.C.: U.S. Dept. Interior.
Rostrup-Nielsen, J. R. *J. Catal.* **31**, 173 (1973).
Sabatier, P., and B. Sonderens. *C.R. Acad. Sci.* **134**, 514 (1902).
Salinas-Martinez de Lecea, C., M. Almela-Alarcon, and A. Linares-Solano. *Fuel* **69**, 21 (1990).
Schlesinger, M. D., J. J. Demester, and M. Greyson. *Ind. Eng. Chem.* **48**, 68 (1956).
Schlinger, W. G. *California Industrial Associates Conf., April* (1967).
Schmid, B. K., and D. M. Jackson. *Proc. 3rd Ann. Conf. on Coal Gasification & Liquefaction, Univ. Pittsburgh, PA* (1976).
Schoenemann, K. U.S. Pat. No. 2,060,447 (1936).
Schora, F. C., B. S. Lee, and J. Huebler. *Proc. 12th World Gas Conf., Nice, France* (1973).
Schrider, L. A., J. W. Jennings, C. F. Brandenburg, and D. D. Fischer. *Proc. 49th Fall Mtg., AIME Soc. Pet. Eng., Houston TX* (1974).
Schultz, J. F., F. S. Karn, and R. B. Anderson. *U.S. Bur. Mines Rept. Invest. 6974* (1967).
Sherwin, M. B., M. E. Frank, and D. G. Blum. *Proc. 5th Synth. Pipeline Gas Symp., Chicago, IL* (1973).
Siemens, William. *Trans. Chem. Soc. (London)* **21**, 279 (1868).

Simone, A. A. *Combustion Magazine,* May (1976).
Spiro, C. L., D. W. McKee, P. G. Kosky, and E. J. Lamby. *Fuel* **62**, 180 (1983).
Stephens, D. R., F. O. Beane, and R. W. Hill. *Proc. 2nd Ann. u/g Coal Gasif. Symp., Morgantown, WV* (1976).
Sze, M. C., and G. J. Snell. U.S. Pat. Nos. 3,852,182 and 3,856,675 (1974); U.S. Pat. Nos. 3,932,266 and 3,974,073 (1976).
Tarrer, A. R., J. A. Guin, W. S. Pitts, J. P. Henley, J. W. Prather, and G. A. Styles. *ACS Div. Fuel Chem., Prepr.* **21**(5), 59 (1976).
Thornton, D. P., D. J. Ward, and R. A. Erickson. *Hydrocarb. Process.* **51**, 81 (1972).
Topsoe, H. *J. Inst. Gas Engrs.* **6**, 401 (1966).
Tsonopoulos, C., J. L. Heidman, and S-C. Hwang. *Thermodynamic and Transport Properties of Coal Liquids,* Exxon Monograph, 1986. New York: Wiley & Sons.
Uhde, F. French Pat. No. 800,920 (1936).
UOP (Universal Oil Products). *The Environmental Fuels Processing Facility, Company Report,* March (1973).
U.S. Dept. of Commerce. *NTIS Rept. PB-234 203* (1974).
U.S. Dept. of Energy. *Clean Coal Technol. Progr. Rept. DOE/FE-0092* (1987).
Voodg, J., and J. Tielrooy. *Hydrocarb. Process.* **46**, 115 (1967).
Wensel, W. *Angew. Chem. B* **21**, 225 (1948).
Whitehurst, D. D., M. Farcasiu, T. O. Mitchell, and J. J. Dickert. *EPRI-AF 480, Res. Proj. 410-1 Final Rept.* (1977).
Wilson, W. G., L. J. Sealock, Jr., F. C. Hoodmaker, R. W. Hoffman, and D. L. Stinson. *ACS Adv. Chem. Ser.* **131**, 203 (1974).
Wolk, R., N. Stewart, and S. Alpert. *EPRI J.* **1**(4), 12 (1976).
Yarzab, R. F., P. H. Given, A. Davis, and W. Spackman. *Fuel* **59**, 81 (1980).
Zielke, C. W., W. A. Rosenhoover, and E. Gorin. *ACS Div. Fuel Chem., Prepr.* **19**(2), 306, (1976); *ACS Adv. Chem. Ser.* **151**, 153, (1976).

CHAPTER 11

Environmental Aspects

1. THE FOCI OF PUBLIC CONCERN

Qualitatively viewed, the production, processing, and end uses of fossil hydrocarbons generate substantially similar waste streams, and remedial measures that prevent or remedy their societal and environmental ill effects differ primarily in whether they center on solid, liquid, or gaseous wastes. Only in a quantitative sense do each of the fossil hydrocarbons generate characteristic wastes, and it is therefore not surprising that public perceptions of the environmental penalties incurred by development of fossil hydrocarbon resources focus on four or five broad topics:

1. Deteriorating air quality as reflected in urban air pollution, in atmospheric warming by so-called "greenhouse gases" such as CO_2 and CH_4, and in putative dangerous thinning of the stratospheric ozone layer by, among others, chlorofluorocarbons
2. Ecological damage to soils, animal habitats, and aquifers
3. Unacceptable acidification of lakes and forests by precipitation of SO_x and NO_x in acid rains
4. Groundwater contamination by toxic hydrocarbon wastes
5. Serious environmental damage by spillages of petroleum hydrocarbons

These concerns are addressed in the following pages by considering the nature of the major challenges to ecological integrity from development of fossil hydrocarbon resources, and by outlining the responses—i.e., the means by which environmental ill effects of such development can be prevented or, if unavoidable, acceptably remedied.

2. THE CHALLENGES TO ECOLOGICAL INTEGRITY

For practical purposes, fossil hydrocarbon wastes—unwanted by-products that could damage the environment and the society it sustains—fall naturally into three classes that reflect their physical states.

Solid Wastes

Major sources of solid wastes are exploration programs—especially exploration in virgin forested areas; mining (including recovery of bitumen from surface-mined oil sands [1]); and coal preparation. But if disposed with some care and, in particular, protected against leaching by water, the massive inorganic wastes generated by these activities pose little environmental hazard. Because they are for the most part composed of previously weathered mineral matter, further mobilization of potentially harmful trace elements is unlikely to proceed at rates that cannot be accommodated in the natural environment without harm. Substantial damage need therefore only be anticipated where freshly exposed pyrite can oxidize to Fe^{2+} and acid runoff generated by the resultant SO_2 and SO_4^{2-} is not fully contained and neutalized before release.

Particulate matter can prove more problematical. Unless discharged in quantities that cause siltation of natural watercourses or seriously impair local air and water quality, *organic* particulates—mainly coal dusts raised during open transport or storage—may be aesthetically objectionable, but constitute no significant health hazard [2] and pose no appreciable environmental threat. They may indeed in some circumstances prove beneficial [3].

More serious potential risks are, however, posed by *inorganic* particulates—in particular, by flyash generated in combustion systems and gasifiers operated in suspension-firing modes. The bulk of such material is made up of partly fused oxides of Al, Ca, Fe, and Si and primarily threatens the environment through the enormous volumes that, if not captured before venting, would be discharged into the atmosphere from large coal-fired power stations [4]; the small, but significant, proportions of *respirable* (<1-μm) particles in flyash ($<0.5\%$) would only pose danger to health if inhaled over long periods. But there are persuasive indications that some components of flyash could slowly hydrolyze when open to surface waters, and then contaminate groundwaters [5], and serious public health hazards could also develop if toxic trace elements entered a biological cycle. How the oxides of such elements are transformed into more dangerous (mostly organic) entities, and the ecological pathways which they then follow, is often still quite unclear [5]. Yet, aside from the well-known toxicity of arsenic and lead, such elements as cadmium, fluorine, and selenium have been unequivocally identified as potent plant poisons, and the lethal effects of organic forms of mercury—notably methyl mercury, $(CH_3)_2Hg$, which seems to develop from previously accumulated *inorganic* mercury in aquatic species—have been dramatically demonstrated by the so-called "Minemata disease" [6].

Table 11.2.1 provides an indication of the magnitude of these potential problems by juxtaposing the mobilization of elements from flyash against mobilization of the same elements by natural weathering of exposed rocks.

TABLE 11.2.1 Potential Mobilization of Trace Elements by U.S. Coal Combustion (10^3 tonnes/year)[a,b]

Element	*a*	*b*	*c*
Aluminum	3,640	64,000	5.7
Arsenic	1.72	8	21
Barium	22.7	480	4.7
Bromine	0.1	34	0.3
Cadmium	0.16	0.6	27
Cesium	0.38	5.4	7
Calcium	1,510	34,000	4.4
Chromium	6.2	90	6.9
Cobalt	1	7.3	13.7
Copper	2.9	28	10
Iron	3,780	35,000	10.8
Lead	1.7	13	13
Magnesium	420	11,000	3.8
Manganese	11.7	770	1.5
Mercury	0.005	0.12	4.1
Molybdenum	3.1	27	11.5
Nickel	5.6	36	15.6
Potassium	536	16,000	3.4
Rubidium	5.3	90	5.9
Scandium	0.77	6.3	12.2
Selenium	0.66	0.5	132
Silicon	8,020	30,500	2.6
Sodium	243	15,000	1.6
Strontium	8	370	2.2
Thorium	0.73	11	6.6
Tin	0.17	5	3.4
Titanium	176	4,500	3.9
Uranium	0.76	0.95	80
Vanadium	9.9	90	11
Zinc	26.8	75	36

[a] Klein *et al.* (1975), [b] van Hook and Shults (1977). *a*, discharged in slags and fly ash; *b*, mobilized by weathering; $c = (a/b)100$.

Liquid Wastes

Toxic liquid wastes that pose potentially serious ecological and public health hazards accrue from a wide variety of operations related to production, preparation, and processing of petroleum hydrocarbons, as well as from production and processing of coal.

Massive liquid effluents are, as a rule, aqueous streams that contain environmentally unacceptable concentrations of dissolved or suspended organic and inorganic matter. Major concerns center, in particular, on the disposition of formation fluids brought up with crude oil and gas; on acid mine drainage and similar runoff from coal stockpiles and waste dumps; on cooling waters from power plants; and on the compositionally still more complex effluents from oil refineries, petrochemical plants, coke ovens, and high-temperature coal conversion operations.

Formation fluids produced with natural gas and crude oils are invariably salty and, if carelessly disposed, would seriously damage and eventually destroy the biological activity of soils [7], and similar concerns have been voiced about contaminated waters accruing from steam-stimulated enhanced production of heavy oil and u/g extraction of hydrocarbons from bituminous sands [8].

Accidental spillages of crude oil and leakages of petroleum hydrocarbons (Wilson *et al.*, 1986) can also cause massive long-term environmental damage. Aside from several widely reported mishaps at sea [9], petroleum-based hydrocarbons have been identified in unconfined aquifers that carry potable water [10]. But detectable contamination, which commonly manifests itself in deterioration of well-water quality, develops very slowly over many years, and when eventually noted, it is more often than not extremely difficult, if not impossible, to remedy.

Acid waste waters that, as noted earlier, form by oxidation of FeS_x and subsequent hydrolysis of the resultant iron sulfates in percolating waters, as in

$$3\,FeS + 6\,O_2 \rightarrow FeO + Fe_2O_3 + 3\,SO_2$$
$$H_2O + SO_2 \rightarrow H_2SO_3;\ 2\,H_2SO_3 + O_2 \rightarrow H_2SO_4$$
$$FeS + 2\,O_2 \rightarrow FeSO_4;\ FeSO_4 + H_2O \rightarrow H_2SO_4 + FeO,$$

almost always also contain Al, Ca, Fe, Mg, SO_4^{2-}, and a broad spectrum of trace elements, and the damage such streams can inflict could therefore be even more serious than damage caused by acid rain [11]. However, volumes of acid runoff depend on coal characteristics, and the distances such runoff can traverse before it is rendered sustantially harmless by interaction with the soil depend on drainage features, topography, and rainfall patterns (Nichols, 1974), as well as on the nature and composition of the soils across which it migrates.

Coals with very low pyrite concentrations pose no hazard from acid runoff because *organic* sulfur is not subject to atmospheric oxidation. But drainage waters from low-rank coals or from the ash of such coals can sometimes pose problems from excessive *alkalinity* [12].

Liquid effluents from refineries, petrochemical plants, coke ovens, and high-temperature coal conversion facilities are more complex. They usually contain dissolved or colloidally dispersed aliphatic and aromatic hydrocarbons, tar components, and processing chemicals, as well as dissolved and suspended inorganic matter, and such streams can pose major hazards. Phenolics, polynuclear aromatic hydrocarbons, and nitrogen heterocycles are ecologically harm-

TABLE 11.2.2 Carcinogens in Coal Gasification and Liquefaction Wastes[a]

Class	Compound
Amines	Diethylamines
	Methylethylamines
Heterocycles	Pyridines
	Pyrroles
Hydrocarbons	Benzene
Phenols	Cresols
	Alkyl cresols
Polynuclear compds.	Anthracene
	Benzo[*a*]pyrene[b]
	Chrysene
	Benzo[*a*]anthracene
	Benzo[*a*]anthrone
	Dibenzo[*a,l*]pyrene
	Dibenzo[*a,n*]pyrene
	Dibenzo[*a,i*]pyrene
	Indeno[1,2,3-*c,d*]pyrene
	Benzoacridine
Trace elements	Arsenic[b]
	Beryllium
	Cadmium
	Nickel
Organometallics	Nickel carbonyl

[a] Bridbord (1976).
[b] Designated as potential mutagen or teratogen by U.S. Natl. Inst. Occupational Safety and Health.

ful and can in some forms also represent serious health risks because they are, or are suspected of being, carcinogenic and/or mutagenic.

In practice, environmental statutes require most liquid effluents to be contained on site and neutralized. But potentially hazardous components are sometimes accidentally released as fugitive vapors that condense on cooling to ambient air temperatures; leakages of tar vapors from HT coke ovens, which could threaten health by skin contact and/or inhalation of fugitive aerosols (Sexton, 1960; Redmond *et al.*, 1972; Mazumdar *et al.*, 1977) are cases in point. Some compounds deemed to be particularly hazardous are listed in Table 11.2.2.

Definitive risk assessments are, unfortunately, very difficult because of a prevailing lack of reliable tolerance limits or threshold values that could serve as data bases for formulating standards. Estimates of risk are, in fact, usually developed from perceived "previous experience," from episodal or isolated epidemiological studies, or from definition of process wastes *expected* to issue from projected conversion plants [13]—and such estimates rarely detail, let alone justify, the assumptions on which they are based.

Nevertheless, waste stream compositions can often serve to pinpoint where problems are *likely* to be encountered. For example, experience of high-temperature coal operations makes evident

1. that health risks lie mostly in exposure to carcinogens, and that gasification would consequently pose less danger than liquefaction;
2. that extensively alkyl-substituted tars, generated below 450°C, are likely to prove fairly innocuous, and are certainly much less hazardous than the highly aromatized tars generated at >450–500°C; and
3. that aerosols containing carcinogens constitute greater threats to occupational and public health than accidental releases of condensible, and therefore much less mobile, vapors

But in addition to carcinogenic and/or mutagenic waste components originating in some HT operations, such processing frequently also generates aqueous waste streams contaminated with harmful inorganic components. An example presents itself in the particularly noxious so-called "weak ammonia liquors" that accrue from coke oven gas scrubbers and that, if improperly disposed, can effectively destroy aquatic life [14].

Gaseous Emissions

Major stationary sources of noxious gases [14] are natural-gas processing plants (H_2S, SO_2); oil refineries (H_2S, SO_2, HCs); thermal power plants (CO_2, CO, SO_2/SO_3, NO_x); and high-temperature coal-processing facilities (SO_2, NH_3, H_2S, HCs). However, potential emission levels are process dependent and often determined by *which* fossil hydrocarbons are being produced. In jurisdictions that host large-scale facilities for natural-gas production, H_2S and SO_2 emissions from gas-processing plants could thus be far greater than the combined volumes generated by other operations. And although volumes of CO_2 per GJ from coal-fired power plants inevitably exceed those from oil-fired facilities (and could be further lowered by moving to natural gas), substitutions can carry unacceptable environmental as well as economic costs: Oil-firing is liable to emit appreciably higher levels of unburned hydrocarbons, and the higher flame temperatures associated with combustion of natural gas would enhance NO_x generation by interaction of atmospheric N_2 and O_2 [16].

But concerns over CO_2 emissions as major causes of global warming ought not to mask the more immediate societal dangers posed by H_2S, SO_2, and NO_x through acid rain [17] and smog formation [18].

TABLE 11.2.3 Toxicity of Various Gases[a]

	s.g.	Threshhold limit[b]	Hazardous limit[c]	Lethal conc.[d]
HCN	0.94	10 ppm	150 ppm/h	300 ppm
H_2S	1.18	10 ppm	250 ppm/h	600 ppm
SO_2	2.21	5 ppm		1000 ppm
CO[e]	0.97	50 ppm	400 ppm/h	1000 ppm
CO_2	1.52	5000 ppm	5%	10%

[a] Longley (1982).
[b] Concentration at which repeated exposure can be tolerated.
[c] Concentration that may cause serious ill effects and death.
[d] Concentration at which brief exposure will cause death.
[e] Releatively brief inhalation can cause anoxia of the brain and result in severe brain damage.

The hazards posed by H_2S are emphasized by comparing its toxicity with the adverse health impacts of SO_2 and other gases associated with natural gas processing (see Table 11.2.3) and are quantified by its physiological effects (see Table 11.2.4). Other material problems arise from facile chemical interaction of H_2S with Fe, which causes embrittlement and can thereby lead to serious

TABLE 11.2.4 Physiological Effects of H_2S[a]

0.000013 vol % (0.13 ppm):
Unpleasant odor, which becomes very noticeable at ~5 ppm; increasing concentrations beyond this limit are accompanied by progressive loss of sense of smell

0.01 vol % (100 ppm):
Coughing, eye irritation, loss of sense of smell after 3–15 minutes; altered respiration, eye pain, and drowsiness after 15–30 minutes; throat irritation after 60 minutes; prolonged exposure increases severity of symptoms

0.02 vol % (200 ppm):
Rapid loss of smell; burning sensation in eyes and throat

0.05 vol % (500 ppm):
Dizziness; loss of sense of balance and reasoning powers; after a few minutes, breathing problems demand prompt artificial resuscitation

0.07 vol % (700 ppm):
Rapid loss of consciousness; breathing will stop within a few minutes

>0.1 vol % (>1000 ppm):
Immediate loss of consciousness; permanent brain damage or death unless victim is immediately artificially resuscitated

[a] American Petroleum Institute (1981).

equipment failure. (An equally facile interaction of SO_2/SO_3 with $CaCO_3$, noted in [17], has already seriously damaged many, often historic, buildings.)

NO_x, on the one hand essential for photo-oxidation of hydrocarbons in the upper atmosphere, will, on the other, promote smog formation as well as visually impair air quality by imparting a light-brown coloration to it at concentrations as low as 0.25 ppm, and NO_2 per se is distinctly toxic: deeply inhaled, it can cause serious bronchial and respiratory damage, and *continuous* exposure to concentrations as low as 1 ppm is considered a cause for some concern.

3. THE RESPONSES: LEGISLATION AND NEW TECHNOLOGY

Increasingly articulated public concerns over environmental depredations caused by resource development have, since the late 1960s, evoked a series of political and technological responses that have gone far to mitigate, and in several instances virtually eliminate, unacceptable impacts of production, processing, and use of fossil hydrocarbons.

The political responses are enshrined in legislation that, among other environmental statues,

1. makes all activities relating to exploration, mining, processing, and use of fossil hydrocarbons subject to permits and control by government agencies with responsibility for environmental protection
2. demands on-site containment, and appropriate remedial treatment, of all noxious solid and liquid wastes before their final release
3. establishes limits on emission of noxious and/or potentially harmful solids, liquids, and gases, and
4. requires satisfactory remediation and restoration of all surface lands disturbed by permitted activity

Compliance with these ordinances is sought by regulations that detail the conditions under which development of fossil hydrocarbon resources is allowable in defined areas, and that specify penalties for noncompliance. Periodic reviews of environmental protection measures endeavor to ensure relevance of these measures in light of experience, better information, and/or availability of improved remedial technology.

However, paralleling legislative action, and directly responding to it, are measures taken by industry, which sought to introduce improved procedures for developing and using fossil hydrocarbons, and invoked some innovative chemistry to prevent or remedy damage from process wastes.

TABLE 11.3.1 Collection Efficiencies for Different Firing Modes[a]

Firing mode	Emision limits (mg/m³)			
	50	100	200	500
Pulverized coal	99.37–99.75	98.75–99.50	97.50–99.00	93.75–97.50
Spreader stoker	99.00–99.50	95.00–98.00	90.00–96.00	75.00–90.00
Grate stoker	95.00–98.30	90.00–96.70	80.00–93.33	50.00–83.33
Cyclone furnace	90.00–96.67	80.00–93.33	60.00–86.67	0–66.67

[a] Klingspor and Vernon (1988).

Disposal of Solid Wastes

With rare exceptions that may require special measures, mineral matter-rich discards from preparation plants and bottom ash [19] from coal-fired combustion systems can be safely disposed in mined-out pits or lagoons where suspended solid matter can settle out before being recovered for disposal. But in either case, *safe* disposal demands careful selection of substantially impervious sites that preclude groundwater contamination by seepage or leaching during long-term containment; and where wastes are ponded, the clarified waters can only be recycled or returned to off-site natural water bodies if free of environmentally unacceptable suspended and/or dissolved matter.

Finely dispersed solids (particulates) can be efficiently captured in collection systems that include wet scrubbers, fabric filters, cyclones, and electrostatic precipitators. All these devices have long performance histories, but each has characteristics that tend to make choices among them contingent on the type and quantity of particulate matter as well as on regional or national ambient-air quality standards (Klingspor and Vernon, 1988). Table 11.3.1 illustrates this with collection efficiencies that different combustion modes must meet in order to comply with statutory emission limits [20]. The efficiency *ranges* in this table reflect generation of particulates in combustion regimes that differ in such details as flame temperature and particle velocity.

Preferred wet scrubbers for particle capture are venturi units (Fig. 11.3.1) whose performance is primarily determined by the pressure drop across the venturi (U.S. Environmental Protection Agency, 1982). Collection efficiencies and operating costs increase therefore fairly rapidly with power input.

Fabric filters or baghouses tend to be disadvantaged by the relatively high pressure drops across them, but are nevertheless acceptable alternatives to electrostatic precipitators (see below). They are usually fabricated of chemically inert fiberglass or Teflon and operated at >150°C to avoid moisture condensation and attendant clogging. Performance depends on the gas flow rate as well as on the pressure drop (which is deemed economically acceptable if <1–1.5

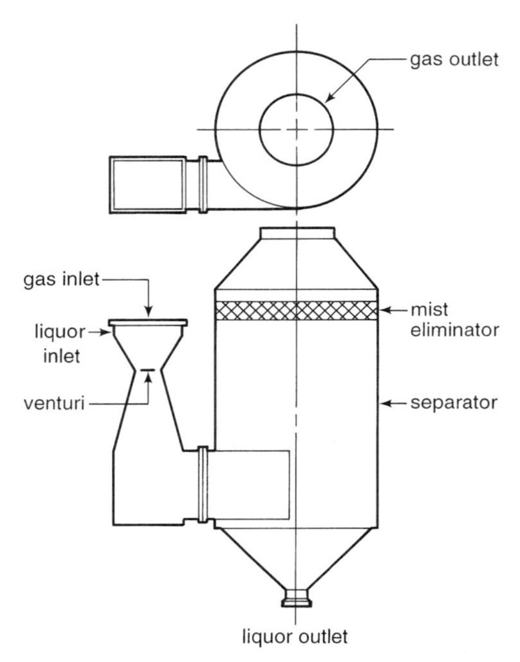

FIGURE 11.3.1 Schematic of a modern venturi scrubber.

kPa). Suitably sized filter bags can capture <1-μm particles with 99.7–99.9% efficiency, and accumulated solid matter can be removed for disposal by reversing the air flow or shaking and deflating the filter bags (Donovan, 1985).

Inertial dust collectors, or cyclones, can in favorable circumstances capture up to 90% of the particulate matter passed through them, but are limited by the fact that collection efficiencies depend on the number of turns of the carrier fluid in the unit. High efficiencies can therefore only be attained with relatively small (<0.75-m diameter) units, and although the overall performance can be improved by using three or more cyclones in series, collection efficiencies fall even then quite dramatically if the particles are smaller than 5 μm. Cyclones are consequently only used for primary particle capture, and are for that purpose deployed ahead of more competent equipment—e.g., in a fluidized-bed combustion system, ahead of a wet scrubber.

Where, as in suspension-fired combustion devices, very large volumes of particulates are generated, the preferred capture system is, for the time being, an electrostatic precipitator. This device consists of negatively charged wire or plate electrodes between positively charged collector plates, and is usually operated with a 30–75 kV potential difference across the positive and negative electrodes. The resultant corona discharge ionizes and thereby electrically charges particulates in the flue gas, and solid matter consequently moves transversely to the collector plates, from which they are later transferred to a disposal hopper by periodically vibrating (or "rapping") the plates. The construction and operating modes of electrostatic precipitators has been described in several publications (Gooch and Francis, 1975; Bump, 1977; Walker, 1977) that note that the size of a unit for a particular installation can, as a first approximation, be determined from the Deutsch equation (Deutsch, 1922):

$$E = 1 - \exp(-WA/V),$$

where E is a dimensionless fractional collection efficiency, A the total collector area (m^2), V the gas flow rate ($m^3\ s^{-1}$), and W the particle migration velocity (m/s). Better size estimates can be obtained by using Matts and Ohnfeldt's (1973) modification of the Deutsch equation, or calculating them from Potter's (1978) formulation, which takes into account such other variables as the applied voltage and apparent particle sizes.

Electrostatic precipitators can capture >99.5% of emitted particulate matter. But the fractional collection efficiency falls sharply when the dust resistivity (ρ) exceeds 10^{11} Ω-cm or is less than 10^4 Ω-cm (McDonald and Dean, 1982). And because ρ is strongly affected by temperatures as well as by the composition of the particles and varies inversely with the sulfur content of the fuel, efficient capture of flyash from *low*-sulfur coal requires operation of the precipitator at 300–450°C rather than at the more usual 125–175°C [21].

Remediation and Disposal of Liquid Wastes

Surface runoff and aqueous waste streams from mining, processing, and large-scale industrial use of fossil hydrocarbons are generally diverted to on-site ponds where suspended solids can settle out and the clarified waters can be neutralized (usually with lime) before being discharged. Unacceptably high concentrations of Ca^{2+}, Cu^{2+}, Fe^{2+}, and other ionic species deemed to be environmentally harmful are lowered to statutorily required levels by precipitation, and the resultant sludges are then dewatered and disposed in safe burial sites. The quality of water is commonly assessed by its hardness, a property ascribed (Brown *et al.*, 1970) to the presence of alkaline earths and mainly associated with Ca^{2+} and Mg^{2+}, but some national standards use different scales for measuring it (see Table 11.3.2).

Environmentally satisfactory action with respect to other potentially hazardous streams is, however, generally much more problematical than acceptable reclamation of waste waters, and is usually determined by what, how much, and where action needs to be taken. So, in particular, in the case of accidental

TABLE 11.3.2 Water Hardness: Definition of °H[a]

International	mval[b]	1 meq/liter
England	1 °H	10 mg $CaCO_3$/liter
		0.8 German °H
France	1 °H	10 mg $CaCO_3$/liter
		0.7 English °H
		0.56 German °H
Germany	1 °H	10 mg CaO (≡ 7.14 mg Ca^{2+})/liter
		1.25 English °H
		1.79 French °H
USA	1 °H	1 mg $CaCO_3$/liter
		0.056 German °H

[a] Angino (1983).
[b] 1 mval = 1 meq/liter of water; proposed as the international unit of hardness.

oil spills and hydrocarbon leakages (even when arising from small mishaps at wells or refineries; Table 11.3.3), where remediation depends on the type of oil that has been spilled, and whether on land or water (see Petroleum Industry Training Service, 1977).

Operations on land center mainly on confining the spill, where necessary by construction of dikes or trenches, before seeking to recover the oil by skimming and/or pumping. At wells and sumps, use is also made of specific "source removal" methods, which include removal of hydrocarbon material by air pumping. Hydrocarbon plumes on aquifers are often amenable to "bioreстoration" (Wilson *et al.*, 1986)—in essence, providing for biodegradation through aeration [22]. And similar microbial degradation of oil to harmless matter is on occasion resorted to when recovery is impractical. As a last resort, but clearly much less attractive, unrecoverable oil is burned off: this is, in fact, routinely done, albeit in incinerators rather than in the open, when the oil is

TABLE 11.3.3 Alberta Oil Spills, 1975–1977[a]

Spilled volume		
m^3	(bbls)	No. of spills
0.2–16	(1–100)	1650
16–80	(101–500)	332
80–800	(501–5000)	85
>800	(>5000)	8

[a] 45% at pipeline facilities, 25% at production facilities, 20% at wells, 10% due to other causes.

heavily contaminated with water and debris, and therefore unsuitable for reprocessing.

However, oil spills on water are often more serious [23] and more difficult to counter. Containment and recovery requires deployment of a variety of booms, skimmers, and/or sorbents [24], but where these are impractical or ineffective, recourse must be made to means of "last resort" that are, as a rule, subject to prior governmental approval. These include the use of (i) surfactants that are substantially nontoxic to aquatic life, concentrate the oil slick, and inhibit evaporation; (ii) fine-grained solids such as gypsum, chalk, or fine sand that adhere to the oil and cause it to sink [25]; and (iii) dispersants that reduce the oil/water interfacial tension, spread the slick over a larger surface of the water, and so accelerate oxidation and biodegradation of the oil [26].

Equally difficult, but different, problems are posed by liquid wastes from HT processing, especially HT coal processing. The usual practice involves extraction of the waste with benzene, tricresyl phosphate, diisopropyl ether, or a light oil—any one of which can remove up to 97–99% of dissolved phenolic matter (see Bond *et al.*, 1974)—and subsequent biological oxidation of the extract (Brinn, 1973). However, even when treating wastes with high concentrations of phenolics (in which case, extraction efficiencies would presumably be maximized), it is frequently not possible to meet environmental guidelines [27] without recourse to secondary treatment with an activated sludge or a strongly oxidizing trickle filter process. Both methods are reportedly able to lower residual phenolic matter to 0.01% (Ashmore *et al.*, 1967, 1968, 1972).

Methods that almost wholly eliminate phenolics can also more or less completely degrade benzenoid hydrocarbons. However, solvent extraction of aromatic bases is usually much less complete: Removal of polynuclear aromatic hydrocarbons by sedimentation and biological oxidation is only 30–90% effective, and heterocyclic compounds are merely *presumed* to be rendered substantially harmless by such waste-stream processing.

The special treatment that must be accorded coke-oven ammonia liquors—which, in addition to NH_3 and NH_4^+, hold cyanides, thiocyanates, sulfides, and thiosulfates—now commonly involves bacterial degradation of noxious components. However, effective treatment is complex because each bacterial strain requires a specific temperature range and/or pH conditions for optimum performance. Much interest has therefore been shown in a so-called "aerated tower biology," which is specifically designed to eliminate N (Pascik, 1982). In a first step, NH_3 is here oxidized to NO_3^+ by

$$2\,NH_4^+ + 3\,O_2 \rightarrow 2\,NO_2^- + 4\,H^+ + 2\,H_2O$$
$$2\,NO_2^- + O_2 \rightarrow 2\,NO_3^+,$$

which is accomplished by *B. nitrosomas* and *B. nitrobacter*, respectively, and this is followed, in a second step, by

$$2\,NO_3^+ + 5\,H_2 \rightarrow N_2 + 2\,OH^- + 4\,H_2O,$$

which is achieved by facultatively anaerobic bacteria in the presence of an H donor such as methanol.

Other processes for rendering ammonia liquors substantially harmless—e.g., oxidation with O_3, H_2O_2, Cl_2 at <45°C/pH 7–10, or chlorine dioxide—are now infrequently used.

Control of Gaseous Emissions

Methods of choice for ensuring environmental acceptability of gas streams emitted into the atmosphere are, first and foremost, precombustion techniques that minimize the input of SO_2 and/or NO_x into flue gas and procedures that hold H_2S concentrations in other HT gas wastes to acceptable limits.

With respect to combustion, this means wherever possible using cleaner (low-sulfur) fuels and/or modified (so-called "advanced") firing methods [28] or, where this is impractical, removing acid components from the waste gas stream by chemical means. For this latter purpose, a number of techniques based on classic inorganic reactions have been proposed (see below). But choices of appropriate stripping methods always depend on the composition of the stream. Some otherwise attractive methods can prove inefficient or excessively costly for removing *trace* components, and others may be unsuitable for high-volume *lean* streams because they impose excessive incremental pressure drops across the capture system.

Techniques for containing H_2S, outlined in Chapter 8 for desulfurizing natural gas or purifying syngas, are in principle also applicable to waste gas streams from coke ovens and coal liquefaction plants. However, in such cases they can only meet air-quality standards if augmented by tail-gas cleanup appropriate for very low concentrations of acid gases—and this does, in fact, offer the most practical means for stripping H_2S, SO_2, and NO_x from flue gases and similar waste streams.

Of several methods for reducing sulfur contents of coal *before* combustion or HT processing (see Chapter 8), separation by gravity can only modestly lower FeS_x concentrations, and chemical techniques that can remove organic as well as inorganic sulfur forms have so far not been taken much beyond bench-scale demonstration [29]. There remain consequently only *in-system* capture methods or *postcombustion* removal of acid gases from flue gas or equivalent waste gas streams.

Effective in-system capture of SO_2 can be achieved either by directly trapping and destroying SO_2 through interaction with lime as in

$$CaCO_3 \rightarrow CaO + CO_2$$
$$CaO + SO_2 \rightarrow CaSO_3,$$

which is accompanied (Hatfield and Slack, 1975) by

$$2\ CaSO_3 + O_2 \rightarrow 2\ CaSO_4$$

or by transiently fixing SO_2 by absorption into an alkaline solution that is then regenerated with simultaneous permanent destruction of SO_2 by reaction with CaO.

Indirect trapping is exemplified by a Na-based double-alkali process (EPRI, 1978) in which the relevant reactions are

$$Na_2SO_3 + H_2O + SO_2 \rightarrow 2\ NaHSO_3$$
$$2\ NaOH + SO_2 \rightarrow Na_2SO_3 + H_2O,$$

accompanieid by

$$2\ NaHSO_3 + O_2 \rightarrow 2\ Na_2SO_4.$$

Na_2SO_3 is then regenerated, and SO_2 simultaneously fixed, by

$$2\ Na_2SO_4 + Ca(OH)_2 \rightarrow Na_2SO_3 + CaSO_3 + 2\ H_2O$$
$$Na_2SO_4 + Ca(OH)_2 + 2\ H_2O \rightarrow 2\ NaOH + CaSO_4 \cdot 2H_2O.$$

Variants of these processes capture SO_2

1. with NH_3 and consequent formation of ammonium sulfite that is subsequently oxidized to the sulfate;
2. with an aqueous solution of $Fe_2(SO_4)_3$ in which

   $$SO_2 + H_2O \rightarrow H_2SO_3$$

   is followed by catalytic conversion of H_2SO_3 to H_2SO_4; or
3. with MgO, in which case the capture and regeneration reactions are

   $$MgO + SO_2 \rightarrow MgSO_3$$
   $$MgSO_3 + SO_2 + H_2O \rightarrow Mg(HSO_3)_2$$
   $$Mg(HSO_3)_2 + MgO \rightarrow 2\ MgSO_3 + H_2O;$$

   and MgO is recovered by thermal decomposition,

   $$MgSO_3 \rightarrow MgO + SO_2,$$

   liberated SO_2 is processed into elemental sulfur.

Unlike fixation of SO_2 in $CaSO_3/CaSO_4$, which is usually discarded as waste, these variants offer some economic advantage by yielding saleable products—i.e., elemental sulfur, H_2SO_4, $(NH_4)_2SO_4$—but this is partly offset by higher operating costs.

The same techniques could, in principle, also be used to control NO_x in flue gas (Slack, 1981). However, *simultaneous* capture of SO_2 and NO_x by them tends to be costly, and greater benefits are consequently seen in three other innovative, but still largely experimental approaches to concurrent in-system capture of SO_2, SO_3, and NO_x (MacRae, 1991).

In one of these, the NOXSO process, flue gas encounters fluidized Na_2CO_3-coated Al_2O_3 particles, which cause SO_2 to be fixed in Na_2SO_4 and NO_x in $NaNO_3$; spent sorbent is continuously withdrawn and regenerated by heating in H_2 or CH_4. The method is claimed to remove 85–90% SO_2 and NO_x.

In another procedure, the SULF-X process, SO_2/NO_x is extracted from the flue gas by contact with a slurry of finely divided Fe^{2+} sulfide, which converts SO_2 to elemental S and NO_x to elemental N_2.

And in a third, the Electron Beam Process, capture is effected by either of two versions. In one, NH_3 is injected into the humidified and cooled flue gas, which is then passed through an electron beam. This converts SO_2 and NO_x to H_2SO_4 and HNO_3, which react with NH_3 to yield $(NH_4)_2SO_4$ and NH_4NO_3, and these are captured in baghouses and disposed as fertilizers. In the alternative lime-based version, the flue gas is humidified and cooled by evaporation of a lime slurry that also reacts with some SO_2 to form $CaSO_3$; the treated gas stream is passed through an electron beam, which converts NO_x and remaining SO_2 to HNO_3 and H_2SO_4; and the process is completed by sending the gas through a fabric filter, where the acids react with $Ca(OH)_2$ to form $CaSO_4$ and $Ca(NO_3)_2$.

In-system capture of NO_x *per se* can also be accomplished by a selective catalytic reduction (SCR) process. In this case, NH_3 is injected into the cooler upper furnace region, where it reacts with NO_x at 350–450°C over an Al_2O_3-supported transition-metal catalyst such as Cu, Cr, Mn, or V. The net reaction is

$$NO + NO_2 + 2\ NH_3 \rightarrow 2\ N_2 + 3\ H_2O.$$

Very similar chemistries are involved in simultaneous removal of SO_2 and NO_x by *scrubbing* the flue gas with slurried CaO or MgO, or with aqueous solutions of NH_3, $NaOH/Na_2SO_3$, or $Fe_2(SO_4)_3$. The scrubbed gas is then stripped of entrained particulates and reheated to prevent condensation of moisture in the stack before releasing it into the atmosphere. Figure 11.3.2 exemplifies operating modes that can remove 70–90% SO_2 and NO_x from a flue gas. The process chemistry is illustrated by scrubbing with NH_3, in which case

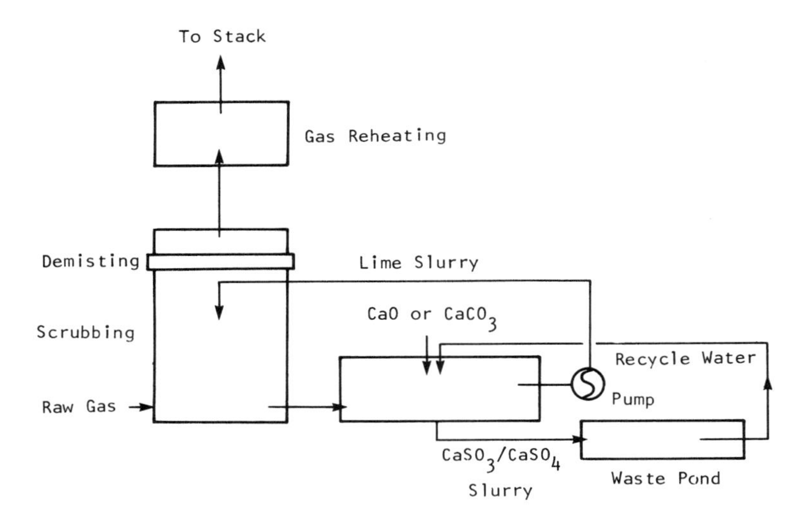

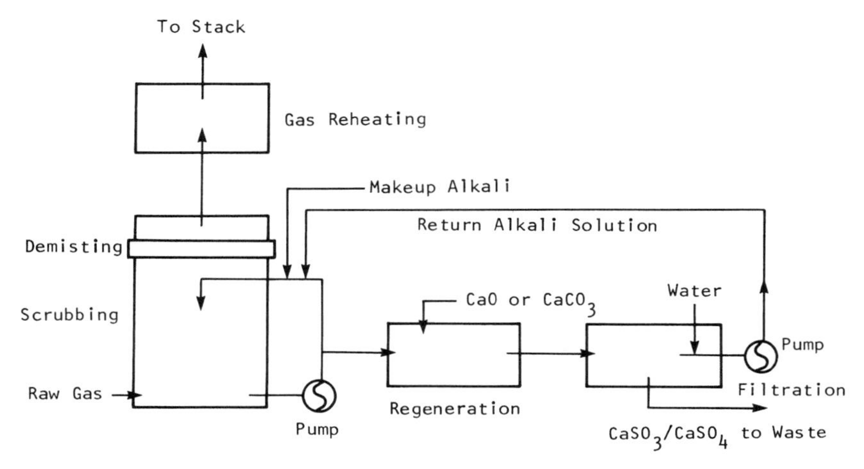

FIGURE 11.3.2 Flowsheets for direct (upper) and indirect (lower) capture of SO_2 and NO_x in flue gas by limestone.

$$2\,NH_3 + H_2O + SO_2 \rightarrow NH_4)_2SO_3$$
$$(NH_4)_2SO_3 + SO_2 + H_2O \rightarrow 2\,NH_4HSO_3$$
$$6\,NH_4HSO_3 + 2\,NO \rightarrow 2\,NH(NH_4SO_3)_2 + (NH_4)_2S_2O_6 + 2\,H_2O.$$

The imidodisulfonate is then hydrolyzed to yield $(NH_4)_2SO_4$, as in

TABLE 11.3.4 Power Plant FGD Technologies[a]

Technology	Status	SO_2/NO_x reduction[b]
Precombustion cleaning		
(Gravity separation)	Commercial	0–30% SO_2
Chemical cleaning	R&D	up to 90% SO_2
Coal–water fuels	Demonstrated[c]	50% pyritic SO_2 some NO_x
Combustion control		
Staged combustion/low-NO_x burners	Commercial	30–50% NO_x
Sorbent injection	Demonstrated[c]	60–75% SO_2
Postcombustion (FGD)[d]		
Wet lime/limestone	Commercial	>90% SO_2
Regenerable FGD	Commercial	70–80% SO_2
Combined SO_2/NO_x FGD	Commercial	90% SO_2/80% NO_x
Selective catalytic reduction	Commercial	30–50% NO_x
Dry sorbent injection	Demonstrated[c]	n.a.

[a] MacRae (1991).
[b] Indicative only.
[c] Used in several small operations.
[d] FGD = flue gas desulfurization; includes NO_x removal.

$$NH(NH_4SO_3)_2 + 2\,H_2O \rightarrow (NH_4)_2SO_4 + NH_4HSO_4$$
$$NH_4HSO_4 + NH_3 \rightarrow (NH_4)_2SO_4.$$

The ammonium dithionate is simultaneously decomposed, as in

$$(NH_4)_2S_2O_6 \rightarrow (NH_4)_2SO_4 + SO_2,$$

and SO_2 is finally converted to elemental sulfur.

An alternative version uses

$$Na_2SO_3 + SO_2 + H_2O \rightarrow 2\,NaHSO_3$$
$$6\,NaHSO_3 + 2\,NO \rightarrow 2\,NH(Na_2SO_3)_2 + Na_2S_2O_6 + 2\,H_2O.$$

and decomposes the sodium imidodisulfonate by interaction with $Ca(NO_2)_2$, CaO, and H_2SO_4, which yield $CaSO_4$, Na_2SO_4, and N_2. The dithionate is decomposed as before.

Table 11.3.4 identifies in summary form the technologies that now present themselves, or are potentially available, for controlling SO_2 and NO_x in flue gas from fossil-fueled generating plants.

To slow global warming, if that is indeed largely due to CO_2 from combustion of fossil hydrocarbons [30], serious thought has been given to lowering CO_2 emission rates by energy conservation, more efficient thermal power generation by combined-cycle operations [31], and greater reliance on nuclear energy and H-rich fuels. Anticipated savings are, however, likely to be offset by energy needs of developing countries, which are expected to account for nearly 60% of global manmade CO_2 emissions by the year 2050 (Environment Canada, 1986; Electric Power Research Inst., 1988; International Energy Agency, 1989). Some attention has been therefore been focused on the technical and economic feasibility of compressing captured CO_2 for injection into suitable u/g caverns. But because nuclear energy generation is beset by problems of its own, and large-scale long-term CO_2 storage is unlikely to prove practical, tangible benefits can ultimately only be expected to accrue from sustained energy conservation and more efficient (combined-cycle) power generation.

NOTES

[1] Exploration, open-pit mining, and other similar surface operations also adversely affect regional soil quality, watershed characteristics, and wildlife habitats. But such damage is always site specific and is not discussed in this survey. Because statutorily required rehabilitation is monitored and adjudicated by public agencies responsible for environment protection, it might reasonably be deemed relatively short-term damage. Public concerns over exploration and mining do, in fact, center mainly on the *feasibility* of acceptable rehabilitation of disturbed lands after operations cease (Doyle, 1976; Downs and Stocks, 1977).

[2] Pneumoconiosis, the infamous "black lung" disease caused by long-term inhalation of fine coal dusts, was for centuries an inescapable fate of miners, but has now been virtually eradicated by safety regulations and dust-control programs in mines. Permissible dust concentrations are set by standards under industrial health and safety ordinances and enforced by government inspectors.

[3] Coal *per se* is not harmful to plants and animals, and may indeed be directly or indirectly beneficial to them. Oxidation during prolonged exposure to air enriches it in soil amendments and organic fertilizers (i.e., humic acids), and its high porosity enables it to sorb odor- and color-forming substances (including bacteria) from water. Coal has therefore often been advocated as a low-cost substitute for activated carbons in waste water treatment (Kahoe, 1967; Biospheric Research Inc., 1969; Perrotti and Rodman, 1973; Bunn, 1974).

[4] Since combustion systems fueled by pulverized coal discharge up to 90% of total ash as flyash, the flue gas emitted by such systems could contain 10–14 g/m^3, and a 1000-MW generating plant burning bituminous coal with 15% ash would discharge some 1200 t/d. If the fuel is a subbituminous coal with 15% ash, the discharge rate would run to ~1800 t/d.

[5] The complexity of such transformations is daunting, but possible pathways are suggested by weathering reactions. Angino (1983) has outlined some examples that illustrate the first steps of reaction sequences to potentially hazardous matter:

1. Congruent dissolution reactions:

Quartz	$SiO_2 + 2\ H_2O \rightarrow H_4SiO_4$
Gibbsite	$Al_2O_3 \cdot 3H_2O + 2\ H_2O \rightarrow 2\ Al(OH)_4^- + 2\ H^+$
Apatite	$Ca_5(PO_4)_3(OH) + 3\ H_2O \rightarrow 5\ Ca^{2+} \cdot 3\ HPO_4^{2-} + 4\ OH^-$

2. Incongruent dissolution reactions:

Magnesite $MgCO_3 + 2\,H_2O \rightarrow HCO_3^- + \underset{\text{brucite}}{Mg(OH)_2} + H^+$

Orthoclase $3\,KAlSi_3O_8 + 2\,H_2CO_3 + 12\,H_2O \rightarrow 2\,K^+ + 2\,HCO_3^- + 6\,H_4SiO_4 + \underset{\text{mica}}{KAl_3Si_3O_{10}(OH)_2}$

Biotite $KMg_3AlSi_3O_{10}(OH)_2 + 7\,H_2CO_3 + 0.5\,H_2O \rightarrow K^{2+}\cdot 3\,Mg^{3+}\cdot 7\,HCO_3^- + 2\,H_4SiO_4 + 0.5\,\underset{\text{kaolinite}}{Al_2Si_2O_5(OH)_4}$

Dolomite $CaMg(CO_3)_2 + Ca^{2+} \rightarrow Mg^{3+}\cdot 2\,\underset{\text{calcite}}{CaCO_3}$

3. Redox reactions:

Hematite $Fe_2O_3 + H_2O + 2\,e \rightarrow 2\,Fe_3O_4(s) + 2\,OH^-$
Galena $PbS + 4\,Mn_3O_4 + 12\,H_2O \rightarrow Pb^{3+}SO_4^{2-} + 12\,Mn^{3+}\cdot 24OH^-$

[6] In the early 1950s, several fishermen and their families on Japan's Minemata Bay died or were permanently disabled by paralysis of their central nervous systems after ingesting Hg-polluted fish. This episode prompted studies of mercury levels in North American and European watercourses, and bans on fishing in lakes in which potentially dangerous Hg levels were indicated. There is, however, some evidence that hazards to mammalian species (including *Homo sapiens*) arise, not from inorganic Hg, which is frequently still used in dental amalgams, but from an *organic* form—methyl mercury, $Hg(CH_3)_2$—which develops during bacterial attack on tissues holding Hg.

[7] Until the early Middle Ages, it was in fact common practice, after razing an enemy city, to ruin the surrounding soil as well as the destroyed city by spreading salt and thereby preventing any form of revegetation for several generations.

[8] Bitumen in strata too deep for surface mining can be recovered by steam injection, steam drive, or wet combustion drive—each conducted in any one of several different modes, and each meeting with a degree of success that depends on the characteristics of the pay zone. Similar procedures are often used in enhanced recovery of heavy oils. Volumes of variously contaminated water generated by these operations can in some cases exceed the volume of recovered hydrocarbons by a factor of 4 or 5.

[9] An extreme instance of accidental crude oil release is a spill of 41,000 m^3 (250,000 bbl) when a large oil tanker, the *Exxon Valdez*, ran aground on a charted reef in Alaska's Prince William Sound. Cleanup costs and settlement of lawsuits filed against Exxon Corporation by the U.S. Federal and Alaska governments totaled over \$3 billion; and a class-action suit brought against Exxon by some 14,000 persons directly harmed by the spill also resulted in Exxon having to pay a further \$5 billion in punitive damages to the plaintiffs. Most spillages are, of course, very much smaller: some illustrative data for a typical two-year period in the mid-1970s, when oil activity in Alberta peaked, are set out in Table 11.3.3.

[10] In the late 1980s, groundwater supplied ~50% of the domestic water demand in the United States, over 25% in Canada, and 70–75% in Western Europe. This water derives mainly from relatively shallow aquifers that contain <1500 mg/liter total dissolved solids, and there is therefore concern over contamination of aquifers by petroleum hydrocarbons (in particular, benzene and toluene). Common causes of such contamination are underground storage tanks for petroleum products.

[11] The severe ecological damage that uncontrolled discharge of such effluents can cause is

depressingly well demonstrated in the older coal-mining areas of Appalachia and parts of Illinois, Indiana, and Kentucky.

[12] Alkaline runoffs have been reported from several mine sites in the Western United States and Western Canada where subbituminous coals with, generally, $<0.5\%$ total sulfur and very little FeS_x, are surface mined. In some places, pH values as high as 11 have been measured.

[13] A case in point is an attempt (Singer *et al.*, 1978) to assess the composition of effluents from the Synthane gasification process (Forney and McGee, 1972; Carson, 1975) at a time when that process was in an early stage of development (and now seems quite unlikely to reach commercial status). Similar attempts have been made to analyze waste streams from a hypothetical 28,000 t/d SRC-II facility. But in this instance, it must be borne that all techniques for converting coal to liquid hydrocarbons promote aromatization, and that liquefaction at 350–400°C might well be expected to generate some potentially carcinogenic and/or mutagenic entities.

[14] Even when containing as little as 2 mg ammonia-N per liter and showing a pH of 7.5–8.0, weak ammoniacal liquors can kill aquatic biota (McKee and Wolf, 1963); at a higher pH, which would arise from shift of the equilibrium reaction $NH_4OH = NH_3 + H_2O$ to the right, this threshold is even lower as NH_3 tends to be much more toxic than its protonated form. According to Adams *et al.* (1975), NH_3 may in fact be the most difficult parameter to control in the treament of coke-oven waste water.

[15] Very large volumes of noxious gases (CO, CO_2, hydrocarbons, etc.) are of course also emitted from *mobile* sources such as automobiles and diesel-fueled trains.

[16] This reaction becomes important at > 1100°C and is followed in the atmosphere by $2\ NO + O_2 \rightarrow 2\ NO_2$. The initial rate of such oxidation increases rapidly with the initial concentration (C_o) of NO, changing from $\sim$0.003 ppm min^{-1} at $C_o = 1$ to 3.0 ppm min^{-1} as $C_o \rightarrow 100$ ppm.

[17] Vented to the atmosphere, SO_2 can remain free for many days and migrate with air currents across several hundred kilometers. Acid rains in Scandinavia have thus been traced to SO_2 emitted from thermal power plants in southern England and continental Europe, and the death of more than 1400 lakes in Ontario, as well as destruction of vegetation over large areas of the Canadian Shield, has been ascribed to SO_2 emissions in the New England states and from nickel smelters at Sudbury (Ontario). It has now become evident that such ecological damage is not necessarily irreparable, and that curtailment of SO_x emissions could serve to repair it over time. But more permanent damage has been inflicted by acid rain on manmade structures. Especially vulverable are limestone structures, which are rapidly corroded by

$$CaCO_3 + H_2SO_4 \rightarrow CaSO_4 + CO_2 + H_2O.$$

Similar damage can be done to metal sheathings, which are also prone to corrosion by acid sulfates.

[18] In November 1952, a dense fog in London, which lasted several days and was attributed to air pollution by coal combustion, was the direct cause of death of more than 4000 persons, mostly from respiratory or heart failures. It was later found tht children and elderly persons are particularly prone to disablement of such smogs.

[19] This is the fused ash or clinker (usually $>$2–3 cm) withdrawn at or near the bottom of the combustion appliance.

[20] The 50 mg/m_3 emission limit applies in the United States and Canada, as well as in several Western European countries. Australia and Britain have apparently not yet adopted it, but are expected to do so.

[21] To reach these temperatures, the precipitator is placed below the air-heater part of the boiler rather than, as is more usual, above it. But because the higher temperatures expand gas volumes, efficient particle capture also demands larger collector plates than would otherwise

be needed. An alternative is to condition the ash by injecting SO_3 or Na^+ (commonly as Na_2CO_3) into the gas stream. This has sometimes been done to avoid costly retrofitting of installations that, to comply with environmental legislation, switched from relatively high-sulfur to cleaner low-sulfur coal. However, because as much as 30% of injected SO_3 passes unchanged into the atmosphere, this is hardly an acceptable practice.

[22] The rate and extent of degradation depends on the nature of the hydrocarbon, on the indigenous microbial population, and, since effective microbial degradation of hydrocarbons is oxidative, on the availability of O_2.

[23] On water, oil spills can have disastrous impacts on the aquatic food chain and/or prove lethal to many marine species: among its effects are fatal liver and kidney poisoning in otters and seals, plugging of dolphin and whale blowholes, and fatal emphysema in other aquatic species through inhalation.

[24] Skimmers are either floating or rotary devices that use impellers, suction, or discs to draw the oil into a collector, or weir skimmers that, in refined versions, are self-leveling in the water and adjustable for skimming depth. Sorbents include natural materials (peat moss, straw, sawdust), mineral products (asperlite, vermiculite, talc), or manufactured sorbents (polyester foam, polystyrene, polyurethane).

[25] Sinking agents are particularly effective when applied to heavy oils, but are only cosmetic and have raised questions about their effect on aquatic life.

[26] Because of accompanying detrimental effects—notably an *increasingly toxic* oil slick—the use of dispersants is discouraged.

[27] Even *current* EPA guidelines, which are certain to become more stringent in the future, require 99.97% removal of phenolic matter.

[28] R & D directed toward development of so-called "clean coal technologies" has successfully tested, and largely perfected, several retrofit techniques that substantially reduce emissions of SO_x and NO_x (Anon., 1991). Examples are

1. TransAlta Resources' LNS (low NO_x/SO_x) burner system—an entrained-flow system for cyclone-fired boilers that uses "staged" combustion: by adding $CaCO_3$ to the coal, the initial combustion stage promotes capture of SO_2, and NO_x is then converted to elemental N_2
2. Babcox and Wilcox's LIMB (limestone injection, multiple burner) system, which injects CaO directly into the boiler at some point above the "low NO_x" burners
3. Georgia Power's deployment of second-generation "low NO_x" burners in conjunction with over-fire secondary air injection
4. The combined use of LIMB and COOLSIDE—the latter developed by the Pittsburgh Energy Technology Center (PETC) and involving injection of a Ca-based sorbent into the cooler upper part of the boiler
5. Babcock and Wilcox's SNRB (SO_x–NO_x–Rox Box) flue-gas cleanup, which removes SO_x by injecting a Na- or Ca-based sorbent into the flue gas as it leaves the economizer, and uses NH_3 to reduce NO_x to elemental N_2 by a selective catalytic reduction process.

[29] Whether, given the alternatives, chemical cleaning methods will *ever* do so is debatable.

[30] Insufficient hard data make predictions of global warming and its effects critically dependent on computer models, and definitive conclusions are, in fact, not expected until well after the turn of the 20th century. Global temperatures appear to have risen ~0.5°C over the past 100 years (MacCracken and Luther, 1985), in part probably due to a 25% rise in atmospheric CO_2 from 270–280 ppmv in the 1850s to ~345 ppmv in 1985 (Trabalka, 1985), and theoretical models designed to generate estimates of resultant climatic change suggest that a doubling of CO_2 concentrations (to ~700 ppmv) could raise average global temperatures by 1.5–4.5°C. This must, however, be viewed against global temperature *fluctuations:* in Devonian times, coal formed, among other places, on Bear island, now within the Arctic Circle; 4000–8000 years ago, global temperatures were several degrees higher than now; over a

10^5-year cycle characteristic of glacial/interglacial periods, average temperatures fluctuated within an 8°C range; and only over short periods were fluctuations limited to 0.5°C (Lamb, 1977; Smith, 1978). In any event, much larger radiative effects than from CO_2 are, because of their IR absorption, likely to accrue from emission of biogenic CH_4 (which is increasing at ~1%/year) and N_2O (which is increasing at ~0.25%/year because of growing use of inorganic fertilizers).

[31] Combined-cycle systems could raise the average efficiency of coal-fired thermal power generation from ~33–34% to ~42–44% and cut CO_2 volumes by 20–25%.

REFERENCES

Adams, C. E., R. M. Stein, and W.W. Eckenfelder, Jr. *Proc. 29th Industr. Waste Conf. Pt. 2,* 865 (1975).

American Petroleum Inst. *API RP 55, Dallas, TX* (1981).

Angino, E. E. In *Applied Environmental Geochemistry* (Iain Thornton, ed.), Chapter 6, 1983. London: Academic Press.

Anonymous. *PETC Review No. 3,* 1991. Washington, D.C.: U.S. Dept. of Energy.

Ashmore, A. G., J. R. Catchpole, and R. L. Cooper. *Water Res.* **1**, 605 (1967); **2**, 555 (1968); **6**, 1459 (1972).

Biospheric Research Inc. *Rept. PB-18455,* 1969. Washginton, D.C.

Bond, R. G., C. P. Straub, and R. Prober. *Waste Water Treatment and Disposal,* Vol. 4, *Handbook of Environmental Control,* 1974. Cleveland: CRC Press.

Bridbord, K. *Proc. Inst. Occupational Health & Safety, Rockville, MD* (1976).

Brinn, D. G. *NTIS-SM/TN/1/25* (1973).

Brown, E., M. W. Skougstad, and M. J. Fishman. *Techniques of Water Resources Invest., U.S. Geol. Surv. Book 5,* 1970. Washington, D.C.: U.S. Govt. Printing Office.

Bump, R. L. *Chem. Eng. News,* Jan 17 (1977).

Bunn, C. O. U.S. Pat. No. 3,798,158 (1974).

Carson, S. E. *Proc. 7th Synth. Pipeline Gas Symp., Chicago, IL* (1975).

Deutsch, W. *Ann. Phys.* **68**, 335 (1922).

Donovan, R. P. *Fabric Filtration for Combustion Sources,* 1985. New York: Marcel Dekker.

Downs, C. G., and J. Stocks. *Environmental Impact of Mining,* 1977. Somerset, U.K.: Halstead Press.

Doyle, W. S. *Strip Mining of Coal: Environmental Solutions,* 1976. New Jersey: Noyes Data Corp.

Electric Power Research Institute. *EPRI Rept. No. FP-713* (1978).

Electric Power Research Institute. The politics of climate. *EPRI Journal,* June (1988).

Environment Canada. *Fact Sheet—The Impact of Global Warming,* 1986. Downsview, ON: Supply & Services Canada.

Forney, A. J., and J. P. McGee. *Proc. 4th Synth. Pipeline Gas Symp., Chicago, IL* (1972).

Gooch, J. P., and N. L. Francis. *U.S. Symp. on Electrostat. Precipitator Control of Fine Particles, EPA-650/2-75-016* (1975).

Hatfield, J. D., and A. V. Slack. *ACS Adv. Chem. Ser.* **139**, 130 (1975).

International Energy Agency (IEA). *Emission Controls in Electricity Generation and Industry, OECD, Paris* (1989).

Kahoe, A. J. U.S. Pat. No. 3,300,403 (1967).

Klein, D. H., A. W. Andren, and N. E. Bolten. *Water Air Soil Poll.* 5(1), 71 (1975).

Klingspor, J., and J. L. Vernon. *Particulate control for coal combustion, IEACR/05, IEA Coal Research, London* (1988).

Lamb, H. H. *Climate: Present, Past and Future,* 1977. London: Methuen.

Longley, M. S. *Hydrogen Sulfide in Production Operations,* 1982. Austin, TX: Division of Continuing Education, Univ. Texas at Austin.
MacCracken, M. C., and F. M. Luther (eds.). *DOE/ER-0235; DOE/ER-0237,* 1985. Washington, D.C.: Dept. of Energy.
McDonald, J. R., and A. L. Dean. *Electrostatic Precipitator Manual,* 1982. New Jersey: Noyes Data Corp.
McKee, J. E., and H. W. Wolf. *Water quality criteria,* 2nd ed. Water Resources Control Bd., California, Publ. No. 3-A (1963).
MacRae, K. M. *New Coal Technology and Electric Power Development, Study No. 28, Can. Energy Res. Inst., Calgary, AB* (1991).
Matts, S. A., and P. P. Ohnfeldt. *Gas Cleaning with SF Electrostatic Precipitators,* 1973. Vaxjo, Sweden: Flakt Industrie AB.
Mazumdar, S., C. Redmond, W. Sollecito, and N. Sussman. *J. Air Pollut. Control Assoc.* **25**(4), 382 (1977).
Nichols, C. R. *Development document for effluent limitation guidelines, EPA, Washington, D.C.* (1974).
Pascik, I. *Hydrocarbon Process.* **61**, 80 (1982).
Perrotti, A. E., and C. A. Rodman. *Chem. Eng. Progr.* **69**(11), 63 (1973).
Petroleum Industry Training Service. *Oil Spill Containment & Recovery—A Primer, 1977.* Edmonton, AB.
Potter, E. C. *J. Air Pollution Contr. Assoc.* **28**(1), 40 (1978).
Redmond, C., A. Ciocco, J. W. Lloyd, and H. W. Rush. *J. Occup. Med.* **14**, 621 (1972).
Sexton, R. J. *Arch. Environ. Health* **1**, 181 (1960).
Singer, P. C., F. K. Phaendler, J. Chincilli, A. F. Maciorowski, J. C. Lamb III, and R. Goodman. *EPA-600/7-78-181, NTIS, Springfield, VA* (1978).
Slack, A. V. In *Chemistry of Coal Utilization,* Suppl. Vol. (M. A. Elliott, ed.), 1981. New York: Wiley.
Smith, I. *ICTIS/ER 01, IEA, London* (1978).
Trabalka, J. (ed.). *DOE/ER-0239,* 1985. Washington, D.C.: Dept. of Energy.
U.S. Environmental Protection Agency. *EPA-450/3-81-005A* (1982).
van Hook, R. I., and W. D. Shults. *Workshop on Trace Contaminants from Coal Combustion, ERDA-77-64, Knoxville, TN* (1977).
Walker, A. B. *Proc. American Power Conf., Chicago, IL, 1974* (publ. 1977).
Wilson, J. T., L. E. Leach, M. Henson, and J. N. Jones. *Groundwater Monitor. Rev.* **6**, 56 (1986).

INDEX

I

K

L

M

U

V

W

X

Z